First Conference of the International Water Association IWA for Young Scientist in Poland 2017

Water, Wastewater and Energy in Smart Cities

E3S Web of Conferences Volume 30 (2018)

Cracow, Poland
12 - 13 September 2017

Editor:

M. Cimochowicz-Rybicka

ISBN: 978-1-5108-5723-0

Printed from e-media with permission by:

Curran Associates, Inc.
57 Morehouse Lane
Red Hook, NY 12571

Some format issues inherent in the e-media version may also appear in this print version.

Printed by Curran Associates, Inc. (2018)

For additional information, please contact EDP Sciences – Web of Conferences at the address below.

EDP Sciences – Web of Conferences
17, Avenue du Hoggar
Parc d'Activité de Courtabœuf
BP 112
F-91944 Les Ulis Cedex A
France

Phone: +33 (0) 1 69 18 75 75
Fax: +33 (0) 1 69 28 84 91

contact-edps@webofconferences.org

Additional copies of this publication are available from:

Curran Associates, Inc.
57 Morehouse Lane
Red Hook, NY 12571 USA
Phone: 845-758-0400
Fax: 845-758-2633
Email: curran@proceedings.com
Web: www.proceedings.com

TABLE OF CONTENTS

edp sciences

E3S Web of Conferences

Statement of Peer review

In submitting conference proceedings to E3S Web of Conferences, I certify to the Publisher that I adhere to the **Policy on Publishing Integrity** of the journal in order to safeguard good scientific practice in publishing.

1. All articles have been subjected to peer review administered by the proceedings editors.
2. Reviews have been conducted by expert referees, who have been requested to provide unbiased and constructive comments aimed, whenever possible, at improving the work.
3. Proceedings editors have taken all reasonable steps to ensure the quality of the materials they publish and their decision to accept or reject a paper for publication has been based only on the merits of the work and the relevance to the journal.

Title, date and place of the conference THE FIRST CONFERENCE OF THE INTERNATIONAL WATER ASSOCIATION IWA FOR YOUNG SCIENTIST IN POLAND „WATER, WASTEWATER AND ENERGY IN SMART CITIES"

Proceedings editor(s): MAŁGORZATA CIMOCHOWICZ-RYBICKA

Date and editor's signature 01/17/2018

Application of Information Technology Solution for Early Warning Systems at Water Utilities

Alicja Bałut[1,*]

[1]Institute of Environmental Engineering, Poznan University of Technology, Berdychowo 4, 60-995, Poznań, Poland.

Abstract. Deployment of IT solutions in water utilities in Poland concerns nowadays lots beyond GIS implementation projects [1]. The scope of modern IT platforms is truly advanced software for complete management of water treatment processes and involved objects, including ranges of various types of equipment. There are multiply factors that disrupt required volumes of supplied water. They are normally classified as natural, accidental and intentional. This paper addresses potential residing in already deployed IT solutions of water utilities in and also in new ones being now developed. Primarily- from the perspective of intentional, terrorist threats. This document depicts operating procedures that are called in case of spotted contamination in a water supply (damage of key elements of the network infrastructure) or in case of an introduction factors. This paper also discusses relevant IT tools with access provided to network operators or water plant owners that are extremely useful in accurate pinpointing the treat and in following relevant operating procedures and related actions.

1 Introduction

Water distribution networks are spatially highly diverse. Their exploitation depends on various external systems, including atmospheric ones. On the way water passes to end clients it meets equipment where various processes occur being a challenge for sewage plant operations. Exploitation may become a real challenge for water supply security, given each network has to meet specific physical and chemical parameters [2].

1.1 Threats

These events can be classified into three basic groups:

• natural (hazardous weather conditions: floods, earthquakes, fires, hurricanes, droughts, also natural contamination of ground water),
• target-oriented (terrorism, contamination with hazardous substances, acts of vandalism),
• incidental (accidents: incidental discharge of sewage or its inflow into protected areas, cracks in water pipes, accidents in areas where clean water and sewage pipelines cross).

* Corresponding address: alicja.balut@put.poznan.pl

The amount of target-oriented (intentional) threat acts on water distribution systems or acts of sabotage of water's supply is relatively small compared to other threats listed.

Especially the following should be noted:

- destruction of water network elements via cyberattacks, mainly targeted at systems and equipment managed by SCADA (Supervisory Control and Data Acquisition),
- water contamination with: chemical substances, pharmaceuticals, microbes, bacteria, toxins. Amongst the most common points of contamination are: water tanks for treated water, main water pipelines [3].

2 Early warning systems (EWS) in water utilities

One of the foundations of an early warning system is its applicability to forecast diffusion of hazardous substances in a water network using a hydraulic model [4]. It is of a particular importance to note that a properly calibrated quality model has all means to become the basis to forecast, simulate diffusion of threats or simulate solutions to crisis situations and thus ensuring the security of water supply.

Above mentioned targets are as well realized by water quality monitoring systems whose purpose is to shorten the time a contaminant remains in water and as well to support the determination of spatial range of the threat. Choice of the best localisation of measuring points is a task based on multiple, often self-excluding tasks. It requires conscious decision making often supported by optimisation tools [5].

Ideal criteria of an early warning system [6],[7]:

- threat is communicated early enough to enable reaction in agreed time,
- affordable price,
- no necessity to train staff (costs),
- inclusion of all threat types,
- threat source identification,
- minimal amount of wrong answers,
- simple interface, long operation with no necessity of changes nor updates.

An early warning system consists of the following basic elements [8]:

- accident / emergency warning systems,
- hazard identification with aid of data in databases,
- hydraulic model that in case of a threat becomes the starting point for further operations,
- local water quality control system, at the source.

As mentioned above early warning systems should be based on measurements points installed at water distribution systems, water intake and measurement data received from SCADA. This information should be integrated and processed in case of threat to authorities and regional institutions [9]. Launch of such a system should be proceeded by an earlier adoption of procedures regulating alarm handling. In general prior to the deployment of an early warning system an in-depth analysis of threats related to the localization and neighborhood of the water network should be carried out [10]. Inventory should also include all available sources of emission, their category, characteristics and utilization methods. It would be also advised to carry out analysis of the risk related to acquisition of raw water.

3 Planning and modelling of threat scenarios

Modelling of water network systems nowadays integrated with GIS (Geographic Information Systems) enables answering key questions related to water quality:

- water sources (wells, tanks) that supply relevant city districts,

https://doi.org/10.1051/e3sconf/20183001001

• spot-measured water origin (percentage by sources), graphic illustration of network fragments that share water from the same source,
• time required to transport water to most remote network sections.

Figures 1,2 and 3 illustrate some screenshot from application of commercial software (InfoWater Protector, Info Works WS Plus, InfoWater) to complete similar tasks also contaminant (multiply species) modelling like [11],[12]:
• disinfection by-products model,
• pathogen inactivation model,
• chloramine decomposition model,
• particle settlement model,
• temperature.

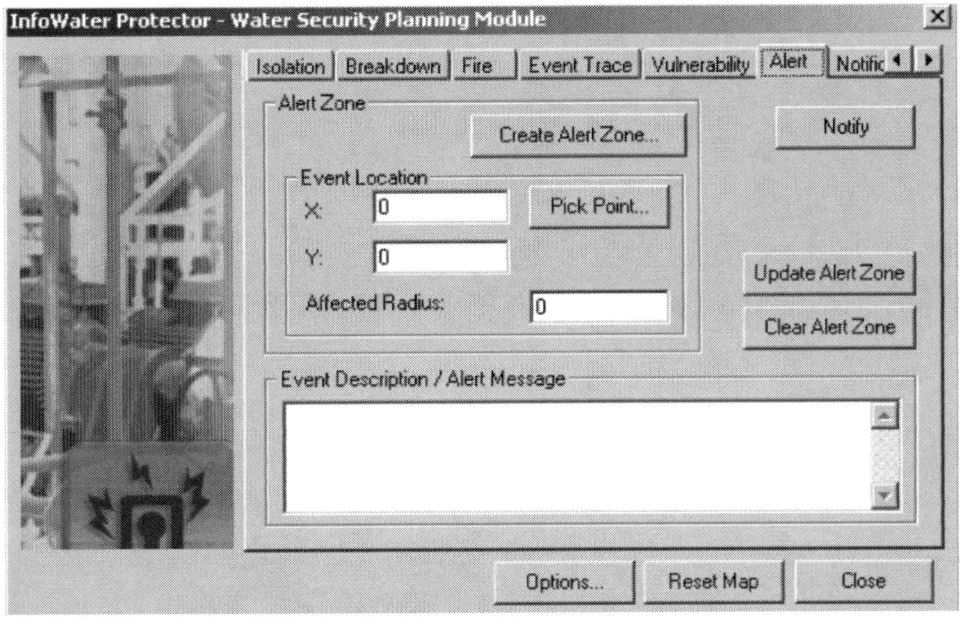

Fig. 1. Screenshot from InfoWater Protector that illustrates The ALERT tab that allows quick reporting and visualization of the extent of an event. It also supports listing of all affected customers for priority notification and sorting them in order of increasing distance (shortest to furthest) from the event location [13].

Unquestionably in case of threat related operating plans a hydraulic model is the best solution of invaluable support. Tasks mentioned above when realized during the modelling process enable generation of multiple scenarios that covered events, strategies and neutralisation techniques of i.e. substances introduced into networks on purpose.

Fig. 2. Screenshot from Info Works Plus, illustrating zones with consumers of contaminated water, during the time of an incident [6].

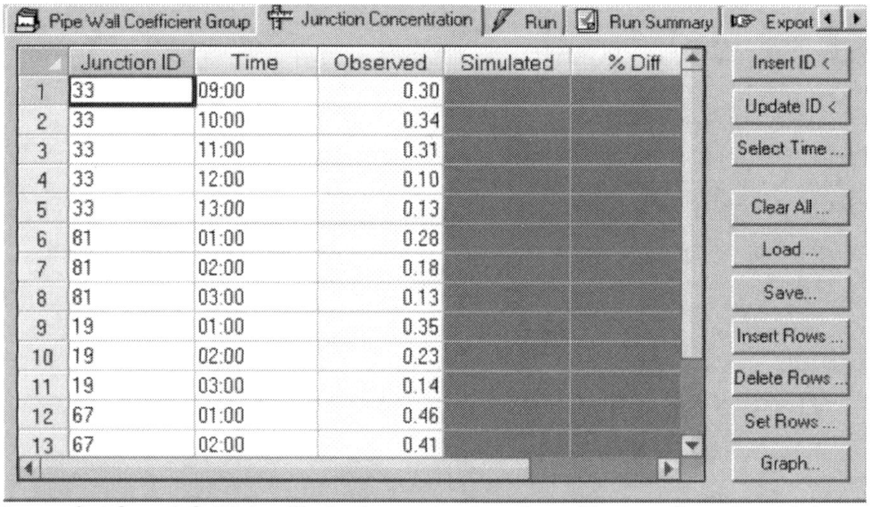

	Junction ID	Time	Observed	Simulated	% Diff
1	33	09:00	0.30		
2	33	10:00	0.34		
3	33	11:00	0.31		
4	33	12:00	0.10		
5	33	13:00	0.13		
6	81	01:00	0.28		
7	81	02:00	0.18		
8	81	03:00	0.13		
9	19	01:00	0.35		
10	19	02:00	0.23		
11	19	03:00	0.14		
12	67	01:00	0.46		
13	67	02:00	0.41		

Fig. 3. Screenshot from Info Water, illustrating contaminant's accident configuration tab in a water network model [11].

Based on this example a graphical display can depict i.e.:
• zones contaminated in a given timeframe, with a given contaminant, with information on water sources,
• spot at the network map where a harmful substance has been introduced. This is available thanks to data on source type from which water in the given spot originates.

3.1 Emergency procedures in case of threats

Thanks to integrated solutions – a model connected with a SCADA system and with measuring point data – automatic notifications on deviations between data computed in the model and real-life data are issued. This enables fast identification of a public security threat (number of end users of contaminated water) or a temporary break in service provisioning. Based on these analyses decisions to isolate endangered areas, their cut-off from further wide spreading of substances, aid to affected population, check of similar places (neighbouring vessels) by collection of probes for lab testing as well as other actions adequate to the threat can be made. Further information on this issue can be found at: [14],[15].

4 Application of modelling software in early warning systems in water utilities

In the last few years companies from the critical infrastructure sector in Poland, including water utilities, have seriously invested in IT solutions, mainly in GIS systems. This has come along with observed rapid development of enterprise management software, particularly oriented at asset maintenance. What is important from engineers' perspective, this also relates to processes 'optimisation and tools' functioning in water and sewage networks. Majority of water network utilities in Poland are in the process of implementing GIS systems. Parallel efforts are made to build hydraulic models based on EPANET.

Water distribution systems security modelling requires continual updating of network data, access and methods of collecting measurement data and taking into account operators' actions when a dangerous substance has been released into the network. In Poland, most of the utilities are facing this challenge.

Despite this article addresses a few examples of applications for modelling water networks its purpose is not to discuss IT tools for water distribution networks' monitoring. This is a spacious theme, already widely quoted in literature, that should be addressed in a separate article [16].

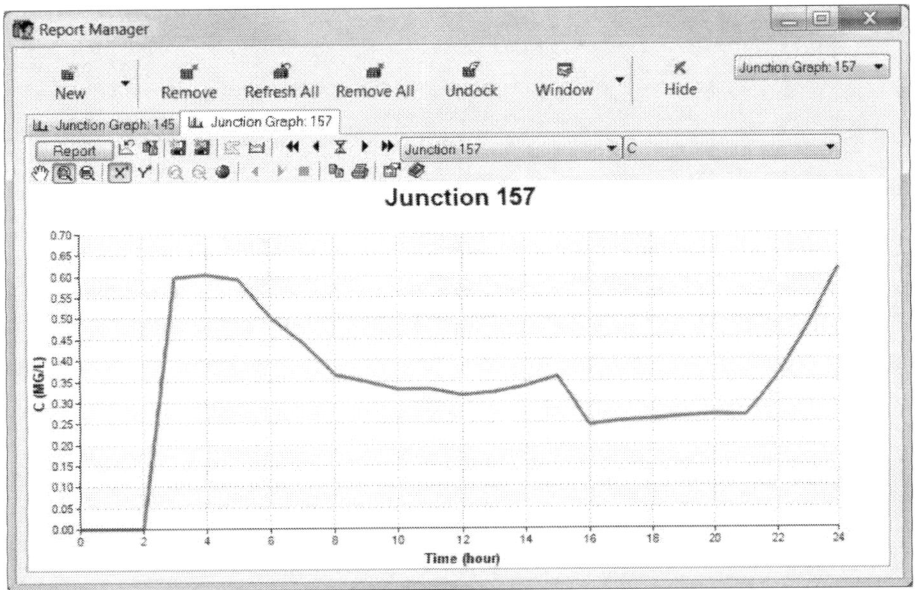

Fig. 4. Screenshot from InfoWater, that presents daily change of the water ratio from reservoir and chlorine concentration at the selected node [17].

Integration of SCADA systems and hydraulic modelling, nowadays called 'online modelling' enables constant modelling of pressure conditions, water flow and chosen water quality parameters in real-time and in any spot of the network [18],[19].

5 Conclusions

Based on briefly presented capabilities of selected IT tools it can be stated that their usability in the management of water distribution systems is sound, especially in case of incidents. Due to modern and integrated GIS, SCADA and online modelling technologies now available real-life and up-to-date data can be analysed. Thanks to these IT tools appropriate decision making can be done fast and relevant authorities/units might receive accurate data and instructions, based on pre-approved Standard Operation Procedures (SOPs). 'Time matters' – right data should be provided in right time to right addressees.

References

1. M. Sobierajski, *Rewolucja informatyczna w wod-kan*, Ochrona Środowiska, no. **3-4/2011**, Wrocław, (2011). pp. 16-18 (in Polish).
2. A. Kuliczkowski, D. Lichosik, *Zagrożenia terrorystyczne związane z systemami zaopatrzenia w wodę*, Instal, **no. 7–8/2015**, (2015), pp.77–81 (in Polish).
3. Y.Y. Haimes, N.C. Matalas, J.H. Lambert, B.A. Jackson and JRF Fellows, *Reducing vulnerability of water supply systems to attack,* Journal of Infrastructure Systems **no. 4(4),** (1998), p.164–167.
4. M. Hosner, *Use of a Model to Help Secure a Water System*, DHI North America Conference, Orlando, USA, June (2002), pp.1-6.
5. R. Murray, R. Janke; W.E. Hart; J.W. Berry; T. Taxon; and J. Uber; *Sensor network design of contamination warning systems: a decision framework*, Journal of American Water Work Association, **100(11),** (2008), pp.97-109.
6. P. Ingeduld, E. Zeman, J.P. Lindgaard, M. Hornser, *Real time analysis for early warning systems based on modelling concept,* Water in Emergency Situations, Springer, USA, (2007), pp. 105-113.
7. [5a] M.T. Brosnan, *Early warning monitoring to detect hazardous events in water supplies,* An ILSI Risk Science Institute Workshop Report, (1999) ISBN: 1-57881-075-2.
8. World Metrological Organization, *Planning of water quality monitoring systems, Technical Report*, Series **no. 1113**, Switzerland, (2013), pp.12-14.
9. EPA-US, *Water Critical Infrastructure and Key Resources Sector-Specific Plan as Input to the National Infrastructure Protection Plan*, USA, (2007), **EPA 817-R-07-001**, pp.25-33.
10. W. Zoleński, *Systemy wczesnego ostrzegania wykorzystujące wiedzę*, Kwartalnik Naukowy, no. **3(11),** Wydawnictwo Politechniki Śląskiej, Gliwice, (2010), pp. 143-161. (in Polish).
11. http://blog.innovyze.com/wp-content/uploads/2012/10/Day-2-1550-Water-Supply-Water-Quality-Modelling-Emily-Stevens.pdf (uploaded: 2012).
12. K.E. Lansey, P.F. Boulos, *Comprehensive Handbook on Water Quality Analysis for Distribution Systems*, MWH Soft, Pasadena, CA, USA, (2005), 448 p.
13. http://www.innovyze.com/products/infowater/protector.aspx

14. A. Bałut, A.Urbaniak, *Model sieci jako narzędzie ochrony system zaopatrzenia w wodę*, Gaz, woda i technika sanitarna, **no. 9**, (2013), pp.359-363 (in Polish).
15. M. Hosner, *Use of a Model to Help Secure a Water System,* DHI North America Conference, Orlando, June, (2002).
16. R.M. Clark, S. Hakim, A. Ostfeld, *Handbook of Water and Wastewater Systems Protection,* Springer, USA, (2011), pp.247-264.
17. *InfoWater User Guide,* Innovyze®Inc, Interlocken Crescent, Suite 200 Broomfield, Colorado 80021 USA, (2009), p.3-100.
18. A. Bałut, A. Urbaniak, *Monitoring system as a tool for risk evaluation in water distribution systems*, in: Proc. of ICCC' 2016, Ivo Petráš - Igor Podlubny - Ján Kačur (eds.) Tatranska Lomnica, Slowakia,(2016), pp. 21-24, ISBN: 978-1-4673-8605-0.
19. P. Ingeduld, G. Turton, *Online analysis of water supply and water distribution systems*, NZWWA Conference, Auckland, New Zealand, (2002).

Sensitivity analysis of water consumption in an office building

Tomas Suchacek[1,*], *Ladislav* Tuhovcak[1] , and *Jan* Rucka[1]

[1]Brno University of Technology, Faculty of Civil Engineering, Institute of Municipal Water Management, Zizkova 17, 60200 Brno, Czech republic, France

Abstract. This article deals with sensitivity analysis of real water consumption in an office building. During a long-term real study, reducing of pressure in its water connection was simulated. A sensitivity analysis of uneven water demand was conducted during working time at various provided pressures and at various time step duration. Correlations between maximal coefficients of water demand variation during working time and provided pressure were suggested. The influence of provided pressure in the water connection on mean coefficients of water demand variation was pointed out, altogether for working hours of all days and separately for days with identical working hours.

1 Introduction

In the Czech Republic, there is currently no legislative regulation defining in a mandatory manner a detailed procedure of calculating water demand related to a public main supplying water to a service area of a town or village type. Traditionally, a combination of empiric experience and partial legislative regulations is employed to determine the volume of water needed for the specific types of buildings or drinking water consumers in units of volume of water per calendar year. Nevertheless, when designing a water supply network, it is necessary to meet the legislative requirements which, in fact, define the hydraulic parameters of the water network. Given the defined design flow-rate it is necessary to ensure the required range of operating pressures in those parts of the water network where consumers are connected. If the water supply is also used for fire-fighting purposes, the required pressure and flow-rate must be ensured for fire water supply. For comprehensive water supply systems, the design or redesign is often based on mathematical modelling and simulation methods [1], where the hydraulic calculations use as inputs also the daily and hourly variation coefficients (demand coefficients). The value of these coefficients is not set in legislation of the Czech Republic, not even with the implementation of legislative norms on the EU level (e.g. ČSN EN 805). It is important to define (design flow) for separate sections or whole networks. In the Czech Republic, the design of water distribution network is calculated with maximal water consumption per hour. The method of calculating maximal water consumption per hour is anchored in legislation, in which the coefficient of maximal variation per hour figures. The value of this coefficient is thus, extremely

* Corresponding author: suchacek.t@fce.vutbr.cz

important for the design of hydraulic capacity of water distribution network. According to legislation, its value can only vary with the number of supplied residents, therefore the focus is set on the maximal coefficients of water demand variation during simulated changes of pressure conditions in the network.

Absence of the coefficients of water demand variation prove to be problematic mainly while designing new networks, when recommended and established values of hourly coefficients of variation are used. But even if the demand coefficients are specified (e.g. in historical and obsolete legislative), the fact that the values of these coefficients might be modulated by various pressure conditions is not taken into consideration. Therefore, they are considered constant, in spite the fact that pressure influences the water consumption. For example, according to [2], water demand is divided for the purposes of simulating water demands changes in relation to pressure changes into "inside the house" and "outside the house". Both these parts of demand have a different coefficient expressing the dependency on pressure while the "outside the house" demand is much more influencable by pressure. Subsequently, the average coefficient is calculated expressing the dependency of water demand on pressure for the entire building. This coefficient is then implemented in the FAVAD equation [3]. The coefficient value for the "inside the house" demand was set for example in [4] for the student campus in Johannesburg at 0.2. According to [5], the dependency of demand on pressure was also proved for pressure toiler flushers in Great Britain and the coefficient value was set at 0.07 and 0.025. For example, for the "outside the house" consumption, the value of the coefficient according to [6, 7] was set to approximately 0.5, with domestic irrigation systems.

Another way of consumption division for the prediction is, for example, the division according to [8] into a part dependent on pressure and a part dependent on volume. Subsequently, depending on the ratio of both parts of the consumption, dependency is determined for the change of consumption with the change of pressure. According to [9] it is possible to describe a pressure dependence using pressure threshold values. These threshold values and the reference pressure are minimal, and it is assumed that when the pressure drops below the minimal threshold value, the consumption is zero, on the other hand, when the reference value is exceeded, the consumption is constant and with increasing pressure.

The pressure conditions in the water distribution network or the pressure is the most important operating factor that affects water consumption, but on the other hand, the water consumption is influenced by other factors of different kinds. For example, according to [10], the consumption is characterized based on the building characterization. Another group is climatological factors, especially the influence of temperature and precipitation as well as partly humidity, for example according to [11, 12].

For these reasons, attention is also paid to coefficients of water demand variation during working hours in response to simulated modulation of pressure conditions.

The aim of this study was to study change in variation of water demand with the variation of pressure in the water connection, to which the building is connected by installing a pressure reducing valve on this water connection and to observe and evaluate water consumption, or its variation, respectively.

2 Methodology

The primary requirement for calculating the influence of pressure conditions on the water consumption, and especially its variation is the modification of the pressure conditions. In this case, the modification has not been applied in the whole water distribution network, but by installing a pressure reducing valve onto the water connection. The valve simulated the regulation of pressure conditions in the whole network. The output value of the pressure reducing valve was randomly selected by random number generator. Restricting conditions on the minimal output pressure were applied, so that no legislative regulations concerning minimal hydrodynamic pressure would be violated, ensuring usual working flow of the respective building.

For execution of measuring campaign, so called measuring cycles were defined. Settings of the pressure reducing valve would remain constant for the duration of a cycle. Duration of a single cycle was defined to approximately 14 days, which usually represents 10 working days in a single cycle.

In order to be able to register potential decreases in hydrodynamic pressure under the actual value of output pressure and other eventual anomalies in the network, a pressure record interval of 15 second was chosen. For recording volume of water per time unit, a water meter with pulse generator, whose value was set to 1 litre so that water categorization into shorter time intervals would be possible. The value of the water meter was recorded every 15 seconds.

As the campaign lasted 1 year, it was necessary to take into account or exclude long-term impacts and trends in water consumption. In the Czech Republic, there has been a downward trend in water consumption since 1989 but in the last 3 years consumption can be considered more or less stable [13]. The effect of long-term trends could thus be excluded. For exclusion of extremely non-standard or anomalous measurements during working days, number of workers in the building during working hours was monitored. From the statistical data set, days during which the number of workers dropped under 10% of the total amount were excluded. Similar days could discredit the results of the study.

Data results were assessed using two methods. Firstly, maximal data set was used in the statistical file by including all days without taking the length of working time into consideration. In order to be able to make use of data from all days, only shared working hours among all days were taken into account. The second method of result assessment was done separately for days with identical working hours and coefficient was determined for the whole duration of the working time. Evaluation and determination of coefficients were done by the following equations:

Flow in the time step i, in day j is determined as average flow in day j multiplied by coefficient in the time step i during day k.

$$Q_{ij} = Q_{MEAN_j} c_{ij} \tag{1}$$

where Q_{ij} [l.hour^{-1}] represents flow, Q_{MEANj} [l.hour^{-1}] is mean flow, c_{ij} [-] is coefficient of water demand variation, i is the label of the time step and j is the label of the day.

Coefficient of variation after modification is determined as follows:

$$c_{ij} = \frac{Q_{ij}}{Q_{MEAN_j}} \tag{2}$$

Mean coefficient for each time step in pressure category k is determined as the sum of all coefficient values in time step i, in the pressure category k.

$$c_{MEAN_{ik}} = \frac{\sum_{j=1}^{m} c_{ijk}}{m_k} \tag{3}$$

where c_{MEANik} [-] is mean coefficient, k is pressure category (as seen below), m_k [-] is the total number of days in category k.

$$c_{MAX_k} = \max\{c_{ijk}\} \tag{4}$$

where c_{MAXk} [-] is the maximal coefficient in k category.

Categorization of measured mean pressures was done with regard to balancing values in the respective categories, as well as with regard to size of intervals of the categories. Too delicate division would mean relatively small, or even no representation in some pressure categories.

$$\varepsilon_k = \sum_{i=1}^{n}(c_{ik} - 1)^2 \tag{5}$$

where ε_k [-] represents square deviation of coefficients from optimal value in each time step.

3 Case study

Case study was conducted on an office building, which is a relatively common type of building in a consumption area up to the size of a regional city. This building has three above-ground floors and the maximal number of workers is 35. The workers are practically evenly divided into each of the above-ground floors, in which identical equipment is provided. For anomaly determining in measured statistical data set, the number of persons in the building was monitored during the working days. The evaluation of correlation was done only for the working days. The duration of the measuring campaign in the case study was one calendar year. According to survey, it was found out that in the sense of [2], all measured water was consumed "inside the house".

There are three working time shifts in the studied building: 7:00 to 17:00 on Mondays and Wednesdays, 7:00 to 15:00 on Tuesdays and Thursdays and 7:00 to 13:00 on Fridays. For the first method of evaluation, shared working hours (from 7:00 – 13:00) from all days were taken into account. For the second method, working hours of each respective day were assessed separately.

The following values of water consumption were characteristic for the studied building.

Table 1. Characteristic values of water consumption during working hours

Water consumption in working hours	Minimal consumption	Mean consumption	Maximal consumption
Day (working hours)	**(litres)**	**(litres)**	**(litres)**
Monday + Wednesday (7-17)	608	892	1307
Tuesday + Thursday (7-15)	481	710	1252
Friday (7-13)	307	623	927
All days (7-13)	333	545	982

For this case study, the intervals of pressure conditions categories and values in individual categories is shown in Figure 1. The value of pressure for one day was determined as arithmetic mean during the working hours of the respected day.

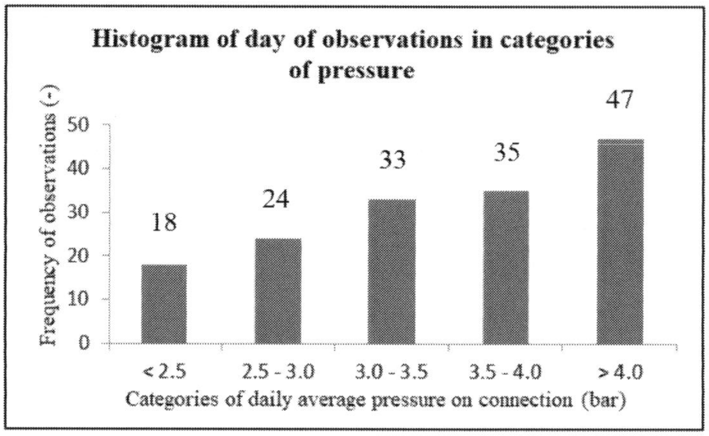

Fig. 1. Histogram of measured values in pressure categories

As a part of this case study, a sensitivity analysis of duration of a time step for maximal and mean coefficient of water demand variation was conducted. Time step durations fluctuate about the most frequent duration of a time, which is 60 minutes, or 1 hour. In this case, correlations for the time step with duration of 10, 20, 30, 40, 60, 90 and 120 minutes were studied.

4 Results

As stated above, in the first study method, shared working hours, from 7:00 – 13:00, of all days were taken into account (with the exception of anomalous values). From the coefficients for individual time steps in each pressure category, mean coefficients for the respective time steps were determined. Using equation (5), square deviation from optimal value was determined. For results presentation, pressure categories 2.5-3.0 bar and 3.5-4.0 bar were selected, and time step durations of 10, 60 and 120 minutes were used. Coefficients of water demand variation were as follows:

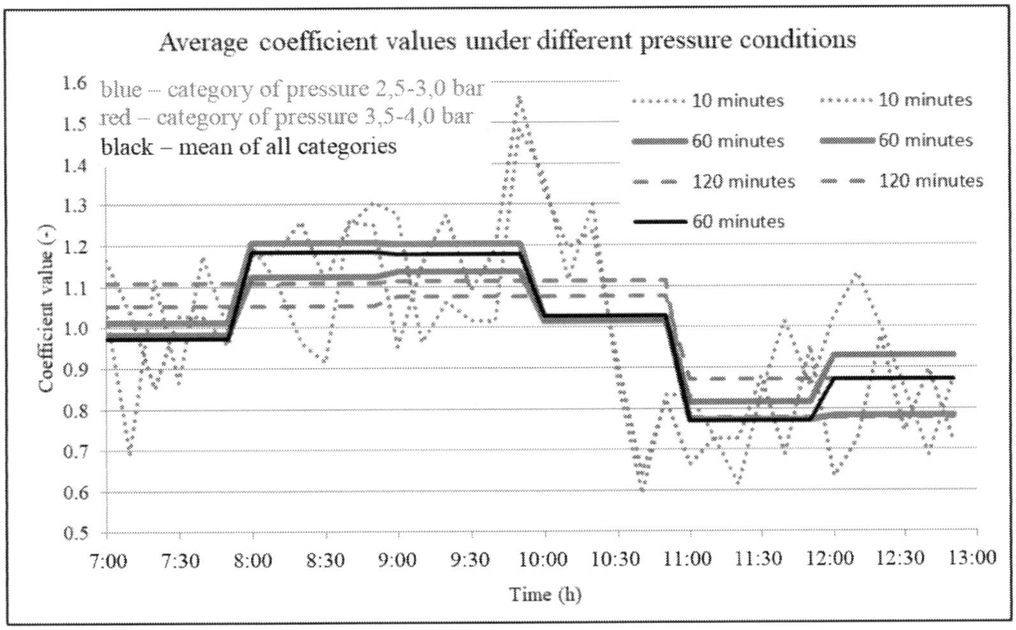

Fig. 2. Mean coefficients in relation in correlation to pressure – all days

Calculated square deviation values for selected categories are presented in the following table:

Table 2. Square deviation values - all days

Time step	10	20	30	40	60	90	120
Category of pressure (bar)	(min)	(min)	(min)	(min)	(min)	(min)	(min)
2.5 – 3.0	1.599	1.086	0.975	0.762	0.444	0.306	0.300
3.5 – 4.0	1.908	1.589	1.487	1.343	1.103	0.845	0.890

In the first study method, maximal coefficients of water demand variation were determined by equation (4). Evaluation of correlation between maximal coefficients and pressure are presented in Figure 3.

For the time step duration of 60 minutes, the maximal coefficient value in the lowest hydrodynamic pressure category was determined as 1.73, while the maximal coefficient value in the highest hydrodynamic pressure category was determined as 2.14, which signifies fluctuation of 26%.

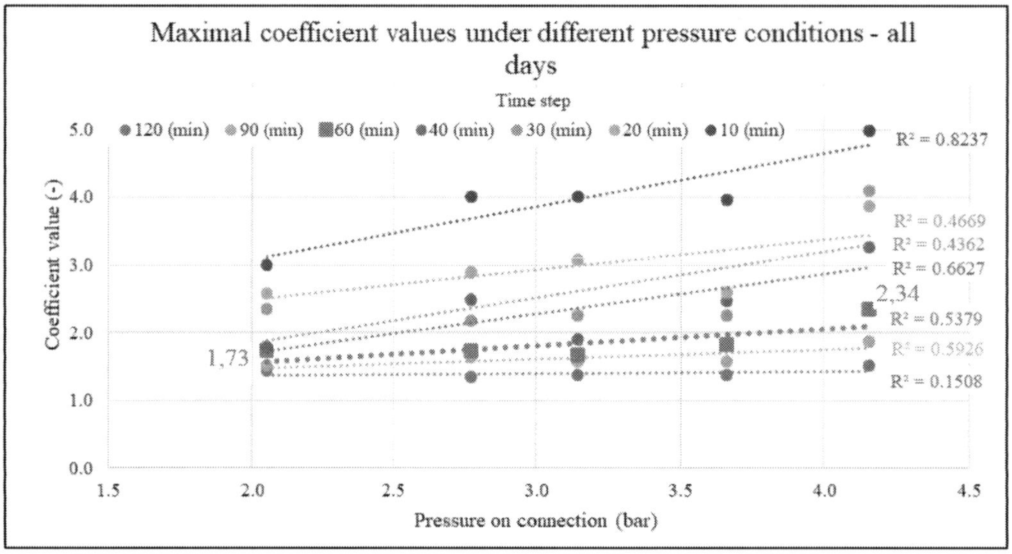

Fig. 3. Maximal coefficients of water demand variation – all days

In the second study method, only days with identical working hours were considered out of the statistical data set of evaluated data, and the whole of working time was taken into account. For results presentation of the second assessment method, pressure categories of 2.5-3.0 bar and 3.5-4.0 bar were selected, and time step durations of 30, 60 and 90 minutes were used. The days selected for results presentation were Tuesday and Thursday (working time from 7:00 to 15:00). Coefficients of water demand variation were as follows:

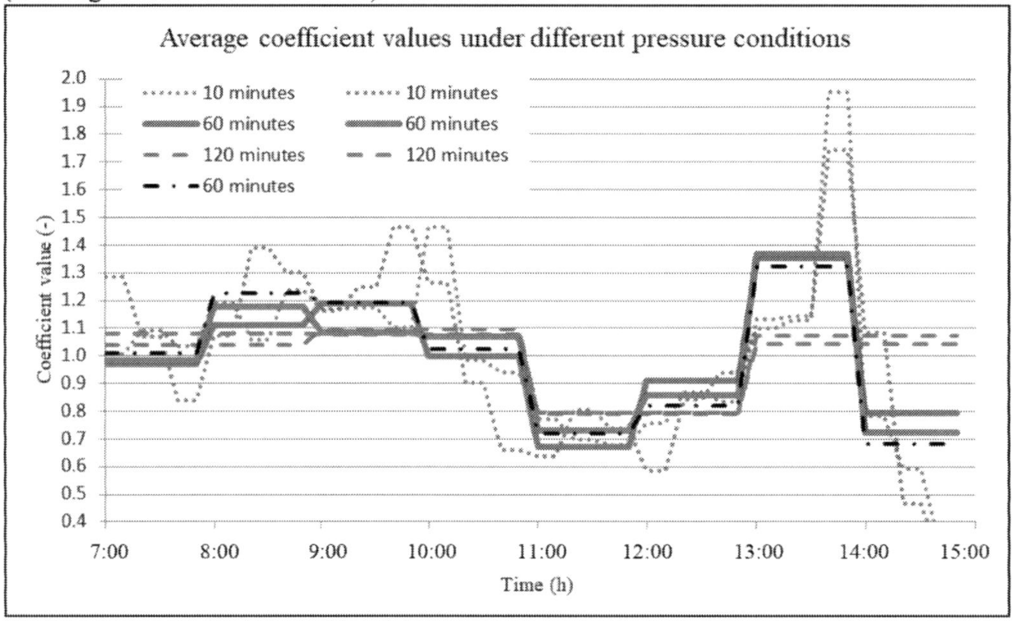

Fig. 4. Mean coefficients in relation to pressure – Tuesday and Thursday

In the second assessment method, square deviations were not determined, since the statistical data set was significantly smaller than in the first method. However, maximal coefficients of water demand variation were determined, as seen in Figure 5.

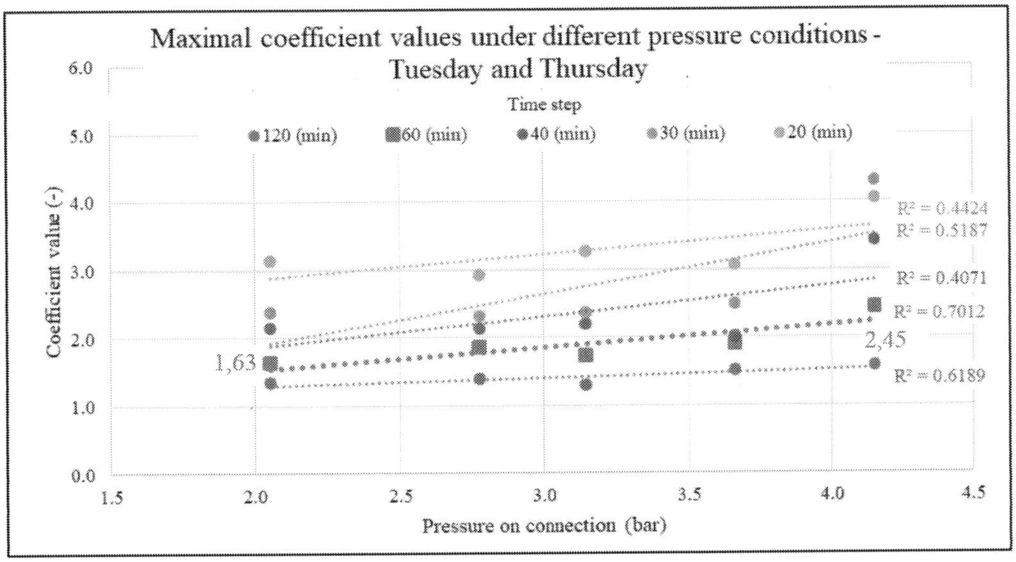

Fig. 5. Maximal coefficients of water demand variation – Tuesday and Thursday

For the time step duration of 60 minutes, the maximal coefficient value in the lowest hydrodynamic pressure category was determined as 1.63, while the maximal coefficient value in the highest hydrodynamic category was determined as 2.45, which signifies fluctuation of 33%.

5 Discussion

Nowadays, there are only few real studies that deal with the issue of uneven water distribution depending on the different pressure conditions. Also, the value of the maximum hourly coefficient is important, according to which distribution networks are proposed, for example, in the Czech Republic. Operational control of pressure conditions could be used as a tool to reduce the uneven distribution from the distribution network, but for the proper management a large number of real studies are needed to be carried out to capture the characteristics of different types of customers. The problem of conducting realistic studies is the need to change the pressure conditions, which can cause customer complaints at the lower and upper limits of the allowed pressure range in the distribution network even if the legal requirements are met.

Although the area of water consumption dependence on pressure ratios and, respectively, unevenness of this consumption is still not fully documented, there are many possibilities to carry out more real measurements in the near future. This option is based on the use of SMART metering, which will allow to collect water consumption information at short time intervals for different types of customers. From these data, it would be possible in the future to be able to compile complex dependences of the unevenness of the distribution for the whole distribution network, not just for individual types of customers.

6 Conclusions

As seen from the aforementioned, in both methods of assessment, a certain correlation between pressure and variation of water demand is observed. This is apparent with both maximal and mean variation coefficient values, as presented in Table 2, where it is clear that fluctuation is higher with higher simulated water pressure in the water distribution network. It is then apparent that when predicting response to water distribution network behaviour, the pressure management of the networks should also take fluctuations of water demand time characteristics into account. Reducing pressure in the network has impact on water losses respectively on their reduction and the consumed volume of water. The influence of pressure in the water connection on the variation of water consumption in the studied building was proved. In this case, it was confirmed that the difference between maximal measured coefficients of water demand variation for lowest and highest categories was 26% at the most common time step duration of 60 minutes for all working days, and 33% for Tuesdays and Thursdays. Findings from this real study for a single water connection should then be implemented, with certain restrictions, for entire water distribution networks, or consumption areas, respectively. It is thus apparent that a detailed analysis of water consumption in time in correlation to pressure conditions, and their following prediction, is a contemporary issue.

Acknowledgements

This research work is funded by the Brno University of Technology in the frame of the research projects titled FAST-J-17-4348 Influence of boundary conditions on optimization of pressure conditions in water supply systems and FAST-S-17-4643 Modelling of selected indicators and processes in the supply of drinking water to users. Thanks to the town of Valasske Klobouky.

References

1. J. Kovář, J. Ručka and O. Andrš, *Proceedings of the 16th International Conference on Mechatronics – Mechatronika 2014*, 697-700 (Brno, 2014)
2. A. Lambert and M. Fantozzi, *Water Loss 2010* (Sao Paolo, 2010)
3. J. May, W. Wat. Env. Eng. (1994)
4. L. Bartlett, *B.Eng Final Year Project Report. Pressure dependent demands in Student Town Phase 3* (Dept. of Civil and urban Engineering, Rand Afrikaans University, Johannesburg, 2004)
5. J. Thorton and A. Lambert, *Leakage 2005* (Halifax, 2005)
6. Bamezai, Lessick, *Water Conservation through System Pressure Optimization in Irvine Ranch Water District* (2003)
7. R. Cullen, *Pressure vs Consumption Relationships in Domestic Irrigation Systems* (2004)
8. V. Kanakoudis and K. Gonelas, Proc. Eng., **89**, 318-325 (2014)
9. J. M. Wagner, U. Shamir and D. H. Marks, J. Wat. Res. Plan. and Man., **114**, 276-294 (1988)
10. W. A. Hussien, F. A. Memon and D. A. Savic, Wat. Res. Man., **30(9)**, 2931-2955 (2016)
11. M. Xenochristou, Z. Kapelan, C. Hutton and J. Hofman, *CCWI 2017 – Computing and Control for Water Industry* (Sheffield, 2017)
12. T. Suchacek, L. Tuhovcak, J. Rucka, *CCWI 2017 – Computing and Control for Water Industry* (Sheffield, 2017)
13. J. Duda, O. Lípa, T. Petr and V. Skácel, *VODOVODY KANALIZACE ČR 2015*, in Czech (Department of Agriculture, Prague, 2016)

https://doi.org/10.1051/e3sconf/20183001003

Drinking Water Supply without Use of a Disinfectant

Marketa Rajnochova[1,*], *Ladislav* Tuhovcak[1] , *Jan* Rucka[1]

[1]Brno University of Technology, Faculty of Civil Engineering, Institute of Municipal Water Management, Zizkova 17, Brno, Czech republic

Abstract. The paper focuses on the issue of drinking water supply without use of any disinfectants. Before the public water supply network operator begins to consider switching to operation without use of chemical disinfection, initial assessment should be made, whether or not the water supply system in question is suitable for this type of operation. The assessment is performed by applying the decision algorithm. The initial assessment is followed by another decision algorithm which serves for managing and controlling the process of switching to drinking water supply without use of a disinfectant. The paper also summarizes previous experience and knowledge of this way operated public water supply systems in the Czech Republic.

1 Introduction to the issue

Objective of the public water supply systems operators is to supply drinking water to the consumers in the required quantity and quality to meet the requirements stipulated by the Decree No. 252/2004 Coll., which sets drinking water limits in the Czech Republic. By applying the chlorine-based chemical disinfection and maintaining sufficient residual chlorine concentration in the distribution network, combined with other measures and actions, drinking water supply in the required microbiological quality should be ensured. However, research and experience from practice bring many new insights into functioning of the chlorine-based chemical disinfection. They show its negative aspects, mainly undesirable affection of taste and smell and occurrence of the so-called disinfection by-products. These disinfection by-products and the negative public approach to chlorine in water have prompted several countries, in particular the Netherlands, Switzerland and Germany, to gradual transition to drinking water supply without use of a chlorine-based disinfectant that remains in water as the residual disinfectant.[1, 2, 3, 4] In these countries, public water supply network operators rely on the adequate number of security barriers protecting the system from microbial contamination from the outside environment. If the number of barriers is sufficient, water can be distributed without any disinfectant. [1, 3, 4]

* Corresponding author: rajnochova.m@fce.vutbr.cz

1.1 Multi-barrier approach

Concept of more barriers (multi-barrier approach) ensures safety in case of failure of any of these barriers. The barriers include protection of the water source, multiple steps in water treatment and proactive management of the distribution system. When it comes to the question whether or not chlorine is irreplaceable in drinking water for human health protection, it can be noted that there are relatively little data confirming that the residual disinfectant in the distribution network would prevent outbreak of infection from drinking water. [2]

The text below describes individual problems that may arise at the water source, in the water treatment plant and in the distribution network that could prevent operation without a disinfectant. This may include, for example, insufficient water source protection, inadequate water treatment due to raw water quality, poor technical condition of the water treatment facility (unprotected air filtration, absence of thresholds under the storage chamber door, etc.), water leakage in combination with vacuum events, etc. The process of initial assessment of the water supply system is shown in the form of a decision algorithm, see Fig. 1. [1, 3, 5] The assessment also includes simulation of chosen indicators of drinking water quality in the water supply network. [6] The initial assessment is followed by the transition strategy extended by defining possible undesired events and their factors which can occur after immediate discontinuing chlorination. Three major undesired events have been defined. For determination of the undesired events and factors, the WaterRisk general methodology [7] has been used.

2 Advantages and disadvantages of drinking water supply without a disinfectant

2.1 Advantages of drinking water supply without a disinfectant

Advantages of operation of the water supply network without use of chemical disinfectants are as follows:

1. By-products of disinfection, such as carcinogenic trihalomethanes (THM), do not occur in drinking water. Halogen phenols, resulting in customer complaints about chlorine flavour and chlorine odour of water, do not occur either. For example, omission of chlorine dosing in Amsterdam led to reduction of THM value below the limit of detection. [1]
2. Masking of microbial contamination of the drinking water cannot appear, as documented during operation with application of a disinfectant. When the disinfectant is used, even very low concentrations of free chlorine cause inactivation of the indicator coliform bacteria and E. coli, as these organisms are very sensitive to chlorine. For instance E. Coli is reduced by 99 percent as early as at the concentration of 0.05 mg $\cdot$ l^{-1}. [8, 9] After addition of the disinfectant, these indicator organisms may be inactivated during the contamination event, whilst the other, more resistant pathogens may still be present. [1]
3. The disinfectant can react with biofilm (material, biofilm material), but cannot react with bacteria. The disinfectant residue can support re-growth, namely by reacting with organic compounds contained in water and producing more readily biodegradable compounds. [1] During common operational conditions the biofilm is stable and the biomass exchange between the biofilm and flow of water is balanced. If the equilibrium is violated due to e.g. rapid changes in the hydraulic

conditions or disinfectant concentration change in the network, the biomass may be revealed from the biofilm which increases the amount of biomass in the drinking water. [10, 11]

2.2 Disadvantages of drinking water supply without a disinfectant

Disadvantages of operation of the water supply network without use of chemical disinfectants are as follows:

1. Presence of adequate concentration of disinfectant may reduce the risk of secondary microbial contamination, which can penetrate into the network through the failures, repairs or water backflow. It may affect re-growth of mainly coliform bacteria and colony counts [1]
2. Presence of disinfectant residue may indicate a system failure. If the network is monitored for concentration of the disinfectant residue, measurement of the reduced concentration of the disinfectant residue may be a signal for a contamination event that has occurred. [1]

Whether or not and to what an extent advantages or disadvantages of the disinfectant are reflected in drinking water depends mainly on the site of dosing and on the amount of the dispensed disinfectant. When considering use of a disinfectant, there is no simple answer to the question of whether the residual chlorine is necessary and beneficial for health and healthy quality of drinking water. [1] The answer naturally also depends on water quality entering the network, on hydraulic and building integrity of the network, and on the ability to apply good hygiene practices for the supplied drinking water. [1,12]

2.3 Experience of water management systems with supplies of drinking water without use of disinfectant

This work encompasses a detailed survey performed in total in nine different water management systems in the Czech Republic which are operated without use of the disinfectant. Most of them utilizes a UV lamp to ensure hygienic water supplies. Experience of transition and subsequent operation without any chemical disinfection is mostly positive. For example, the water management system of the city of Mladá Boleslav, supplying ca 60,000 inhabitants, has faced masking of the system integrity disruption. Discontinued chlorine dosing resulted in appearance of coliform bacteria in drinking water. Occurrence of the coliform bacteria was caused by failed horizontal seal of the storage chamber of one reservoir. [13]. The water supply network in Přelouč, supplying ca 19,000 inhabitants, experienced reduction of nitrites. When the chlorine-based disinfectant was used, increased nitrite concentrations occurred in drinking water. In case of full absence of chlorine, i.e. water was hygienically protected by a UV lamp only, nitrite concentrations decreased. [14] A smaller Vrbka municipal water supply network, supplying 77 inhabitants, faced the problem of increased iron content in the network before chlorination was discontinued. When sodium hypochlorite dosing was shut down and the UV lamp put into operation, iron concentrations were reduced to the standard values ($0.20\ mg \cdot l^{-1}$). [15]

It is possible to summarize that all examined water supply networks meet the microbiological requirements stipulated by the Decree No. 252/2004 Coll. The water treated without any chemical disinfectant is evaluated positively by the consumers, in particular its taste and smell, and there is no concern about disinfection by-products. In some systems the costs for drinking water production without chemical disinfectant are lower, and in some systems they are higher. It depends on technical condition of the system and its size. Obviously the technical condition of the water supply infrastructure affects the drinking water quality significantly. Thus the local conditions needs to be analysed first to estimate the potential water quality deterioration and also to determine the risk of such an event. [11, 16]

3 Initial assessment regarding transition to drinking water supplies without any disinfectant

By discontinuing chemical disinfection, the water supply system loses one of its security barriers. The other barriers must therefore be reliable and functional. It is important to assess whether or not the system is suitable for operation without any disinfectant. The assessment is particularly important for small water supply systems operated by a small operator or a municipality. It has to be considered what defects in water quality ("the undesired events") may occur when switching to supplies of drinking water free from any disinfectant, and what factors affect them. The effort is to cope with this process of transition so that possible disruptions in water quality may be avoided. Before the operator begins to consider water supply without any disinfectant, the water supply system should be subjected to the initial overall analysis of whether it is suitable for this type of operation at all. The water source, the water treatment plant (if included) and the distribution network with its structures should be reviewed. [5]

3.1 Water source

An underground source not affected by surface water is most suitable for operation without a disinfectant. Compared to surface water, the groundwater has an almost constant temperature throughout the year. Although the water temperature can change also during the distribution in the pipes [17, 18]. Groundwater does not need not be treated complexly, because the soil and the rocks above it fulfil the function of a filter and therefore in the majority of cases the water meets the microbiological and biological quality (according to the Decree No. 252/2004 Coll.) and the related overall biological stability. Coliform bacteria or E. coli present in groundwater, or a higher content of organic substances in water, indicate penetration of surface water or pollution that is usually of anthropogenic origin. It is important for non-disinfectant operation that the water source may contribute to the multi-barrier approach to protection of network operation and that its barrier may be effective. Proper barrier functions are achieved if, for example [5]:

1. Functional protective zone of degree I, which prevents unauthorized persons and animals from entering, is achieved by high-quality fencing, by secured access to the source (hood, etc.). Protective zone of degree II, where the requirements of § 30 of the Act No. 254/2001 Coll., on waters, (as amended), are observed, minimization of livestock grazing and farming round the source, etc.
2. Appropriate protection of the water source and its collection, well-made back-filling of the collecting devices, no turbidity events occur.
3. Sufficient depth of the collecting device guaranteeing that water quality may not be affected by surrounding soil loss by splash (surface water effect), confined groundwater level.
4. Good civil part condition, rodents, earthworms, insects, fallen leaves, etc. may not penetrate.

If the source meets these conditions, its functional barrier protecting it from contamination may be assumed. Functionality of the barrier should be confirmed by comparing with results of microbiological analyses performed during the last 2 to 3 years. If affection of the

https://doi.org/10.1051/e3sconf/20183001003

underground source by surface water (wash during rainfalls) is confirmed and the operator is interested in operation of the water supply network without chemical disinfection, replacement of chemical disinfection for UV radiation can be considered.

3.2 Water treatment plant (WTP)

If a water treatment plant (WTP) is the integral part of the water supply system, this plant should also have a functional barrier and contribute by its reliability to the multi-barrier protection of operation. A reliable barrier of the water treatment plant and its storage facilities is created by the following combination [5]:

1. Protection of WTP from penetration of unauthorized persons, game, rodents, insects, etc.,
2. Regular cleaning of all areas (disinfection of floors, removal of webs, etc),
3. Installation of doorsills to prevent dust, pollen, plaster and other impurities from entering the free level of the storage tanks filters, resulting in formation of biofilms and sediments in the storage facilities,
4. Water circulation, avoiding dead corners,
5. Regular cleaning and disinfection of storage tank walls, filters and their refilling,
6. Good technical condition of the storage tank, especially its sealing, preventing leakage of surface water in case of rainfalls,
7. Reliably working vent holes preventing mould formation, protection of these vent holes, especially in the storage and filter zone, from insects, pollen and dust,
8. Avoiding penetration of solar radiation into the filter and storage zones,
9. Filter washing should be determined, based on hydraulic losses on the filter,
10. Filtering should be done under simultaneous measurement of the water turbidity at the outlet,
11. Staff training, good hygienic habits of the staff (clean clothing and footwear to avoid bringing impurities from the outside), state of health of the staff in terms of infectivity, etc.

If the water treatment plant and storage facilities meet these conditions, a functional barrier protecting them against contamination can be assumed. Functionality of the barrier should be confirmed by comparing with the results of microbiological analyses performed during the last 2 to 3 years.

3.3 Distribution network

Distribution network with relevant structures, such as water reservoirs (WR) and pumping stations (PS) is another part of the multi-barrier protection approach. Reliable barrier of the distribution network and its structures is created [5]:

1. Adequate overpressure in the network (higher than 0.05 MPa) without any significant fluctuations,
2. Low failure rate of the network (less than 0.2 failures / km / year),
3. Low water loses,
4. Proper operation of the distribution system by the qualified staff,
5. Adequate pipeline speed ($> 0.5\ m \cdot s^{-1}$) preventing stagnation of water, formation of biofilm and development of microorganisms.

The same requirements, like for WTP with storage facilities, see above, are applied for the network structures such as WR and PS. According to the previous German field study, the internal distribution line of the consumer is the final place, where quality of transported

https://doi.org/10.1051/e3sconf/20183001003

water may be deteriorated [19]. If the aforementioned conditions are met, it can be assumed that the system is operated with the multi-barrier protection approach and is suitable for non-disinfectant operation, which fact should be confirmed by comparing with the results of microbiological analyzes of the last 2 to 3 years. A detailed description is contained in the German strategy for transition to water supplies without any disinfectant. [20]

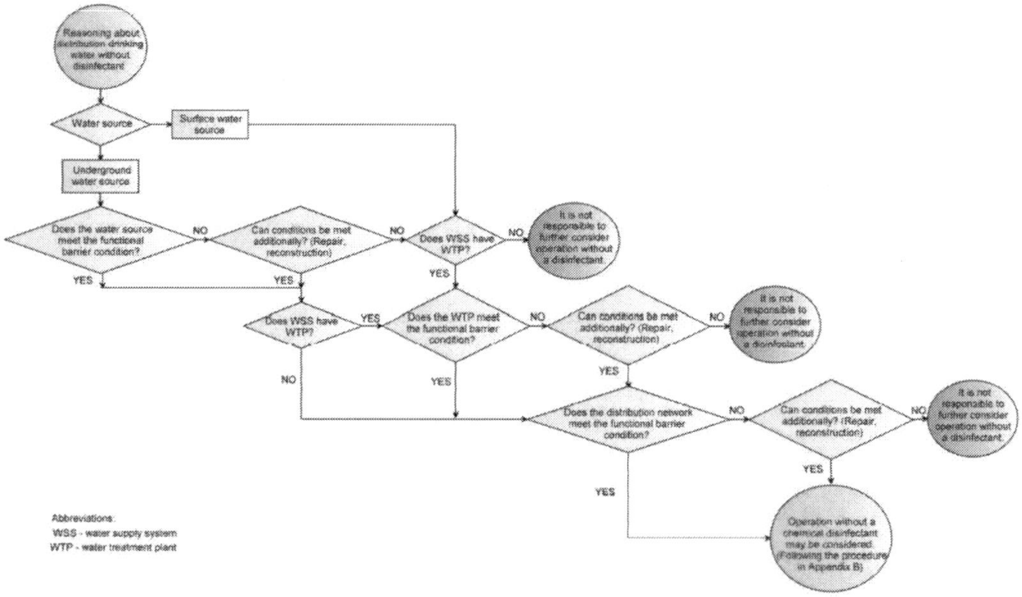

Fig. 1. Decision algorithm of first assessment for transition to supply of drinking water without disinfectant. [5a]

4 Process of transition to supply of drinking water without disinfectant

The decision algorithm of the first assessment is linked with the algorithm controlling the process of transition to the operation without use of any disinfectant, and the latter is extended by undesired events and factors that can occur immediately after discontinued chlorination. Under the undesired event (UE) [7] we shall understand the condition when the system (structure, system element, product) loses its desired feature or ability to perform the requested function. The undesired event is accompanied by occurrence of undesired consequences and is always defined for a particular element of the system. The undesired event is, in principle, a defect, i.e. water quality defect in this case. The general WaterRisk methodology and methodology of Water Safety Plans elaboration were used to describe (determine) UE and factors. [7] Three main undesired events that may occur have been defined for the process of transition to supplies of drinking water without use of any disinfectants, namely as follows:

1. UE 1 - Increased count of colonies at 22°C and 36°C,
2. UE 2 - Occurrence of coliform bacteria,
3. UE 3 - Occurrence of E. Coli bacteria.

4.1 UE 1-Increased count of colonies at 22°C and 36°C

The count of colonies is understood a group of the so-called heterotrophic (organotrophic) bacteria, needing a carbon source from organic substances for their growth. It is an indicator of microbiological recovery of drinking water. This indicator does not provide any direct evidence of pathogenic germs presence and its increased value in water is not connected directly with threats to human health. [21]

The factors that can affect the increased count of colonies are as follows [5]:

1. F 1.1 – Formation of biofilm
2. F 1.2 – Biological stability of water
3. F 1.3 – Water hold-up time in the network
4. F 1.4 – Presence of sediments at the bottom of the pipeline
5. F 1.5 – Concentration of free chlorine before discontinued dosing
6. F 1.6 – Water temperature

4.2 NS 2 – Occurrence of coliform bacteria

Total coliform bacteria are the bacteria commonly present in an environment, such as soil and vegetation. They are generally harmless. If presence of only the total coliform bacteria is detected, the natural environment is most likely their source. Faecal contamination is improbable in this case. However, if natural pollution could penetrate into the system, pathogens could appear here as well. It is therefore important to find and investigate the source of contamination.

The factors that may affect occurrence of coliform bacteria are as follows [5]:

1. F 2.1 – Impairment of physical and hydraulic integrity of the water supply system
2. F 2.1 – Water temperature

4.3 NS 3 – Occurrence of E. Coli

The E. coli indicator bacteria is a subset of coliform bacteria. In most cases it is completely harmless. However, there are also pathogenic strains (e.g. Escherichia coli O157: H7) [22] Presence of E. coli in drinking water is an indicator of fresh fecal contamination. E. coli is very sensitive to chlorine, even a very low concentration of residual chlorine can inactivate it. [8] The factor influencing occurrence of E. coli is as follows:

1. F 3.1 - Impairment of physical and hydraulic integrity of the water supply system.

5 Control algorithm of the process of transition to operation without use of a chemical disinfectant

The algorithm, controlling transition to supply without use of a chemical disinfectant, consists in assessment of water analyzes for the last 2 to 3 years. If microbiological abnormalities did not occur in the analyses in the past, a preliminary survey programme lasting for at least 2 months will be commenced. If microbiological abnormalities do not occur in the preliminary survey programme, dose of the disinfectant will be decreased gradually by ca 0.1 to 0.15 mg · l^{-1}. The surveillance programme then follows for 2 months as a minimum. Occurrence of water quality defects is monitored at this stage, see the undesired events shown above. If undesired events do not occur, dosage of the disinfectant can be discontinued completely. And again, it is necessary to perform the surveillance programme for at least 2 months. If neither any microbiological abnormalities nor undesired events occur at this stage, it can be stated that disinfectant dosing can be discontinued permanently. [5, 20]

The decision algorithm, controlling transition to supplies of drinking water without use of disinfectants, is based on the German strategy, stipulating 3 possible methods of system operation. The first method is drinking water supply in absolute absence of disinfection, both chemical and physical. This is possible only if a well-protected underground source is available. The assumption is that water enters into the distribution system complying with the microbiological and biological requirements stipulated by the Decree 252/2004 Coll., and the network does not provide suitable conditions for re-growth. The second option is to eliminate additional chlorination. This can only be realized if water after disinfection meets the microbiological requirements, and the network has such conditions regarding water flow rate, amount of sediments, concentration of biodegradable carbon, etc., that there is no need to maintain the disinfectant residue. The last option is to replace chemical disinfection for the UV disinfection. [20]. This solution is most frequently chosen in the Czech Republic.

6 CONCLUSION

The paper is focused on the possibility to operate drinking water distribution without application of any chemical disinfectant. Examples of the systems using chemical disinfectants neither for production nor for distribution of drinking water have been described and documented.

The fact that foreign chemicals of disinfectants, which - under some circumstances - may create toxic by-products of disinfection, are not brought into drinking water is the undisputed advantage of operation without a disinfectant. Reaction of chlorine can result in conversion of high-molecular natural organic compounds into simpler substances serving as a substrate for bacteria. No serious microbiological water quality defect has occurred in absence of a disinfectant in any of the systems that have discontinued chemical disinfection. Only a short-term increase in colony count due to biofilm re-formation was observed. Reduced iron concentration in the network, reduced nitrite content and positive evaluation of water smell and taste by the consumers is the positive experience of the water management system of this operation.

The paper also describes the process controlling transition to drinking water supply without a disinfectant, based on the German strategy. However, before the operator begins to consider switching to operation without use of chemical disinfection, initial assessment should be made, whether or not the water supply system in question is suitable for this type of operation. Individual problems that may arise at the water source, in the water treatment plant and the distribution network, which could prevent operation without use of a disinfectant, have been mentioned here. This may include, for example, insufficient water source protection, inadequate water treatment due to raw water quality, poor technical condition of the water treatment facility (unprotected air filtration, absence of thresholds under the storage chamber door, etc.), non-observance of hygiene standards by the operators, water leakage in combination with vacuum events, etc. The initial assessment procedure was displayed through the decision algorithm. The initial assessment is linked with the above-mentioned German strategy complemented by definition of possible undesired events and their factors which may occur immediately after discontinued chlorination. Three major undesired events have been defined: UE 1 - Increased count of colonies at 22°C and 36°C, UE 2 - Occurrence of coliform bacteria, and UE 3 - Occurrence of E. coli. Occurrence of coliform bacteria and E. Coli has been broken down into two undesired events, because the coliform bacteria are restricted by the Decree by the limit value only, i.e. be exceeding it, water loses the respective quality, but water supply is not interrupted, whereas if the limit value is exceeded in case of E. Coli, which is subject to the highest limit value, water supply must be interrupted unconditionally and an alternative drinking water supply source must be ensured.

Acknowledgements

This research work is an output of the research project "Modelling of selected indicators and processes in drinking water supplies", reg. no.: FAST-S-17-4643 and also a project "Influence of hydraulic conditions in the water supply network on selected indicators of drinking water quality", reg. no.: FAST-J-17-4475. Both of them was supported through the internal grant agency of the Brno University of Technology.

References

1. Kooij, D. *Microbial Growth on Drinking Water Supplies*, London: IWA Publishing, (2014)
2. Rosario-Ortiz, F., Rose, J., Speight, V., Gunten, U. and Schnoor, J. How do you like your tap water?. Science, 351(6276), pp. 912-914. (2016)
3. Smeets, P. W. M. H., G. J. Medema a J. C. van Dijk. The Dutch secret: how to provide safe drinking water without chlorine in the Netherlands. In: Drinking water engineering and science [online]. Copernicus publications. Avalilable at: http://www.drink-water-eng-sci.net/2/1/2009/dwes-2-1-2009.pdf, (2009)
4. Rosario - Ortiz, Fernando a Vanessa Speight. Can drinking water be delivered without disinfectants like chlorine and still be safe? In:The conversation [online]. UK Dostupné z: http://theconversation.com/can-drinking-water-be-delivered-without-disinfectants-like-chlorine-and-still-be-safe-55476, (2016)
5. Rajnochová M. Drinking water supply without use of chemical disinfection. Diploma thesis. Brno University of Technology, Faculty of Civil Engineering, Institute of Minucipal Water Management. (in Czech) (2017)
6. Kovář, J., Ručka, J., Andrš, O. Simulation Modelling of Water Supply Network as Mechatronic System. In Proceedings of the 16th International Conference on Mechatronics – Mechatronika 2014. 1. Brno: VUT v Brně, 2014. s. 697-700. (2014)
7. Tuhovčák, L., Ručka, J., Kožíšek, F., Pumann, P., Hlaváč J. and Svoboda M., et al. Risk analysis of public water supply systems, Brno: Akademické nakladatelství CERM, s.r.o. Brno. ISBN 978-80-7204-676-8. (In Czech) (2010)
8. Payment, P., Poor efficacy of residual chlorine disinfectant in drinking water to inactivate waterborne pathogens in distribution systems. Canadian journal of microbiology. 45(8), 7. (1999)
9. Vostrčil, J. Tuhovčák L. Water supply systems: WAQUAL. Methodology development for water and transportation systems managing. s. 35. (In Czech) (1999)
10. Albert, M., W. Hijnen, J. Vossen a M. Blokker. Modelling bacterial biomass in a non-chlorinated drinking water distribution system. WDSA XVIII, 8 [cit. 2017-11-25]. [online]: http://www.sciencedirect.com/science/article/pii/S187770581731367X (2016)
11. Fish, K., A.M. Osborn a J.B. Boxall. Biofilm structures (EPS and bacterial communities) in drinking water distribution systems are conditioned by hydraulics and influence discolouration. ScienceDirect [online], In: . s. 10 [cit. 2017-11-26]. [online]: https://ac.els-cdn.com/S0048969717306988/1-s2.0-S0048969717306988-

main.pdf?_tid=da63c9de-d2a8-11e7-a19a-00000aacb35e&acdnat=1511701011_52461c8347a49b38ef7836f7ead087dd (2017)

12. Potho, M. a M. Blokker. Dynamic hydraulic models to study sedimentation in drinking water networks in detail. Drinking Water Engineering and Science, [online]: https://www.drink-water-eng-sci.net/5/87/2012/dwes-5-87-2012.pdf (2012)
13. Žitný, T. Practical experience of the water utility with chemical disinfection termination. SOVAK [online]: http://www.szu.cz/uploads/documents/chzp/voda/pdf/bez_dezinfekce/Zitny_konference_SOVAK2014.pdf (in Czech) (2014)
14. Hampl, S. Ten years of experience with the operation of the Přelouč waterworks with UV protection. In: Pitná voda 2010,. Č. Budějovice: W&ET Team, 2010, s. 6. ISBN 978-80-254-6854-8 (in Czech) (2010)
15. Růžička, J. Change in the disinfection system in the Vrbka water supply system. In: Pitná voda 2008 [online]. České Budějovice: W&ET Team, 2008, s. 2 [cit. 2016-10-29]. ISBN 978-80-254-2034-8. [online]: http://www.szu.cz/uploads/documents/chzp/voda/pdf/bez_dezinfekce/Ruzicka_Pitna_voda_Tabor_2008.pdf (in Czech) (2008)
16. Douterelo, I., S. Husband a J.B. Boxall. The bacteriological composition of biomass recovered by flushing an operational drinking water distribution system. Water Research. (54), 15 [cit. 2017-11-25]. [online]: http://www.sciencedirect.com/science/article/pii/S004313541400089X (2014)
17. Blokker, M. a I. Pieterse-Quirijns. Modeling Temperature in the Drinking Water Distribution System. Research Gate, [online]:https://www.researchgate.net/publication/262728472_Modeling_Temperature_in_the_Drinking_Water_Distribution_System (2013)
18. Pieterse-Quirijns, I. a M. Blokker. Early warning systems to predict temperature in the drinking water distribution network. ScienceDirect [online].8 [cit. 2017-11-26]. [online]: https://ac.els-cdn.com/S187770581400006X/1-s2.0-S187770581400006X-main.pdf?_tid=177485b6-d281-11e7-8326-00000aacb361&acdnat=1511683934_20dd3cf7b06c1f49af06bde4327f3972 (2013)
19. Flemming, H.-C., B. Bendinger, M. Exner, J. Gebel, T. Kistemann, G. Schaule, U. Szewzyk a J. Wingender. The last meters before the tap: where drinking water quality is at risk. In: Researchgate [online]: https://www.researchgate.net/publication/288023712_The_last_meters_before_the_tap_where_drinking_water_quality_is_at_risk, (2014)
20. Korth A., Wrickle B. Entwicklung und Überprüfung einer Strategie zur Ablösung der chemischen Desinfektion bei Sicherung mikrobiologisch stabiler Verhältnisse im Leitungsnetz. Abschlussbericht zum DVGW-Forschungsvorhaben W 18/00. (in German) (2004)
21. Kožíšek, F., Pumann, P., Šašek J., Methodical recommendations for assessment of the microbiological indicators of drinking water. National Reference Center for the Drinking Water of the National Institute of Public Health. In: National Institute of Public Health, Praha [online]: http://www.szu.cz/uploads/documents/chzp/voda/pdf/Metodicke_doporuceni_NRC_pocty_kolonii.pdf (in Czech) (2014)
22. Kožíšek, F., Kos, J., Pumann, P. Hygienic minimum for water workers. Praha. (in Czech) (2006)

Changes of physicochemical and microbiological parameters of infiltration water at Debina intake in Poznan, unique conditions – a flood

Sylwia Kołaska[1,*], *Joanna* Jeż – Walkowiak[1], and *Zbysław* Dymaczewski[1]

[1]Poznan University of Technology, Faculty of Civil and Environmental Engineering, Berdychowo 4, 60 – 965 Poznan, Poland

Abstract. The paper presents characteristics of Debina infiltration intake which provides water for Poznan and neighbouring communes. The evaluation of effectiveness of infiltration process has been done based on the quality parameters of river water and infiltration water. The analysed water quality parameters are as follows: temperature, iron, manganese, DOC_{KMnO4}, TOC, turbidity, colour, dissolved oxygen, free carbon dioxide, conductivity, total hardness, carbonate hardness, pH, heavy metals, detergents and microorganisms. The paper also includes an assessment of the impact of flood conditions on the quality of infiltration water and operation of infiltration intake. In this part of the paper the following parameters were taken into account: iron, manganese, DOC_{KMnO4}, TOC, turbidity, colour, dissolved oxygen, free carbon dioxide, conductivity, total hardness, the total number of microorganisms in 36°C (mesophilic), the total number of microorganisms in 22°C (psychrophilic), coli bacteria, Clostridium perfringens, Escherichia coli, Enterococci. Analysis of the effects of flood on infiltration process leads to the following conclusions: the deterioration of infiltration water quality was due to the deterioration of river water quality, substantial shortening of groundwater passage and partial disappearance of the aeration zone. The observed deterioration of infiltration water quality did not affect the treated water quality, produced at water treatment plant.

1 Introduction

Infiltration intakes have more than hundred years of tradition [1]. Their implementation were strictly related to decrease of groundwater resources [2]. Among Polish cities which draw water from infiltration intake are Poznan, Warsaw, Cracow, Wroclaw and Legnica – fed directly from river bed (riverbank infiltration) and infiltration ponds (artificial infiltration) [3].

Water, after the infiltration process, is transformed into water with the parameters characteristic for groundwater. Its temperature is stable throughout the year, quality

* Corresponding author: sylwia.kolaska@put.poznan.pl

parameters are improved and their fluctuations which cause difficulties in operating of water treatment plants are smaller (buffering properties of the soil). Technology of a treatment plant employed for the infiltration water is usually the same as for groundwater. In most cases technology is focused on the removal of excessive amounts of iron and manganese from the water and preventive disinfection, however the infiltration process can be less effective and there are also water treatment plants which use technology typical for surface water treatment, for example coagulation [2]. Considering that most of infiltration intakes were created as a result of the development of groundwater intakes, these adjustments in the treatment technology has eliminated a need to build the new water treatment plants, technologically adapted for surface water treatment.

The research presented in this paper is focused on artificial infiltration. In this process water from infiltration ponds goes through a four stage purification process. The first stage occurs in the infiltration pond, where the water quality indicators are averaged. Partial oxidation of ammonia and some organic substances extends in the infiltration pond. Phosphates and nitrates are eliminated at a certain range. Developing algae partly promote decarbonisation of water. In the second stage of water purification, water filters through the natural mineral – biological membrane formed at the base of the pond. Flowing through the membrane and ground layer at the bottom of the pond, water is purified by mechanical filtration, ion exchange, adsorption and biological processes. In this phase the most of heterogeneous impurities are stopped. The next two steps of purification take place in aeration and saturation zones. In the aeration zone as a result of biochemical process, ion exchange or adsorption, contaminants and micro – contaminants are retained. Bacteria and soil fungi, using the oxygen contained in the water, mineralize adsorbed organic pollutants. As water flows to the deeper layers, it is in contact with the increasingly impoverished bacterial environment – the organic substrate is used and the oxygen concentration drops. Products of organic compounds are oxidized by autotrophic bacteria. Carbon dioxide, which is one of the mineralization products, reacts with carbonates in the ground or remains in the water as free/aggressive carbon dioxide. [4]

Presented analyses refer to changes in water quality occurring in Debina intake in Poznan.

2 The brief description of Debina intake

Debina infiltration intake covers 247 – 251 km of the Warta River. Originally formed in1902, initially it was only groundwater intake. In 1911 a series of 50 wells located along the Warta River were initiated, what started the process of drawing water from riverbank. Due to the increasing demand of water the infiltration ponds were started to be built in 1924. The construction of the ponds were completed in 1964. At that time the capacity in wet years was over 100 000 m^3/d and in dry years was about 80 000 m^3/d. Due to the construction of A2 motorway passing through the infiltration intake its capacity has decreased by approximately 20%. The balance of water extraction is: about 70% – water from infiltration ponds, about 20% – water from riverbank, about 10% – groundwater. [4]

3 Materials and methods

The paper presents data from 5.5 years of infiltration intake operation. During the period of investigation two floods occurred. To characterise the impact of floods on infiltration water quality the analysis include two years before and two years after the high river level.

Samples of the Warta River water were taken once a month. As a routine every value of each water quality parameter was determined twice.

Infiltration water was analysed once a week in the period of 5.5 years. Each parameter was analysed two times. Average values of two results were taken into account for the purpose of the further calculation, fatal errors were rejected. Maximum, average and minimum values of water quality parameters were determined.

The values of water quality parameters, physical, chemical and microbiological, were determined according to standard methods. The analyses were performed by certificated laboratory of waterworks.

4 The quality of river and infiltration water

To determine the changes occurring in water quality during the infiltration process, a comparative analysis of river water quality and infiltration water has been done. There are selected parameters analysed.

4.1 Temperature

As a result of the analysis of temperature data, some characteristic properties of the infiltration process have been observed. This is a reduced amplitude of the infiltration water temperature relative to the amplitude of the water in the Warta River. In the analysed data from 2008 to 2013 the maximum amplitude (measured as the difference of the highest and the lowest temperature) of the Warta River water is 23.5 K, while the amplitude of infiltration water is only 13.5 K [5]. As was expected the temperature value of infiltration water is stabilised in comparison with the river water. A study done at other infiltration intakes proved the similar trend. Kennewick by the Columbia River has similar trend of change. The maximum amplitude of the Columbia River is 17.8 K, the amplitude of infiltration water is 3.0 K [6]. Stabilization of water temperature is a beneficial modification that makes process of water treatment easier to lead. Both too high and too low temperature are not good for efficiency of water treatment and cause operational difficulties.

4.2 Physicochemical parameters

The following were taken into account: iron and manganese concentration, DOC_{KMnO4}, TOC, turbidity, colour, dissolved oxygen, free CO_2, conductivity and total hardness collected in the table 1.

Table 1. Quality parameters of the Warta River and infiltration water

parameter	unit	Warta River water			infiltration water		
		max	min	average	max	min	average
iron	mg/dm^3	0.896	0.206	0.508	2.100	0.032	0.434
manganese	mg/dm^3	0.420	0.001	0.110	0.429	0.015	0.298
DOC_{KMnO4}	mgO_2/dm^3	15.10	3.70	6.79	5.50	1.10	3.05
TOC	mg/dm^3	36.90	4.80	8.71	7.11	2.30	4.81
turbidity	NTU	21.0	2.6	6.8	10.8	0.5	1.9
colour	$mgPt/dm^3$	70.0	10.0	23.1	65.0	7.5	11.7
DO	mg/dm^3	14.7	3.4	8.0	11.0	0.9	3.6
free CO_2	mg/dm^3	17.6	0.0	4.6	53.5	3.0	21.3
conductivity	µS/cm	742.0	506.0	604.7	713.0	578.0	644.5
total hardness	$mgCaCO_3/dm^3$	322.0	180.0	245.0	311.0	226.0	265.0

Analysing the data summarised in the table 1 it is surprising that the average concentration of iron in infiltration water is less than in the Warta River water. Despite the fact that maximum concentration of iron in infiltration water exceeds more than two times the maximum concentration of iron in the Warta River water, the average result indicates the decreasing tendency in the iron concentration after the infiltration process. This may occur due to quality of underground water mixed with infiltration water and the geological conditions in Debina intake. As was expected – the average concentration of manganese increases 2.7 times after the infiltration process. This indicates a supply of groundwater reach in manganese and the location of infiltration intake on the soil which is reach in that compound [7]. The DOC_{KMnO4} of infiltration water decreases more than twice comparing to the DOC_{KMnO4} of the Warta River water and finally achieves quality requirements regarding the water intended for consumption, defined at 5 mgO_2/dm^3 (except the time when it reaches maximum values). The TOC value drops almost twice after the infiltration process and also achieves requirements for water intended for human consumption (5 mg/dm^3) for the average value. The turbidity of the water after the infiltration process drops more than 3.5 times. Finally, the average value exceeds the permissible turbidity of drinking water almost twice (max 1 NTU). The colour is reduced twice after the infiltration process. Finally the colour in its average value falls within the standards, but its maximum value exceeds the level of 15 $mgPt/dm^3$ (currently the acceptable colour value is defined as acceptable for a consumer). Infiltration water, through the flow in the reducing ground environment, is deoxidized and enriched by free CO_2. Significant increase in carbon dioxide in infiltration water is a sign of the intensity of chemical and biological processes in infiltration water. The Warta River water is characterized by a greater amplitude of conductivity variations, but average conductivity value is higher in infiltration water. Water conductivity about 600 µS/cm is appropriate for drinking water which should range from 300 to 1200 µS/cm. The average value of total hardness increases by 20 $mgCaCO_3/dm^3$ after the infiltration process. The appropriate level of total hardness for drinking water is between 60 – 500 $mgCaCO_3/dm^3$, infiltration water is in this range. [5]

4.3 Heavy metals in the Warta River water and infiltration water

Heavy metals concentration such as lead, copper, chromium and nickel were analysed. The data are presented in fig. 1.

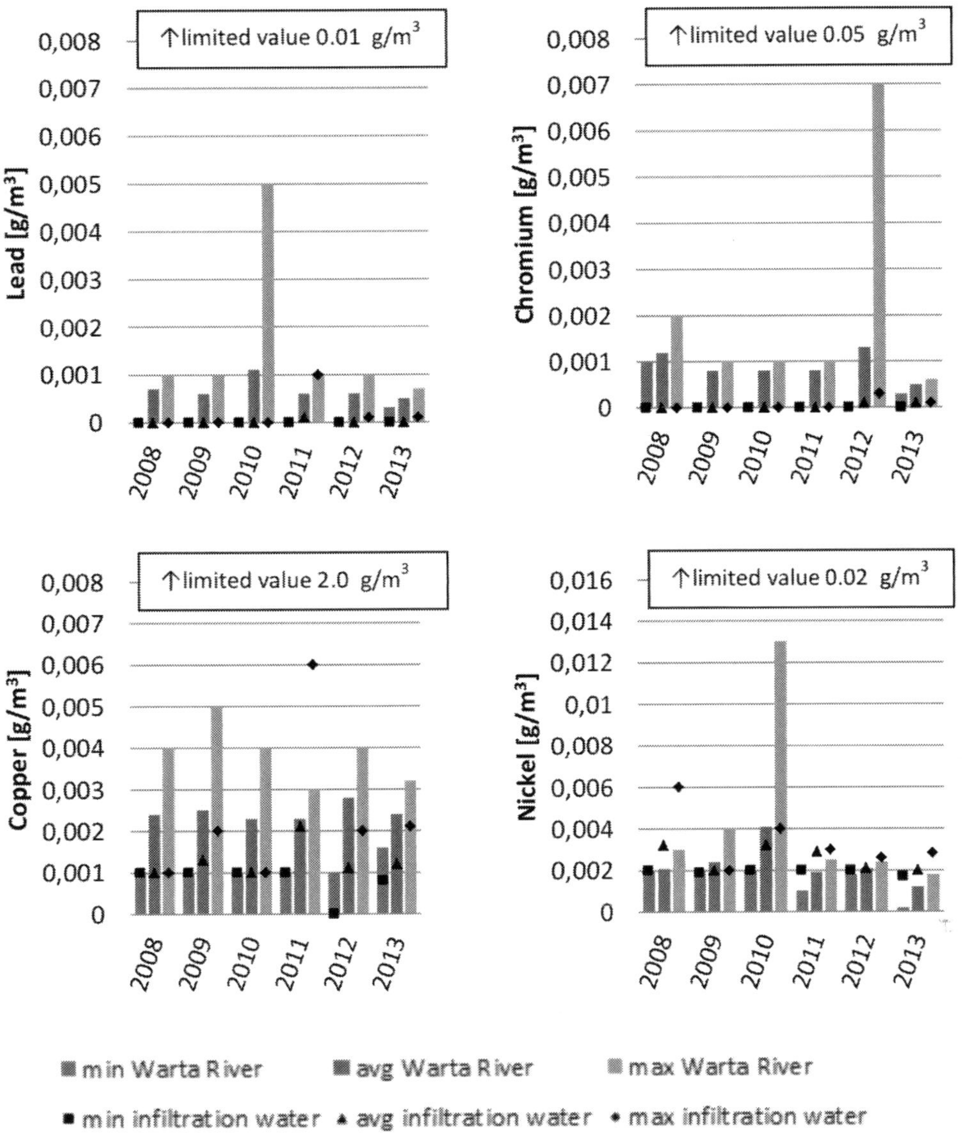

Figure 1. Heavy metals concentration parameters in the Warta River water and infiltration water

The samples were taken usually once a month. In addition to the average value, the minimum and maximum values were also determined. The limited values for drinking water are also presented on the graphs for selected heavy metals. The efficiency of infiltration process was calculated based on the average values, according to the formula 1:

$$E_i = \frac{(c_{in})_i-(c_{out})_i}{(c_{in})_i}\cdot 100 \qquad (1)$$

c_{in}, c_{out} – concentrations of i – heavy metal in the river and after infiltration process, g/m^3.

In the period of research the concentration of heavy metals were low in river and infiltration water, below the limited values. In 2008 – 2013 the lead average concentration in the Warta River had been consistently low at around 0.0006 g/m^3. Only in 2010 that concentration had exceeded the average value of 0.001 g/m^3, what is still almost ten times lower than allowed by the standards regarding drinking water. The efficiency of the infiltration process ranged from 82 – 100%, thus giving a satisfying result showing the ability of infiltration process to reduce the lead concentration.

In 2008 chromium average concentration in the Warta River was 0.0012 g/m^3. In 2009 – 2011 the chromium concentration did not exceed 0.0008 g/m^3. In 2012 it had temporarily increased to 0.0013 g/m^3 and a year later dropped to 0.0005 g/m^3. The effectiveness of the process during the first four years was 100%. In the next two years, it slightly decreased to 91% and 83%, but did not exceed the limit value for drinking water which is 0.05 g/m^3.

In 2008 – 2013 the average concentration of copper in the Warta River water fluctuated around 0.0025 g/m^3, whereas in infiltration water about 0.001 g/m^3, resulting in the efficiency of the process from 42% to 60% excluding 2011 year, where the efficiency value was only 6.2%. The permissible concentration of copper in drinking water is 2 g/m^3.

The average concentration of nickel in the Warta River water in 2008 – 2013 was in the range of 0.0012 g/m^3 to 0.0041 g/m^3. In 2009 and 2010 years there had been the elimination of the metal at about 20%, while in 2008 and 2011 – 2013 an increase in the concentration of nickel in infiltration water occurred. The permissible concentration of nickel in drinking water is 0.02 g/m^3 and even under the influence of the increase in nickel concentration after infiltration process, the values are still ten times less than permissible.

The highest concentration of lead, chromium and nickel occur in individual cases and slightly increase the average value for the year.

During analysis of metal concentration in infiltration water both tendencies – for decrease and increase of metal concentrations in infiltration water – can be noticed. The increase in metal concentration is probably caused by the change of steady water parameters, such as pH, which releases prematurely adsorbed particles from the ground or to exceptional conditions in the Warta River [8]. Despite the occasional increase in concentration of some heavy metals, in any of the analysed cases the concentration of metals in infiltration water did not exceed the limit value specified in the Regulation of the Minister of Health in Poland.

4.4 Detergents in the Warta River water and infiltration water

In table 2 the concentration of detergents is presented.

Table 2. The detergents concentration parameters in Warta River and infiltration water

year	detergents in Warta River water [g/m^3]			detergents in infiltration water [g/m^3]			infiltration efficiency [%]
	max	min	average	max	min	average	
2008	0.280	0.140	0.197	0.100	0.060	0.077	61.1
2009	0.350	0.110	0.195	0.100	0.070	0.080	59.0
2010	0.360	0.070	0.204	0.160	0.070	0.123	40.1
2011	0.460	0.140	0.360	0.400	0.050	0.313	13.2
2012	0.650	0.210	0.436	0.420	0.420	0.420	3.7
2013	0.420	0.160	0.320	0.230	0.050	0.140	56.3

Since 2011 there had been a noticeable increase in the concentration of detergents in the Warta River water. Concentration had increased about twice in comparison to the three previous years (2008 – 2010). Starting with average value of about 0.2 g/m^3 it finally oscillated around 0.4 g/m^3 and more. This is probably due to the fact that population is using more and more surfactants, which are contaminating ground water. The effectiveness of the process in the first two years after the increase in concentration of detergents in the Warta River significantly decreased – from values close to 60% in 2008 and 2009 to 3.7% in 2012. However, a year later the efficiency again increased to 56.3%, as well as surfactants concentration in the Warta River water slightly decreased.

4.5 Chemical stability of the Warta River water and infiltration water

Parameters related to chemical stability of water are presented in table 3. The infiltration water is characterised by higher value of carbonate hardness (alkalinity) than the Warta River water. The pH value of infiltration water is lower than the Warta River water.

To evaluate the chemical stability of the Warta River water and infiltration water the **LSI** – Langelier Saturation Index was calculated and presented in table 3 as well. The value of LSI was calculated according to the formula 2:

$$LSI = pH - pH_{stability} \tag{2}$$

where:
$pH_{stability}$ is a saturation pH.

According to presented data the river water is supersaturated with respect to calcium carbonate ($CaCO_3$) and scale forming may occur (LSI>0). After the infiltration the water reaches the calcium-carbonate equilibrium (LSI=0). [9]

Table 3. Chemical stability parameters of the Warta River and infiltration water

parameter	unit	Warta River water					infiltration water				
		avg. spr.	avg. sum.	avg. aut.	avg. win.	avg. wh. per.	avg. spr.	avg. sum.	avg. aut.	avg. win.	avg. wh. per.
carbonate hardness	$mgCaCO_3/dm^3$	171	163	175	178	171	191	191	187	187	189
pH	-	8.1	8.1	8.0	7.9	8.0	7.4	7.4	7.4	7.4	7.4
$pH_{stability}$	-	7.5	7.5	7.5	7.4	7.5	7.4	7.4	7.4	7.4	7.4
LSI	-	0.6	0.6	0.5	0.5	0.5	0	0	0	0	0

spr. – spring; **sum.** – summer; **aut.** – autumn; **win.** – winter; **wh. per.** – whole period

4.6 Microbiological parameters

The following groups of bacteria were taken into account: coli bacteria, *Escherichia coli*, *Clostridium perfringens* and *Enterococci*. The article presents results of the removal of *Escherichia coli* (table 4), but when it comes to the other kinds of bacteria the efficiency of the process is comparable, always almost 100%.

Table 4. The *Escherichia coli* concentration parameters in the Warta River and infiltration water

year	*Escherichia coli* in Warta River water [MPN in 100 ml]			*Escherichia coli* in infiltration water [MPN in 100 ml]			infiltration efficiency [%]
	max	min	average	max	min	average	
2008	5000	140	1431	0	0	0	100
2009	5172	270	1402	0	0	0	100
2010	1401	100	598	9	0	0	99.960
2011	2600	110	663	3	0	0	99.978
2012	2610	55	642	0	0	0	100
2013	4500	246	1516	1	0	0	99.997

Between 2010 and 2012 there was more than twice decrease in *Escherichia coli* concentration in the Warta River. Unfortunately, in 2013 the number of *Escherichia coli* had returned to the level noticed before 2010. Worse parameters of the infiltration water quality in 2010 and 2011 were related to summer flood in 2010 and winter flood in 2010/ 2011. After the infiltration process *Escherichia coli* is reduced almost in 100%.

5 River and infiltration water quality during floods

Two periods of alarm condition in the Warta River were taken into account. The first is the summer flood from May to June 2010 and the second is the winter flood 2010/ 2011. The parameters were the same as in the list of parameters for the river and infiltration water quality (table 5).

Table 5. Quality of the Warta River and infiltration water during flood periods

parameter	unit	summer flood 26.05. – 18.06.2010 Warta River water	summer flood 26.05. – 18.06.2010 infiltration water	winter flood 24.11. – 24.02.2011 infiltration water
iron	mg/dm^3	0.454 – 0.762	0.338 – 0.691	0.184 – 0.604
manganese	mg/dm^3	0.056 – 0.250	0.300 – 0.410	0.280 – 0.370
DOC_{KMnO4}	mgO_2/dm^3	12.20 – 15.10	3.10 – 5.50	3.21 – 4.06
TOC	mg/dm^3	15.76 – 17.16	5.97 – 7.11	5.76 – 6.22
turbidity	NTU	2.60 – 4.91	1.22 – 1.74	0.95 – 3.19
colour	$mgPt/dm^3$	60 – 70	7.5 – 25	10 – 20
DO	mg/dm^3	4.6 – 5.5	1.9 – 2.6	3.3 – 5.7
free CO_2	mg/dm^3	7.48 – 17.60	23.54 – 26.18	24.20 – 53.46
conductivity	µS/cm	532 – 556	630 – 678	641 – 691
total hardness	$mgCaCO_3/dm^3$	248 – 249	265 – 277	274 – 311

The river water and infiltration water quality during the flood was deteriorated. To describe the intensity of the water quality deterioration the instability coefficient – IC was introduced. The value of IC was calculated according to the formula 3:

$$IC = \frac{AF}{AV} \tag{3}$$

where:
AF is an average value of analysed parameter during the flood time,

AV is an average value of analysed parameter during 5.5 years of intake operation without flood period.

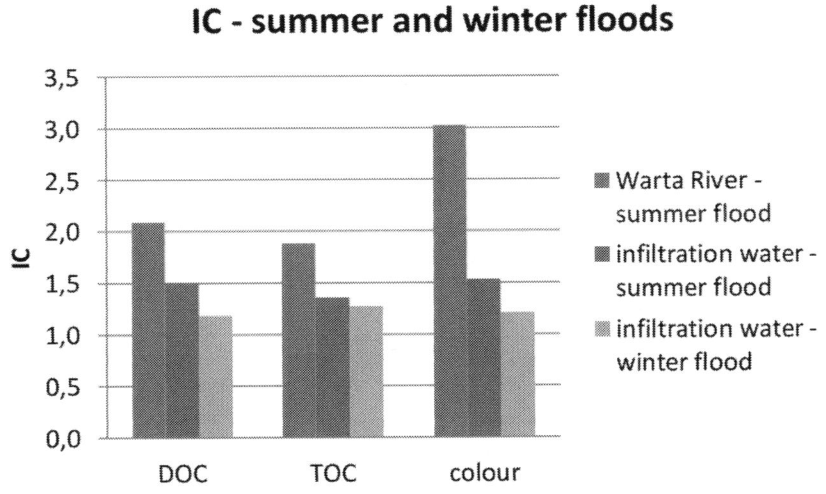

Figure 2. Instability coefficient for DOC_{KMnO4}, TOC and colour

The results of IC values for selected important parameters are presented in figure 2. Comparison of IC values in river and infiltration water allowed to evaluate the buffering properties of infiltration ground passage for chosen parameters. During summer flood IC for DOC_{KMnO4} in the river is 2.09 compare to 1.5 in infiltration water. IC for TOC in the river is 1.88, in infiltration water is 1.36. For colour IC in the river is 3.02, in infiltration water 1.53 – almost twice times more. During winter flood only IC for infiltration water were taken into account. In the winter flood instability of water parameters is lower: $IC_{DOCKMnO4}$ is 1.18, IC_{TOC} is 1.27 and IC_{colour} is 1.21.

$IC_{DOCKMnO4}$, IC_{TOC} and IC_{colour} accurately reflect changes during the floods. In the summer flood water's pollution is higher than during the winter flood. More variation in stability parameters are in the Warta River water than in the infiltration water – the soil has buffering properties.

Referring to the data from 5.5 years of exploitation of the infiltration intake, including flood periods, it can be noticed that during floods:

- in the case of summer flood parameters of the Warta River water quality are usually above average, reaching their maximum levels in the case of DOC_{KMnO4} (15.10 mgO_2/dm^3), colour (70 $mgPt/dm^3$) and free CO_2 (17.6 mg/dm^3);
- the infiltration water during summer flood also reaches its maximum values for such parameters as: DOC_{KMnO4} (5.5 mgO_2/dm^3) and TOC (7.11 mg/dm^3);
- during winter flood water quality parameters associated with contamination are also above average level;

In addition, the microbiological parameters of the Warta River water and infiltration water were analysed. The noticed regularities were presented in the summary table 6.

Table 6. Microbiological parameters during flood periods

parameter	Warta River water	infiltration water	
	summer flood	summer flood	winter flood
Total number of microorganisms at 36°C after 48 h	no data	4 – 263 CFU/1ml	range: 5 – 170 CFU/1ml average: 76 CFU/1ml
Total number of microorganisms at 22°C after 72 h	no data	2 – 1510 CFU/1ml	range: 6 – 590 CFU/1ml average: 164 CFU/1ml
Coli bacteria colilert method	1600 – 4350 MPN in 100 ml	0 – 613 MPN in 100 ml	range: 0 – 11 MPN in 100 ml average: 4.4 MPN in 100 ml
Escherichia coli colilert method	100 – 166 MPN in 100 ml	0 – 9 MPN in 100 ml	range: 0 – 3 MPN in 100 ml average: 0.5 MPN in 100 ml
Clostridium perfringens	140 – 200 CFU/100 ml	0 – 1 CFU/100 ml	range: 0 – 1 CFU/100ml average: 0.3 CFU/100 ml
Enterococci (Streptococci) enterolert method	120 – 192 MPN in 100 ml	0 – 22 MPN in 100 ml	range: 0 – 1 MPN in 100 ml average: 0.3 MPN in 100 ml

The results from tables 5 and 6 show the potential threat to the stability of the infiltration water parameters during floods. Deterioration of the infiltration water parameters is caused by the flooding of the infiltration area – shortening the infiltration zone. In case of wells leakage, contamination of captured water may occur and in the result the intake may stop to function. However on Debina intake there were no major problems with drinking water production during the flood periods due to the high efficiency of water treatment technology applied for infiltration water treatment.

Current modernization of Debina intake aims to protect infiltration water from flood by elevation of the ground level and hermetic closure of the wells.

6 Conclusions

The infiltration process stabilizes temperature of water and simplifies the further processes on infiltration water treatment plant. The parameters of infiltration water are similar to the parameters of groundwater, so treatment technology typical for groundwater treatment can be effective. Colour and turbidity – characteristic water quality parameter of surface water, are significantly reduced after the infiltration. Concentration of manganese increased after infiltration process. [10]

In the analysed period at Debina intake problems with exceeding the concentration of heavy metals in the river and infiltration water were not reported. Periodically there is a tendency to elution of some heavy metals from the ground.

The infiltration process effectively eliminates the microbiological contaminants from water. In the most of analysed samples no bacteria were detected and the reduction of the bacteria always was more than 99%. Two floods had occurred in the analysed period. First in May – June 2010 and the second one from November 2010 to February 2011. During floods, the parameters of the river water and infiltration water were worse. Especially during the summer flood unfavourable changes were more intensive. The calculated instability coefficients (IC) for selected parameters: DOC_{KMnO4}, TOC and colour, show the intensity of water quality changes during flood periods. They indicate the buffering properties of the infiltration process, minimizing the effect of river water deterioration on infiltration water quality during the flood. After analysis of the data, it can be clearly stated that infiltration intakes seem to be a perfect system for natural and ecological water treatment, which replaces complex process with the usage of chemicals in surface water treatment plants.

References

1. W.D. Gollnitz, J.L. Clancy, B.L. Whitteberry, J.A. Vogot, Journal American Water Works Association, **95/12**, 56 - 66 (2003)
2. A. L. Kowal, Ochrona środowiska, **2**, 5 – 7 (1983)
3. M. Sozański (ed.), *Wodociągi i kanalizacja w Polsce, tradycja i współczesność* (Polska Fundacja Ochrony Zasobów Wodnych, Poznań- Bydgoszcz, 2002)
4. A. Bartosik, I. Chomicki, T. Jankowski, Ochrona środowiska, **3**, 51 – 54 (2007)
5. S. Kołaska, I. Chomicki, Instal, **7/8 (364)**, 66 – 69 (2015)
6. S. Mikels, Journal American Water Works Association, **84/9**, 77 – 84 (1992)
7. A. L. Kowal, M. Świderska – Bróż, Ochrona środowiska, **2**, 5 – 7 (1983)
8. M. Błażejewski, Ochrona środowiska, **2**, 7 – 13 (1983)
9. Falewicz P., Drela I., Kuczkowska S., Ochrona przed korozją, **4/5**, 187 – 191 (2008)
10. A. L. Kowal, M. Świderska – Bróż, *Oczyszczanie wody* (Wydawnictwo Naukowe PWN, Warszawa – Wrocław 2000)

Innovative solutions in monitoring systems in flood protection

Klaudia Sekuła[1,*], *Marzena* Połeć[1], and *Aleksandra* Borecka[1]

[1]AGH University of Science and Technology, Department of Hydrogeology and Geology Engineering, Krakow, Poland

Abstract. The article presents the possibilities of ISMOP – IT System of Levee Monitoring. This system is able to collecting data from the reference and experimental control and measurement network. The experimental levee is build in a 1:1 scale and located in the village of Czernichow, near Cracow. The innovation is the utilization of a series of sensors monitoring the changes in the body of levee. It can be done by comparing the results of numerical simulations with results from installed two groups of sensors: reference sensors and experimental sensors. The reference control and measurement sensors create network based on pore pressure and temperature sensors. Additionally, it contains the fiber-optic technology. The second network include design experimental sensors, constructed for the development of solutions that can be used in existing flood embankments. The results are important to create the comprehensive and inexpensive monitoring system, which could be helpful for state authorities and local governments in flood protection.

Keywords: ISMOP, flood protection, levees, levee monitoring system

1 Introduction

Climate change and increasing human influence in the natural ecosystems of rivers pose a greater risk to flooding in areas near the riverbeds. This causes the public a greater interest in river embankments, their technical state and the threats they carry.

ISMOP – IT System of Levee Monitoring, which was created in cooperation with AGH University of Science and Technology and companies: SWECO Hydroprojekt Kraków and Sp. z o.o., Neosentio, has the task of supporting local authorities in flood protection and also for the development of an early warning network. The result of project will be the separation of recommended sensors and controls, which can be implemented in flood protection systems. Based on sensor data and analysis, the system will send a warning signal to the crisis management panel. Similar projects were carried out in Denmark, the Netherlands [1-2] and France [3], but the ISMOP project is unique in scale. The 1:1 experimental levee was built for this purpose, and an extensive network of sensors not

*Corresponding author: klaudia.sekula@agh.edu.pl

previously used jointly in the single hydrotechnical object was installed in its body. The data provided in this way is compared with the results of numerical simulations. This is an innovative project that combines many areas of science to achieve a common goal of creating a reliable flood warning system by developing methods and tools for forecasting floodplains especially during floods [4-6].

2 Characteristics of the object

The experimental flood embankment is located in Czernichów, about 30 km west of Cracow. It consists of two parallel levee segments in the shape of an ellipse. It was built on a scale 1:1, with a length of 200 m and a width of 53 m. It is capable of holding 28 500 m^3 of water, the height of the levee is 4,5 m and the maximum height of the flood wave is 4,0 m. Levee was constructed of soils characterized by variable filter coefficients in the range from 10^{-5} m/s to 10^{-8} m/s (Figure 1.) with a built-in control and measurement networks.

https://doi.org/10.1051/e3sconf/20183001005

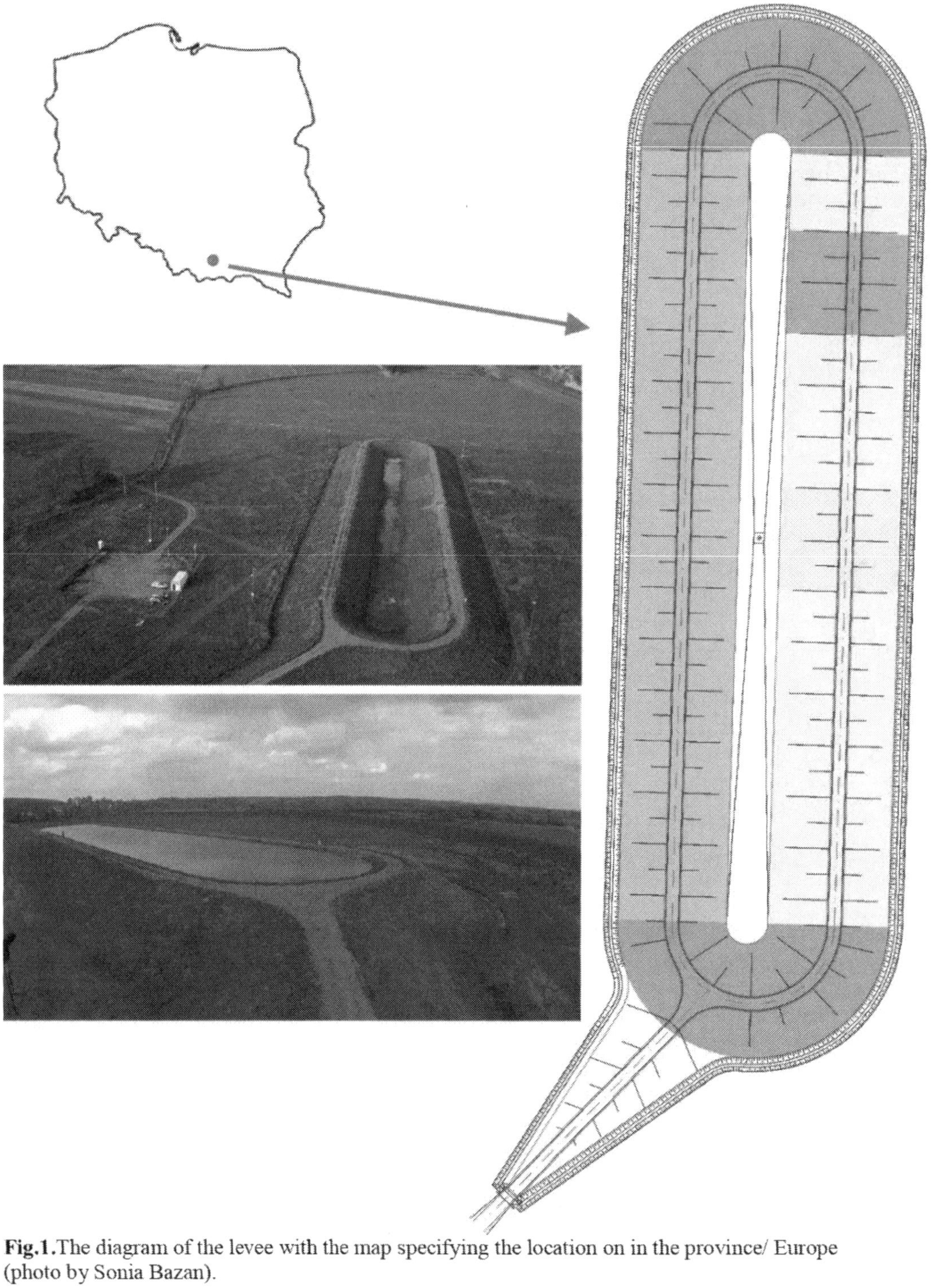

Fig.1. The diagram of the levee with the map specifying the location on in the province/ Europe (photo by Sonia Bazan).

The control and measurement system consists of two elements: the control and measurement equipment built into the experimental levee and the automatic measurement system that collects, processes and visualizes the data obtained from the experimental measurements.

The control and measurement equipment includes the reference network and experimental network. The reference control and measurement network based on specific historically proven technologies and consists of 35 sensors for spot measurement of pore pressure and temperature, and 6 sensor for measurement ground pressure located in measuring sections. There is also a fiber optic sensor cable for the linear measuring of the temperature gradient, inbuilt on two levels of the downstream slope (Figure 2.). Sensors for pore pressure, temperature and ground pressure measurement are located in three sections of levee: southern, central and northern. The experimental control and measurement network based on the alternative technological solutions and on the distributed measurement using the MESH topology proposed by NeoSentio Sp. z o.o. There are about 1000 sensors of the experimental control and measurement network (Figure 3.) that measure pore pressure and temperature [7].

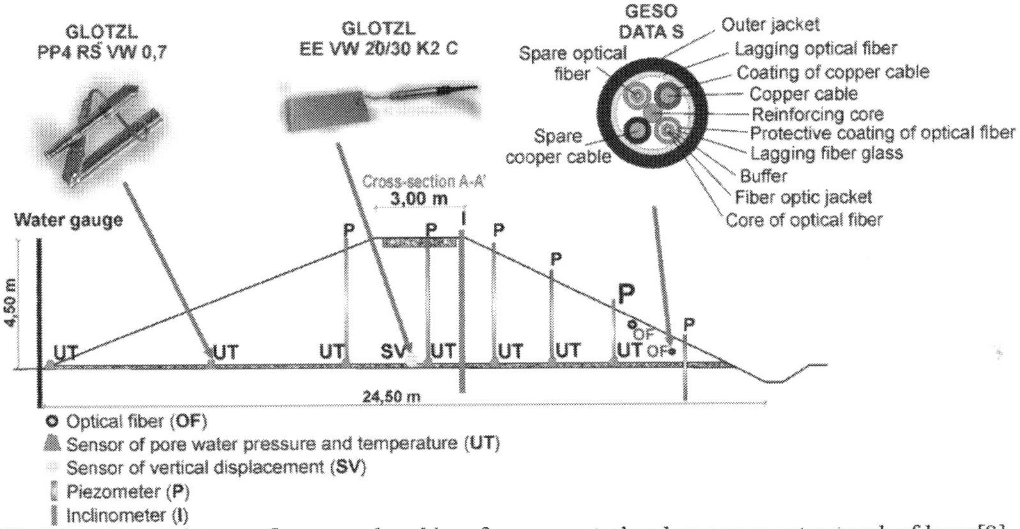

Fig.2. Location and types of sensors placed in reference control and measurement network of levee[8].

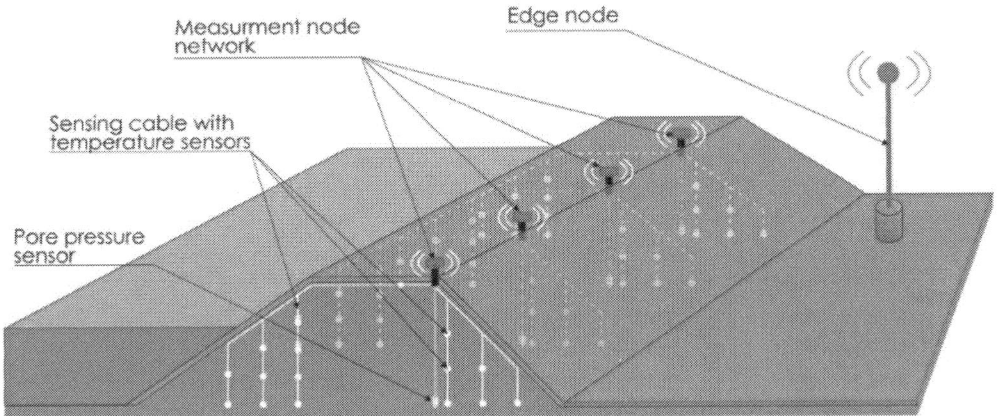

Fig.3. Schematic illustration of the idea of experimental measuring network – a matrix of sensors, a number of measuring nodes and an external base station from the edge node[7].

Mechanical and physical parameters measured by sensors are recorded and analyzed by Monitoring & Modeling Center. It is system which is based on the incoming sensor data and numerical models will perform during the current risk assessment. In addition, the system is enable data visualization on synoptic maps (meteorological data), and models of the monitored sections of the shaft (data from telemetric network) and view historical data. Based on the analysis of the measured parameters will be triggered alarm, providing the occurrence of a crisis situation [9]. Exemplary screens from Monitoring & Modeling Center are presented on Figure 4.

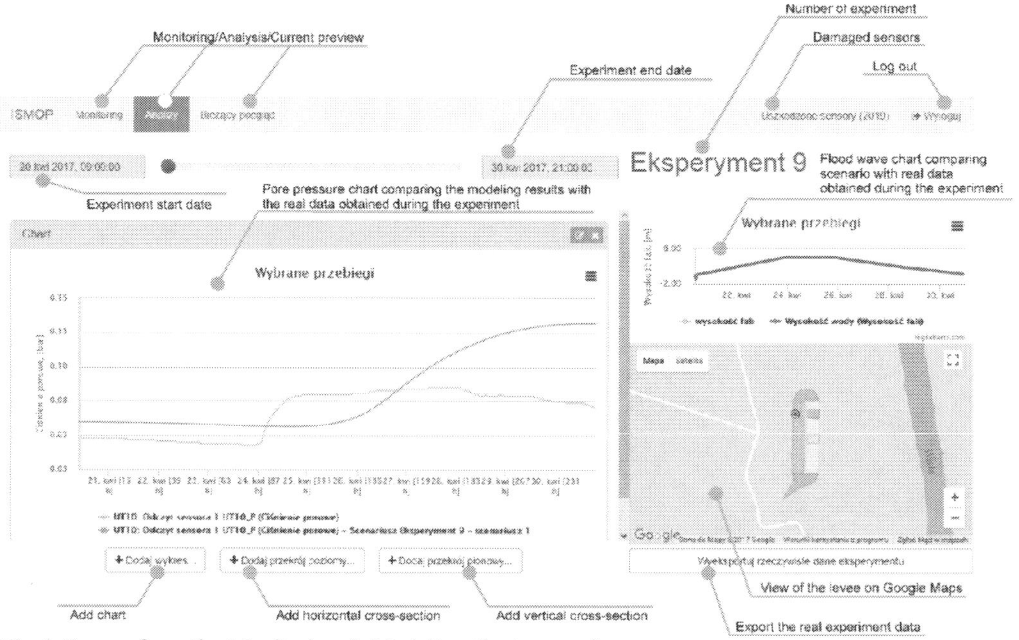

Fig.4. Screen from the Monitoring & Modeling Center panel.

What is one of the most important aspect, the experimental levee in Czernichów reflects the majority of flood embankments in Poland. It has been designed with a number of technical solutions which have a task to imitate the real problems occurring in the existing hydrotechnical contractions. They are, for example, horizontal drainage (north-west section) designed to imitate animal burrows or a sand lens (south-west section) designed to imitate frequent interlaces of good permeability materials encountered in older embankments. The experimental levee was also covered with vegetation, which was supposed to stabilize the slopes and reflect the conditions that prevail in such hydrotechnical constructions in the country.

In addition, it is carried out the monitoring of levee surface using geodetic network to register displacements and deformations of levees based on classical methods (total stations, possibly GPS). The monitoring also using measurements provided by ground radar interferometer (IBIS-L). During the experiments were also used infrared camera – FLIR T620. On the outer parts of the levee there is conducted geophysical measurements with the use of geoelectrical tomography [4].

3 Experiments

The experiments consist in simulating the static flood wave using the pump system located in the Vistula River, which filling the tank of levee to the appointed water level. Up to this article, 10 experiments were conducted (as of 10.06.2017). Each of the experiments was characterized by an individual flood scenario, namely the time of arrival of the flood wave, its height and duration (filling the tank phase, plateau phase and emptying the tank phase). Automated system for conducting experiments on the experimental levee allows for manual and automatic (remote) control. Based on the flood scenarios, the flood embankment is filled and the sensor data is monitored and made available on a special panel of Monitoring and Modelling Center with access for the researchers involved in the project (Figure 4). In next step these data are analyzed by comparing the results obtained with the numerical modeling effect. An important aspect of the functioning of the panel is the archiving of data from previous experiments. Obtained data are used to create further simulations and are helpful in forecasting the effects of floods.

4 Preliminary results

Measurement experiments are still during implementation and detailed analysis of the results obtained. The preliminary results of these were presented in publications Chuchro et al. [6], Borecka et al. [10]. One of their most important issues was enabled the control of the operation of the automated pump system, and the identification of factors that could interfere with measurements, as well as to determine the correctness of the experiment, and to prepare instructions for subsequent simulation experiments.

In each experiment, temperature, pore pressure and deformation were measured. Data obtained from sensors measuring pore pressure and temperature allow to plot changes in the water table level over time. In next step it was able to determine how deep the water infiltrate into the levees body and allows the earlier detection of damage that it can cause.

The first of the experiments allowed only for observation of levee behavior. As the duration of the flood wavelength in next experiments, changes of parameters values made it possible to define a part of the levee that has leaked, what was confirmed in fact.

5 Summary

Collected data from the sensors will become the basis for analyzing the real dangers of leaks or loss of slopes stability in the levees. Thanks to these data, it will be possible to identify areas that are particularly susceptible to damage and rapid response of emergency services and their modernization after the flood wave has subsided. Ultimately, this system is intended to assist local governments and emergency situations to send an alarm if anomalous readings occur or the limit values are exceeded. Designed from the ground monitoring system for flood embankments is applied in the centers of crisis management, which allows for faster and more effective response, eg. the evacuation of the population from the area exposed to flooding as a result of floodbank failures. Simulations that can be made based on the data of the experiments show the changes that occur in the embankments as a result of water soaking, which will allow for predictions of flood effects in future based on real time analyzes. This is an innovative approach to the issue of flood protection. In Europe, similar projects were created, but they were not carried out on such a large scale and with such an extensive network of measurement.

Acknowledgments. The project is financed under grant No. PBS1/ B9/18/2013 awarded by the National Centre for Research and Development within the framework of the Applied Research Programme.

References

1. J.R. Courivaud, P. Pinettes, C. Guidoux, J.J. Fry, Y.L. Beck, *Fiber Optic based monitoring of levees and embankment dams*, 21st Century Dam Design – Advances and adaptations, 31st Annual USSD Conference, 11–15 April, San Diego, 1560–1577 (2011)
2. M.A. Van, C. Zwanenburg, A.R. Koelewijn, H. Van Lottum, *Evaluation on full scale levee stability tests at booneschans and coresspondig Centrifuge Test,* 17th ICSMGE, 5–9 October 2009, Alexandria, Egypt, 2048–2051 (2009)
3. O. Artieres, Y.L. Beck, A.A. Khan, P.Cunat, J.J. Fry, J.R. Courivaud, C. Guidoux, P. Pinettes, *Assessment of dams and dikes behavior with a fibre optics based monitoring solution*, 2nd Conference on dam rehabilitation and Maintenance, 23–25 November, Taylor & Francis Group, London, 79–86 (2010)
4. J. Stanisz, K. Korzec, A. Borecka, Geol., Geophys. and Envi. Protec., **41(1)**, 137-139 (2015)
5. A. Borecka, J. Stanisz, K. Korzec, A. Leśniak, *Smart levee in Poland,* FLOODrisk 2016, 258 (2016)
6. M. Chuchro, A. Franczyk, B. Bukowska-Belniak, A. Leśniak, *Flood risk reducing using Computer System for Monitoring River Embankment*, Technologie informatyczne w ochronie i kształtowaniu środowiska: I konferencja naukowo-techniczna: 17 January 2017, Warszawa (in Polish) (2017)
7. K. Sekuła, A. Borecka, D. Kessler, P. Majerski, Comp. Sci., **18(4),** 357-284 (2017)
8. A. Borecka, K. Korzec, J. Stanisz, *Full-scale monitoring system of levees,* CGW Workshop'16: proc., 49–50 (2016)
9. J. Stanisz, A. Borecka, A. Leśniak, K. Zieliński, Prz. Geol., **62**, 699–703 (in Polish) (2014)
10. A. Borecka, K. Korzec, B. Bukowska-Belniak, M. Dwornik, A. Leśniak, *Smart Levee in Poland – preliminary results of experiment*, Problemy hydrotechniki: XI konferencja naukowo-techniczna, 10–12 maja 2017, Świeradów-Zdrój, Poland (2017)

 https://doi.org/10.1051/e3sconf/20183001006

Assessment the impact of samplers change on the uncertainty related to geothermal water sampling

Katarzyna Wątor[1*], *Anna* Mika[1], *Klaudia* Sekuła[1] and *Ewa* Kmiecik[1]

[1]AGH University of Science and Technology, Department of Hydrogeology and Engineering Geology, Krakow, Poland

Abstract. The aim of this study is to assess the impact of samplers change on the uncertainty associated with the process of the geothermal water sampling. The study was carried out on geothermal water exploited in Podhale region, southern Poland (Małopolska province). To estimate the uncertainty associated with sampling the results of determinations of metasilicic acid (H_2SiO_3) in normal and duplicate samples collected in two series were used (in each series the samples were collected by qualified sampler). Chemical analyses were performed using ICP-OES method in the certified Hydrogeochemical Laboratory of the Hydrogeology and Engineering Geology Department at the AGH University of Science and Technology in Krakow (Certificate of Polish Centre for Accreditation No. AB 1050). To evaluate the uncertainty arising from sampling the empirical approach was implemented, based on double analysis of normal and duplicate samples taken from the same well in the series of testing. The analyses of the results were done using ROBAN software based on technique of robust statistics analysis of variance (rANOVA). Conducted research proved that in the case of qualified and experienced samplers uncertainty connected with the sampling can be reduced what results in small measurement uncertainty.

1 INTRODUCTION

All operation performed in the laboratory are biased by the systematic and random errors and all of them have contribution in the total uncertainty. It is estimated that during sampling process and samples transport to the laboratory almost 30% of errors arise. The process of samples treatment before analysis causes next 60% of errors, whereas the analysis is a source of about 10% of total sum of errors [1]. There are two methods of measurement uncertainty estimation. The first one, theoretical (bottom-up), relies on the quantitative assessment of selected sources of uncertainty one by one and in the next stage, their connection to total uncertainty. The second method, empirical (top-down), uses different parameters of analytical method determined during method validation or the results of analyses of control samples [2]. The empirical method is much common and is used more often. In this approach, four methods are used to measurement uncertainty estimation [2]:

– collection and analysis of duplicate samples (only random errors connected with sampling and analysis can be estimated);

* Corresponding author: wator@agh.edu.pl

- use of different sampling protocols (random and systematic errors connected with sampling and random errors arising from analysis can be estimated);
- interlaboratory comparisons in sampling (both random and systematic errors connected with sampling and analysis can be estimated);
- proficiency testing in sampling (both random and systematic errors connected with sampling and analysis can be estimated).

The simplest and probably the cheapest method of uncertainty estimation is duplicate samples method. It can be carry out using balanced design (fig. 1) or its simplest form:

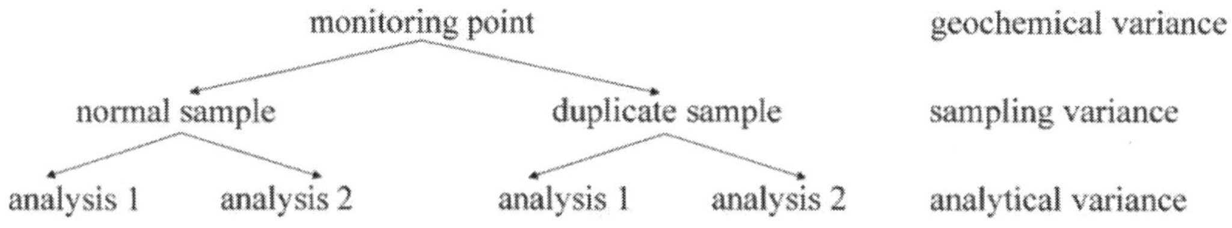

Figure 1. Balanced design schema for duplicate samples [3].

The main advantage of the use of the empirical method is that it takes into account all sources of uncertainty without necessity of their identification. This is universal, fast and easy method. However, it gives an approximate value of uncertainty [3]. The study was conducted to present how the uncertainty of geothermal water sampling changes in case of different samplers. The influence of sampler experience on the total uncertainty may e.g. change the water classification during assessment of curative water potential as it was indicated by Wątor et al. [4].

1.1 Characteristic of the research area

The research was carried out in southern Poland (Podhale region) in Małopolska province. Investigated geothermal water is classified as sulfate-chloride-sodium-calcium water (according to Szczukariew-Prikłoński classification). Mineralization of these waters is about 2.5 g/L, including specific component such as a metasilicic acid [5-8] and a temperature is between 82.9-86.3°C [9].

2 MATERIALS AND METHODS

Sampling was conducted between October 2014 to March 2016 in two-week intervals by two qualified samplers (tab. 1). Samples of geothermal water were collected directly on the wellhead according to procedure described in ISO 5667-11 standard [9] and methodology proposed by Witczak et al. [10] and Korzec et al. [11].

During sampling not only normal but also duplicate samples were collected. Control/duplicate samples were taken as duplicates of normal samples (8 pairs in each series) using the same technique.

Before the samples were collected, the unstable parameters (pH, electrical conductivity and temperature) were measured in online system. Samples of geothermal water were filtered in the field with the use of microporous filters with pores diameter 0.45 μm and also they were preserved with nitric acid (V). Then they were cooled below 4°C and delivered to the laboratory in polyethylene, disposable containers.

Table 1. Scheme of sampling to assess the influence of sampler change on the uncertainty connected with sampling

Sampler/Series	**S1**	**S2**
Number of samples pairs (normal+duplicate)	8	8
Period of samples collection	10.2014-03.2015	10.2015-03.2016
Method of analysis	ICP-OES	
Limit of silicon; determination (LOD)	0.1 mg/L	
Measurement uncertainty declared by the laboratory* (k=2, 95%)	12%	

*including sampling

The chemical analysis were performed in the Hydrogeochemical Laboratory of the Hydrogeology and Engineering Geology Department at the AGH University of Science and Technology in Krakow (Certificate of Polish Centre for Accreditation No. AB 1050). Concentration of silicon was determined using ICP-OES method (inductively coupled plasma optical emission spectrometry — tab. 1) with the use of OPTIMA 7300DV spectrometer of PerkinElmer company according to PN-EN ISO 11885:2009 standard [12], and balanced design (fig. 1).

The measurement uncertainty declared by the laboratory (tab. 1) was estimated during validation of analytical procedure and the confirmation of their fitness for selected purposes. This uncertainty involves several factors arising from all steps of analytical procedure and vary dependent on the type of measure and the field of application [13]. Laboratory took into account all uncertainty sources — both random (precision) and systematic (accuracy) joint with sampling and analysis — during the combined standard uncertainty calculation.

There are known different methods of uncertainty estimation. The most common and probably the cheapest one is empirical approach (top down) based on analysis of control field samples and certified reference materials. In this paper the influence of sampler change on the uncertainty connected with geothermal water sampling is considered. In the literature it is indicated that the experience of sampler can affect the groundwater sampling uncertainty [11, 14-16].

Assessment of uncertainty can be conducted for particular measuring series by ROBAN software using robust ANOVA statistics, which allow 10% of outliers in analyzed database. The software calculates (on the base of results in normal and duplicate samples) three independent components of total variance: geochemical, sampling and analytical and their percentage (1):

$$ {}^{2}_{total} = {}^{2}_{geochemical} + {}^{2}_{sampling} + {}^{2}_{analytical} \quad (1) $$

Figure 2 shows limit values of relative parts of particular variances parts in total variance [17]. Percentage of geochemical variance should not be less than 80% and analytical variance should not be higher than 4% of total variance. Measurement variance – as a sum of analytical variance and sampling variance – should be lower than 20% [17].

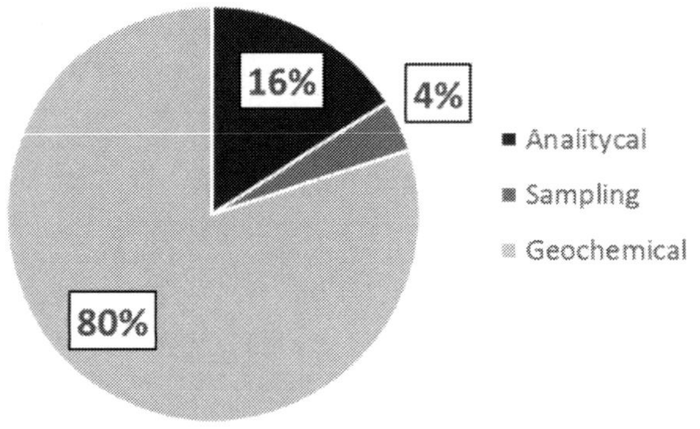

Figure 2. Percentage of particular variances in total variance ([17]– modified).

Additionally standard uncertainty (u) connected with this factor can be calculated:

$$u_{\text{geochemical}} = s_{\text{geochemical}} \quad (2)$$

$$u_{\text{sampling}} = s_{\text{sampling}} \quad (3)$$

$$u_{\text{analytical}} = s_{\text{analytical}} \quad (4)$$

$$u_{\text{measurement}} = s_{\text{measurement}} \quad (5)$$

To estimate expanded uncertainty (U) at 95% confidence level, standard uncertainty supposed to be multiplied by coverage factor which is 2 [18]. The application provides also relative uncertainty (U') which corresponds to mean value ($\bar{x}$) of analyzed indicators in normal and duplicate samples. The detailed information about calculation of expanded and relative uncertainty can be find in many publications, e.g. [2, 3, 15, 19].

When all samples are collected by the same person with the use of the same sampling protocol, systematic errors are minimised and the estimated uncertainty arising from sampling include only random effects. In this case, systematic errors should be additionally estimated, for example by using sampling reference materials or by taking into account results from proficiency testing in sampling.

3 RESULTS AND DISCUSSION

The silicon content in samples was recalculated to the concentration of the metasilicic acid and all further analysis and calculations were done for metasilicic acid.

In table 2 the values of particular elements in total variance are presented. For sampler S1 measurement variance exceeds almost 16% of total variance. Analytical variance in case of set S1 is around 8.75% of total variance whereas in series S2 it is 11.95%, so in both cases it exceeds 4% showed on figure 2.

In investigated geothermal water samples geochemical variance, which is connected with natural variability of metasilicic acid concentration, is the most visible. In both cases it is around 84–87% of total variance. Sampling variance is equal more than 7% for sampler S1 and is six times higher than for sampler S2 (1.19%).

Table 2. The variance values estimated based on results obtained from Roban software for metasilicic acid concentration in analysed geothermal water.

Parameter	Calculated values	
	Sampler/Series S1	Sampler/Series S2
Geochemical variance [mg/L]^2	6.25	15.21
Percentage of geochemical variance in total variance [%]	84.05	86.86
Sampling variance [mg/L]^2	0.536	0.209
Percentage of sampling variance in total variance [%]	7.20	1.19
Analytical variance [mg/L]^2	0.651	2.10
Percentage of analytic variance in total variance [%]	8.75	11.95
Measurement variance [mg/L]^2	1.19	2.31
Percentage of measurement variance in total variance [%]	15.95	13.14

The values of uncertainty estimated on the basis of results from ROBAN software are presented in table 3.

Table 3. The results of uncertainty of analysed geothermal water samples (calculated for metasilicic acid analysis) for two different samplers.

Parameter	Value	
	Sampler/Series S1	Sampler/Series S2
$\bar{x}$ [mg/L]	80.02	76.72
u_{total} [mg/L]	2.73	4.19
U_{total} [mg/L]	5.46	8.38
U'_{total} [%]	6.82	10.92
$u_{geochemical}$ [mg/L]	2.50	3.90
$U_{geochemical}$ [mg/L]	5.00	7.80
$U'_{geochemical}$ [%]	6.25	10.17
$u_{sampling}$ [mg/L]	0.73	0.46
$U_{sampling}$ [mg/L]	1.46	0.91
$U'_{sampling}$ [%]	**1.83**	**1.19**
$u_{analytical}$ [mg/L]	0.81	1.45
$U_{analytical}$ [mg/L]	1.61	2.90
$U'_{analytical}$ [%]	2.02	3.78
$u_{measurement}$ [mg/L]	1.09	1.52
$U_{measurement}$ [mg/L]	2.18	3.04
$U'_{measurement}$ [%]	**2.72**	**3.96**

Uncertainty connected with sampling for set S1 is 1.83% (with average metasilicic acid concentration in analyzed samples 80.02 mg/L), and total uncertainty is 6.82% (± 5.46 mg/L). In case of second sampler S2 uncertainty connected with sampling is a little bit lower — 1.19% (with average metasilicic acid concentration 76.72 mg/L), however, total uncertainty is higher 10.92% (± 8.38 mg/L). Due to the fact that the greater part of total uncertainty is geochemical uncertainty, the higher uncertainty for the S2 sampler can be related to the natural variability of the water samples during the study period.

The obtained results indicate that in the case of samples collection by qualified and experienced samplers there is no difference in uncertainty arising from sampling. Also measurement uncertainty for both tested samplers is similar and is much smaller than uncertainty declared by the laboratory. When the uncertainty is estimated using duplicate samples collected by the one person using single sampling procedure, only random effect are included, whereas during uncertainty estimation in the laboratory also systematic errors are considered. The estimation of uncertainty in a reliable manner and then using it during the inference process contribute to increase the reliability of decisions making on the obtained results.

3.1 Conclusion

In this research the influence of the samplers change on the uncertainty of the sampling was presented on the example of metasilicic acid analysis in geothermal water samples from Podhale region, southern Poland. In the assessment of total uncertainty the empirical approach was applied, based on normal and duplicate samples analyses. The samples were collected from the same well by two samplers from October 2014 to March 2016. The uncertainty determination was performed using ROBAN software and the analysis of variance with robust statistics (rANOVA).Uncertainty connected with sampling for set S1 is 1.83% (with average metasilicic acid concentration in analyzed samples 80.02 mg/l), and total uncertainty is 6.82% (± 5.46 mg/l). In case of second sampler S2 uncertainty connected with sampling is lower — 1.19% (with average metasilicic acid concentration 76.72 mg/l), however, total uncertainty is higher — 10.92% (± 8.38 mg/L).

The obtained results indicate that in the case of samples collection by qualified and experienced samplers there is no difference in uncertainty arising from sampling and the value of this uncertainty can be reduced. Therefore it is very important to take into account systematic sampling errors connected with the change of sampler. Problem of sampling uncertainty of hydrogeochemical data was mentioned in publications [3, 14-16].

Measurement uncertainty for both samplers is about fourth times smaller than uncertainty declared by the laboratory. When the uncertainty is estimated using duplicate samples collected by the one person using single sampling procedure, only random effects are included. Estimation of uncertainty in the laboratory takes into account both random and systematic errors. The estimation of uncertainty in a reliable manner and then using it during the inference process contribute to increase the reliability of decisions making on the obtained results.

Acknowledgement
This work was financed by the Polish National Centre for Research and Development, grant no. 245079 (2014–2017).

References

1. Nielsen D.M. (ed.) (1991) Practical handbook of ground-water monitoring. CRC Press
2. EURACHEM (2007) Measurement uncertainty arising from sampling. A guide to methods and approaches. EURACHEM.

3. Kmiecik E. (2018) Analytical procedures for ion quantification supporting water treatment process. [In:] Geothermal water management. J. Bundschuh, B. Tomaszewska (eds.), Boca Raton [etc]: CRC Press. Taylor & Francis Group (Sustainable Water Developments: Resources, Management, Treatment, Efficiency and Reuse; ISSN 2373-7506; vol. 6) ISBN: 978-1-138-02721-3; p. 83-112
4. Wątor K., Kmiecik E., Tomaszewska B. (2016) Assessing medicinal qualities of groundwater from the Busko-Zdrój area (Poland) using the probabilistic method. *Environmental Earth Sciences*, **75**(804), 1–13. DOI: 10.1007/s12665-016-5538-0.
5. Kleszcz A. and Tomaszewska B. (2013) Prognozowanie scalingu na przykładzie wód ujmowanych otworem Bańska PGP-1 [Prediction of scaling phenomenon based on Bańska PGP-1 geothermal well]. *Technika Poszukiwań Geologicznych Geotermia, Zrównoważony Rozwój – Geological Exploration Technology Geothermics, Sustainable Development*, **1**, 115–122 (in Polish).
6. Bujakowski W., Tomaszewska B. and Miecznik M. (2016) The Podhale geothermal reservoir simulation for long-term sustainable production. Renewable Energy **99**, 420-430.
7. Kępińska B. and Ciągło J. (2008) Możliwości zagospodarowania wód geotermalnych Podhala do celów balneoterapeutycznych i rekreacyjnych [Possibilities of use of the Podhale geothermal waters for balneotherapeutical and recreational purposes]. *Geologia,***34**(3), 541–559 (in Polish).
8. Kmiecik E., Korzec K. and Mika A. (2016) Metasilicic acid in thermal water from Bańska PGP-1 well, Podhale. In: EGC 2016, European Geothermal Congress 2016, Strasbourg, France.
9. Korzec K., Kmiecik E., Mika A., Tomaszewska B. and Wątor K. (2016) Metodyka opróbowania ujęć wód termalnych – aspekty techniczne [Methodology of thermal water sampling – technical aspects]. *Technika Poszukiwań Geologicznych Geotermia, Zrównoważony Rozwój – Geological Exploration Technology Geothermics, Sustainable Development*, **1**, 75-87 (in Polish).
10. Witczak S., Kania J. and Kmiecik E. (2013) Katalog wybranych fizycznych I chemicznych wskaźników zanieczyszczeń wód podziemnych i metod ich oznaczania [Guidebook on selected physical and chemical indicators of groundwater contamination and methods of their determination]. Inspekcja Ochrony Środowiska, Warszawa, ISBN: 978-83-61227-13-7, http://www.gios.gov.pl/images/dokumenty/raporty/ebook2_20130422.pdf (in Polish).
11. ISO 11352 (2012) Water quality. Estimation of measurement uncertainty based on validation and quality control data.
12. Kmiecik E. and Korzec K. (2015) Uncertainty associated with the sampling of geothermal water. [In:] World Geothermal Congress. WGC 2015: views from down under – geothermal in perspective. ed. Roland Horne, Toni Boyd. www.geothermal-energy.org/pdf/IGAstandard/WGC/2015/14037.pdf?
13. Kmiecik E. and Podgórni K. (2009) Ocena wpływu zmiany próbobiorcy na niepewność związaną z opróbowaniem w monitoringu wód podziemnych [Estimation of sampler influence on uncertainty associated with sampling in groundwater monitoring]. Przegląd Geologiczny, **436**, 253-260 (in Polish).
14. Kmiecik E. (2018) The Impact of Human Errors on the Estimation of Uncertainty of Measurements in Water Monitoring. In: Boring R. (eds) Advances in Human Error, Reliability, Resilience, and Performance. AHFE 2017. Advances in Intelligent Systems and Computing, vol 589. Springer, Cham doi : 10.1007/978-3-319-60645-3_16.

15. Ramsey M. H., Thompson M. and Hale M. (1992) Objective evaluation of precision requirements for geochemical analysis using robust analysis of variance. *Journal of Geochemical Exploration*, **44**, 23–26.
16. Ellison S.L.R., Rosslein M. and Williams A. (2000) Quantifying Uncertainty in Analytical Measurement. 2nd edition. EURACHEM/CITAC Guide CG4.
17. Kmiecik E., Tomaszewska B. Wątor K., Bodzek M., Rajca M. and Tyszer M. (2017) Implementation of QA/QC program in research related to the membrane processes use in thermal water treatment. *Desalination and Water Treatment* **73**, 339-347. doi:10.5004/dwt.2017.20604
18. ISO 5667-11 (2011) Water sampling. Part 11. Guidance on groundwater sampling.
19. PN-EN ISO 11885 (2009) Inductively coupled plasma optical emission spectrometry (ICP-OES)

Drinking of tap water is smart, but how do it better? – A tap water quality research

Anna Mika[1*], *Klaudia* Sekuła[1], *Marta* Dendys[2], *Weronika* Ptaszek[1], *Adam* Postawa[1]

[1]Department of Hydrogeology and Geology Engineering, AGH University of Science and Technology, Av. Mickiewicza 30, 30-059 Krakow

[2]Mineral and Energy Economy Research Institute, Polish Academy of Sciences, Wybickiego 7 St., 31-261 Krakow

Abstract Drinking tap water has recently become popular. It is a way to fight with the tons of garbage (disposable, plastic bottles). However, many people are afraid of water quality. The research was performed in December 2015 in Krakow, during one week. 56 samples were collected. The samples were taken in different times of the day and in the two types of building (old one with installation from the 80s and new one with installation built in past few years). Samples were taken by two qualified operators. The first sample was collected at the morning at 6 a.m., before anyone uses the tap. The second one after the tap was flushed and then the third one after 30 minutes stagnation. At the evening was taken one sample (after using the tap all day).The aim of the research was to check the quality of drinking water in the end-user. The results show that quality of tap water in Krakow is good, also in the end-user, but the concentration of chemical elements are changing during the flushing and using of the tap.

1 THE QUALITY OF DRINKING WATER IN POLAND

Drinking water in Poland must fulfil the requirements of Drinking Water Directive from 3 November 1998 (and Commission Directive (EU) 2015/1787 of 6 October 2015 amending Annexes II and III to Council Directive 98/83/EC) [1] and also Regulation of Polish Minister of Health of 13 November 2015 on the scope on the quality of water intended for human consumption [2]. The specified parametric values for selected components should not be exceeded both at the initial point of water distribution as well as in the end-user. Many factors related to the water supply network and inner installation in the building influence on the concentration of chemical elements in water [3]. However, water supply companies are not obligated to analyse the water in the end-user. The companies haven't got any influence on the condition of pipe system in the buildings.

1.1 The quality of water in Krakow

The water in Krakow fulfils all the requirements. The quality of water is analysed regularly by Krakow Waterworks Laboratory. The map of the water supply system is available on company website. It is possible to check from which treatment plant is the water

*Corresponding author: annamika@agh.edu.pl

supplying in concrete district. The chemical composition of water (from all treatment plants for actually period) is available online. According to information from the Krakow Waterworks' page [4], the water in Krakow is of the same high quality as in other European cities because the quality requirements laid down in Polish regulations are as stringent as EU requirements, and some indicators must meet standards even more rigorous than those laid down by the EU. Treated drinking water is as good as mineral water from bottles, because tap water contains minerals, including calcium and magnesium. The Krakow Waterworks ensures safety of water consumption, guaranteed by daily quality tests of the potable water supplied to Krakow's residents.

2 THE RESEARCH

The research was performed in December 2015 in Krakow, during one week. The aim of the research was to assess the variability of ions concentrations during the using of the tap. The influence of the pipe system age was estimated. The samples were taken in two different types of buildings, the old one with the inner water supply installation from the 80s and in the new one, with installation built in last few years. The new installation is made from copper-zinc pipes with PCV elements. There is no information about material of the old installation probably it is made partly of iron, lead. . All the samples were collected from the same taps. The taps were normally used during the day. The analysed water was supplied from Rudawa Water Treatment Plant. This knowledge makes possible to compare the chemical composition of water analysed by Krakow Waterworks Laboratory(certificate of accreditation No. AB 776) in the Treatment Plant and in the end-user (analysed in this research in certified Hydro-geochemical Laboratory of Hydrogeology and Geology Engineering Department, AGH, certificate of accreditation No. AB 1050). The chemical composition published online for December 2015 see below in table 1 (selected elements according to Krakow Waterworks statement of water quality for 5.12.2015-29.02.2016) [5].

Table 1. The chemical elements of water from Rudawa Water Treatment Plant [5]

Chemical component	Concentration (mg/l)	Parametric value (according to CD 2015/1787)
Calcium (Ca^{+2})	110	-
Magnesium (Mg^{+2})	11.3	125
Chlorides (Cl^{-})	36.3	250
Sulphates (SO_4^{-2})	58.0	250
Iron (Fe^{+2})	<0.025	0.2
Copper (Cu^{+2})	<0.003	2.0
Chromium (Cr^{+3})	<0.002	0.05
Nickel (Ni^{+2})	<0.0025	0.02

2.1 The sampling

The samples were taken in twotimes of the day and in two different types of building (old and new one). Samples were taken by two qualified operators. To avoid the influence of the tap itself, 1000 (ml) samples were taken. Thus, the influence of whole installation in the building was analysed [3]. The first sample was collected at the morning at 6 a.m., before the tap in thepremise has been used for other purposes and according to Postawa [6] during the over-night stagnation period no water was drawn from any outlet within the property. When the sample was takenthe tap was fully opened or as open as possible without losing sample. The second sample was taken after the tap was flushed and the third one after 30 minutes stagnation. At the evening was taken one sample (after using the tap all day).

This sampling scheme make possible toanalyse the concentrations of metals in water properly and is recommended by experts of metal analysis in drinking water [6]. The samples were described in table 2.

Table 2. Description of samples

Sample	Where was collected?	Description of sample
1	N and O (new and old building)	Sample was taken at 6 a.m., without prior flushing
2	N and O (new and old building)	Sample was taken after the tap was flushed (3 minutes)
3	N and O (new and old building)	Sample was taken after 30 minutes stagnation
4	N and O (new and old building)	Sample was taken after normal using the tap all day (6 p.m.)

2.2 The chemical analyses

Samples were transported at the day of sampling to the certified Hydrogeochemical Laboratory of Hydrogeology and Geology Engineering Department, AGH and analysed using ICP-OES and ICP-MS methods (according to ISO 11885:2009 [7] and ISO 17294-2:2016 [8]). The physical components (pH, conductivity) were measured. The concentrations of hydrogen carbonate and chlorides were determined titrimetrically.The analysis contained all the main ions and metals. That made possible to calculate the ion balance.The correctness of chemical composition analysis was verified by calculate analytical error based on the ion balance. According to Polish technical standard PN-89/C-04638/02 [9] and other guidelines [10] analytical error for this water should not exceed 5%. Average errors of the analyses do not exceed this value.

2.3 The statistical analyses

In Table 3 have been collated the mean concentration of ions. Statistics were obtained with the exploration procedure using PS IMAGO Software based on IBM SPSS Statistics v.24 – http://psimago.pl/. All the box-and-whisker plots were also prepared using IBM SPSS software. Box-whiskers graph presents the quartile distribution. There are values falling within the range between the first and the third quartile, median (the centre line), whiskers (situated about 1,5 the length of the box), outliers and extremes values. The outliers values lies within 1.5-3 the length of the box from lower and top the edge of the box. The extremes values are located over three lengths [11], [12].

Table 3. Mean concentration of selected chemical elements

Sample	Mean concentration of chemical elements (mg/l)								
	Ca	**Mg**	**Cl**	**HCO_3**	**SO_4**	**Fe**	**Cu**	**Cr**	**Ni**	**Zn**
N1	84.0	10.34	37.62	208.4	55.32	0.022	0.020	<0.005	<0.001	1.15
N2	70.1	9.49	31.75	191.9	43.08	0.019	0.004	<0.005	<0.001	0.19
N3	71.7	9.63	32.86	191.6	44.65	0.015	0.007	<0.005	<0.001	0.51
N4	85.3	10.48	40.67	218.6	58.10	0.010	0.007	<0.005	<0.001	0.42
mean N	**77.8**	**9.99**	**35.73**	**202.6**	**50.29**	**0.017**	**0.010**	**<0.005**	**<0.001**	**0.57**
O1	73.0	9.70	34.06	193.5	48.49	0.037	0.005	<0.005	<0.001	0.10

O2	68.9	9.54	32.10	187.0	44.80	0.025	0.002	<0.005	<0.001	0.01
O3	79.1	10.23	36.36	197.8	55.06	0.023	0.004	<0.005	<0.001	0.05
O4	86.3	10.62	39.19	210.8	60.49	0.018	0.003	<0.005	<0.001	0.03
mean O	**76.8**	**10.02**	**35.43**	**197.3**	**52.21**	**0.026**	**0.004**	**<0.005**	**<0.001**	**0.05**

3 THE RESULTS

The studies did not show any concentrations higher than the parametric value. This leads to the conclusion that, based on the analysed elements (no microbiological analyses were performed), the quality of water in Krakow is good, but concentrations of chemical components are different at the initial point of water distribution and by the consumers. The results show also, that concentration of metals and related substances are mostly higher in the samples taken without passing the stagnant water in the water system. However all the measured concentrations were lower than the parametric values. The ions concentrations measured in the certified Hydro-geochemical Laboratory of Hydrogeology and Geology Engineering Department, AGH are different than the values presented by Krakow Waterworks (Table 3). It is very important to take into consideration the uncertainty of measure. All certified laboratories declare their own uncertainty. According to certificates of each laboratory (AB 776 and AB 1050), presented elements were measured with different methods. That also influences on the results.

At figure 1, on the box-and-whisker plot, were presented differences between Cu concentration in old (O) and new (N) buildings, in all samples. The highest concentrations were measured in sample N1 (without passing the stagnant water in the water system). All values are also higher than the concentration of Cu measured in the treatment plant. What is also important, the concentrations in the new building are larger than in the old building. Probably the inner installation in new building is partly made of copper. The parametric value of Cu concentration in drinking water is 2 (mg/l) (CD 2015/1787).

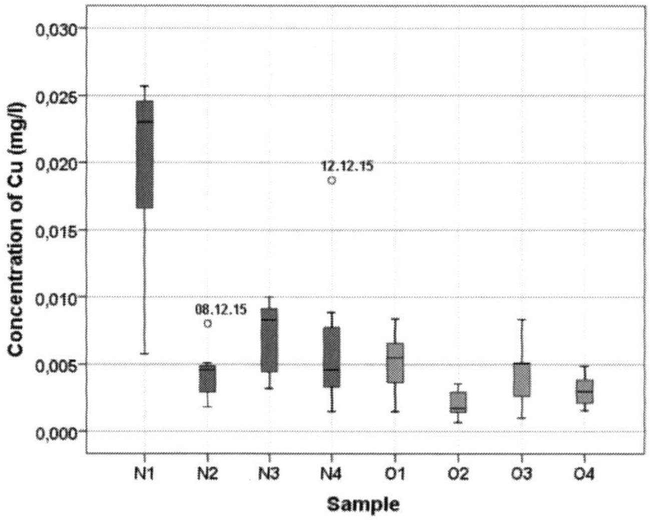

Fig. 1. Concentration of Cu (mg/l) in different samples.

At the figure 2 were presented values of Fe concertation for all samples. The higher concentrations were measured in old building, especially in samples O1 (without passing the stagnant water in the water system). All values are also higher than the concentration of

Fe measured in the treatment plant. In this case, probably the old installation is made partly of iron. The parametric value of Fe concentration in drinking water is 0.2 (mg/l) (CD 2015/1787).

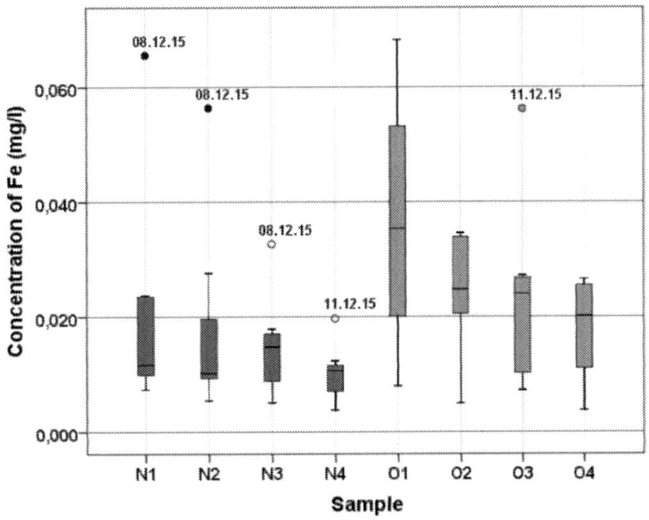

Figure 2. Concentration of Fe (mg/l) in different samples.

At the figure 3 were presented concentrations of Zn for all samples. The higher concentrations were measured in new building, especially in samples N1 (without passing the stagnant water in the pipe-system). The higher concentration could be a result of pipe-system material type. All values are also higher than the concentration of Zn measured in the old building.The concentration of Zn in treatment plant is unknown. In (CD 2015/1787) is no parametric value of Zn concentration in drinking water.

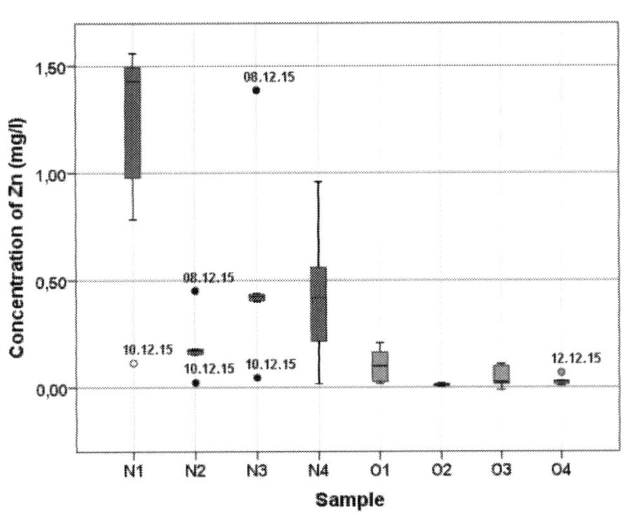

Figure 3. Concentration of Zn (mg/l) in different samples.

At the figure 4 (a) and (b) were presented concentrations of Ca and Mg for all samples. The concentrations of two most important cations in water (calcium and magnesium) measured in the certified Hydro-geochemical Laboratory of Hydrogeology and Geology Engineering Department, AGH are lower than the values presented in Krakow Waterworks statement (dotted line). However, the difference is not so high. For example the uncertainty for calcium and also magnesium measure in AGH Laboratory is 11.7%.

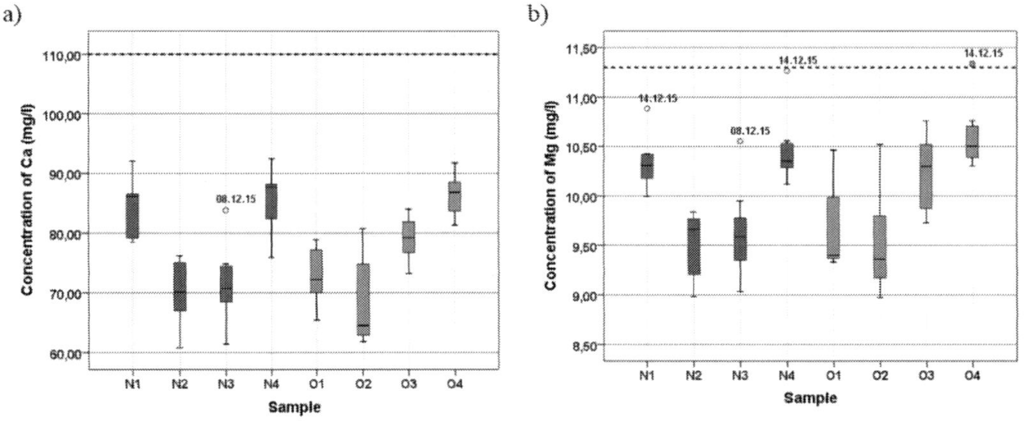

Figure 4. Concentration of (a) Ca (mg/l) and (b) Mg (mg/l) in different samples, dotted lines present the concentrations measured by Krakow Waterworks Laboratory

3.1 Conclusions

The research shows that the quality of water in Krakow is good. The results from certified Hydro-geochemical Laboratory of Hydrogeology and Geology Engineering Department, AGH are different than the concentrations presented in Krakow Waterworks statement. However, the differences are not so high and could be a result of used method or uncertainty. The water in treatment plant but also in the end-user fulfils the requirements of all regulations. The results confirmed the information from Krakow Waterworks about the good quality of water. However, the research shows that the concentrations of selected metals are higher in water collected without flushing the tap.Additionally the material of inner water supply installation influences on the chemical composition of water. But all concentrations of metals are lower than the parametric value include in law regulations [1].The issue is very interesting and consistent with literature, for example this subject was also described in [6],[14] and others publications. The results encourage to expand the research and describe the problem more accurately.

Acknowledgements
The study was partially supported by AGH-UST11.11.140.797.

References

1. Commission Directive (EU) 2015/1787 of 6 October 2015 amending Annexes II and III to Council Directive 98/83/EC on the quality of water intended for human consumption M. Ben Rabha, M.F. Boujmil, M. Saadoun, B. Bessaïs, Eur. Phys. J. Appl. Phys. (to be published) (2015)
2. RMH Regulation of Polish Minister of Health of 13 November 2015 on the scope on quality of water intended for human consumption, Journals of Laws (Dz.U.) No.1989 – RMH 2015 (2015)
3. Postawa A., Witczak S., Kmiecik E., Wątor K. Metals and related substances in drinking water in Poland, AGH Publishing 2011(2011)
4. Krakow Waterworks page - http://en.wodociagi.krakow.pl/about-tap-water.html (2017)
5. Krakow Waterworks– statement of water quality for 5.12.2015-29.02.2016 (2015)
6. Postawa A. Best Practice Guide on the Sampling and Monitoring of Metals in Drinking Water, IWA Publishing 2012 (2012)
7. ISO 11885:2009 Water quality — Determination of selected elements by inductively coupled plasma optical emission spectrometry (ICP-OES) (2009)
8. ISO 17294-2:2016-11 Water quality — Determination of selected elements by inductively coupled plasma mass spectrometry (ICP-MS)(2016)
9. PN–89/C–04638/02 Water and waste water – Ions balance of water – Methods of calculation (in Polish) (1989)
10. Witczak S., Kania J. and Kmiecik E. Catalog of selected physical and chemical indicators of groundwater pollution and methods of marking them. Library of Environmental Monitoring. PIOŚ, Warsaw, 2013. (in Polish) (2013)
11. Kmiecik E. Metodological aspects of assessing the chemical status of grounwater, AGH Publishing 2011 (in Polish) (2011)
12. Szczepańska J., Kmiecik E., Drzymała M. Assesment of stability of curative waters chemical composition from ZdrójGłówny in Krzeszowice. Geological Bulletin of the Polish Geological Institute, 63(10/1):830-833 (in Polish) (2009)
13. Wątor K., Mika A. and Postawa A. Influence of installation materials and water stagnation time on the concentrations of selected metals in drinking water (In:) Water supply and water quality, p. 877–889. (2016)

Application of GPR Method for Detection of Loose Zones in Flood Levee

Tomisław Gołębiowski[1,*], *Tomasz* Małysa[2]

[1]Cracow Uniwersity of Technology, Faculty of Environmental Engineering, Institute of Geotechnics, Department of Geodesy, Geophysics and Engineering Geology, Warszawska st. 24, Cracow, Poland
[2]Firm: MET-GEO, Styczniowa st. 37, Trzebinia, Poland

Abstract. In the paper the results of non-invasive georadar (GPR) surveys carried out for detection of loose zones located in the flood levee was presented. Terrain measurements were performed on the Vistula river flood levee in the village of Wawrzeńczyce near Cracow. In the investigation site, during the flood in 2010, leakages of levee were observed, so detection of inner water filtration paths was an important matter taking into account the stability of the levee during the next flood. GPR surveys had reconnaissance character, so they were carried out with the use of short-offset reflection profiling (SORP) technique and radargrams were subjected to standard signal processing. The results of surveys allowed to outline main loose zone in the levee which were the reason of leakages in 2010. Additionally gravel interbeddings in sand were detected which had an important influence, due to higher porosity of such zones, to water filtration inside of the levee. In the paper three solutions which allow to increase quality and resolution of radargrams were presented, i.e. changeable-polarisation surveys, advanced signal processing and DHA procedure.

1 Introduction

Detection of inner damages in hydrotechnical constructions is very important matter because such damages (e.g. loose zones in flood levees or fractures in concrete dams etc.) may lead to leakage of those construction and in the extreme case (e.g. during the flood) to the destruction of levees and dams. If damages did not appear on the surface of hydrotechnical construction it's impossible to know of its existence. Therefore it is very important to find the methods for examination of inner condition of hydrotechnical construction. Such examination may be carry out on the basis of the boreholes and analysis of material obtained from them or mounting in the constructions different sensors. Both methods weaken the construction, are invasive and expensive and, which is the most important, delivers only punctual information. Solution of described problems might be the application of non-invasive geophysical methods which delivered continuous information between boreholes and sensors. Method presented in the paper may be applied for examination of different hydrotechnical constructions but authors focused only on non-invasive examination of flood levees.

* Corresponding author: goleb@wis.pk.edu.pl, tomaszmalysa1984@o2.pl

There is may geophysical methods which may be applied for examination of flood levees; generally electrical and electromagnetic methods were applied for this purpose, i.e. Electrical Resistivity Tomography (ERT), Electromagnetic Profiling (EMP), Ground Penetrating Radar (GPR) and occasionally Capacitively Coupled Resistivity method [1 – 4]; increasingly Multichannel Analysis of Surface Waves (i.e. MASW technique) is applied for examination of flood levees. Very seldom microgravimetric surveys were applied for location of loose zones in the levees. In the paper authors focused only on the GPR method.

2 Investigation site

GPR surveys were carried out on the flood levee of the Vistula river in the village of Wawrzeńczyce near Cracow (Fig. 1). The levee in the investigation site were built before World War II and renovated in years 1970-72. During the flood in 2010, flood gate was partly destroyed and in the body of the levee leakages were observed (Fig. 1 – places with marks no. I, II, III, IV and V). After the flood, the levee in place no. III (Fig. 1) and flood gate were reconstructed. Other parts of the levee where leakages were observed during the flood were only protected with bags with sand (Fig. 1) and no renovation was carried out. Therefore the aim of GPR surveys was detection of loose zones existed in the levee which were water filtration paths and during the next flood may threaten the stability of the levee.

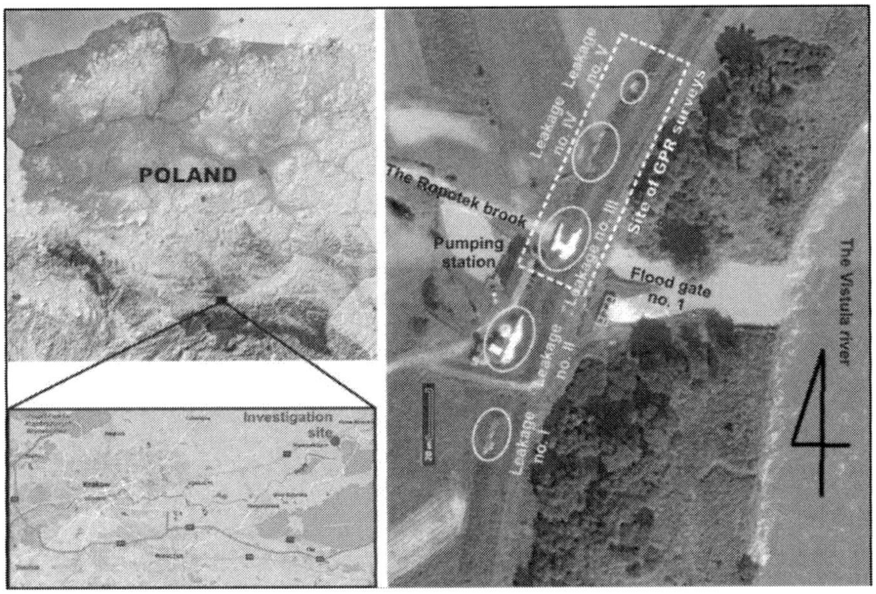

Fig. 1. GPR investigation site on the flood levee of the Vistula river in the village of Wawrzeńczyce near Cracow (base maps: *www.google.pl/maps*).

Reconnaissance geophysical (i.e. ERT and GPR – 100 MHz) surveys and geotechnical (i.e. DLP) investigations were carried in this site in 2014 and the results were presented in scientific conference in Korbielów (Poland), in 2015. High resolution GPR – 300 MHz surveys were performed in 2016 and selected results were presented in scientific conference in Korbielów, in 2017. In this paper authors would show and discuss all results of GPR – 300 MHz surveys and present additional solutions which allow to increase quality and resolution of radargrams recorded on flood levee, i.e. changeable-polarisation surveys, advanced signal processing and DHA procedure. GPR surveys, in 2D mode, were carried out on the top and on the shelfs of the levee (Fig. 2). On the basis of borehole information (Fig. 3) inner construction of the levee in the investigation site was known (Fig. 3). Geotechnical investigations delivered information that body of the levee is built mainly of sand which is compacted and semi-compacted but in several places of the levee interbeddings and accumulations of gravel occurred.

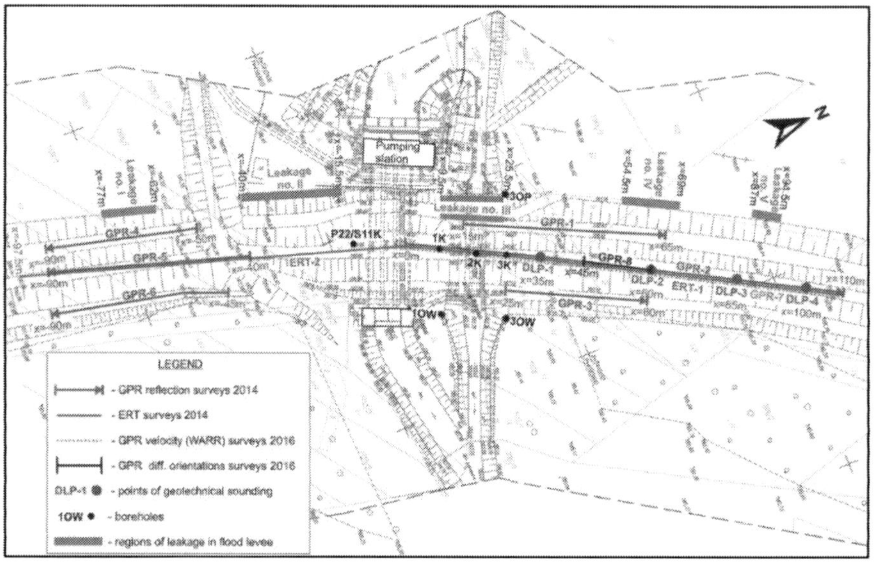

Fig. 2. Project of geophysical and geotechnical investigations (base map: [5]).

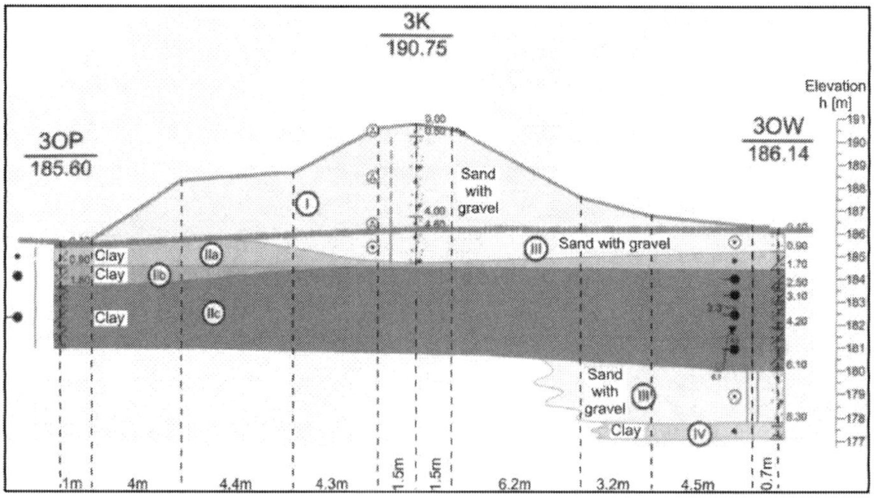

Fig. 3. Geological information from the investigation site [5].

3 Fundaments of GPR method

In the GPR method many different measurement techniques are known, but for reconnaissance surveys the short-offset (occasionally zero-offset) reflection profiling (SORP) technique is used. In SORP technique the transmitter (Tx) and receiver (Rx) antennae move along the profile (Fig. 4A), where the transmitter antenna emits electromagnetic signals at every specified distance interval (Δx). The signals propagate in the form of an electromagnetic (e-m) wave within the geological medium (or examined construction), reflecting from anomalous bodies and geological boundaries (Fig. 4A). The reflected waves are recorded by the receiver antenna and displayed on the computer during the measurement (Fig. 4A) in the form of a radargram (Fig. 4B). The horizontal axis on the radargram is recorded on the length scale, whereas the vertical axis is recorded on the time scale; during the processing of measurement data, the time axis is converted into the depth axis, taking into consideration the information about the velocity of the e-m wave within the studied medium. Depending of type of analysed problem, antennae with different constructions and different frequencies are used. SORP technique is fast, cheap and measurements may be performed by only one person (Fig.4A) but disadvantage of this technique is fact that only qualitative interpretation may be performed. More information concerning SORP technique might be found, between others, in book by [6].

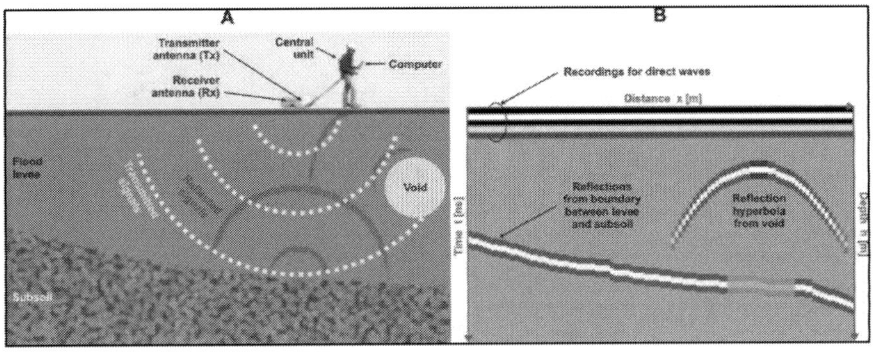

Fig. 4. A) GPR SORP technique; B) Recorded radargram ([7] – changed).

The contrast of values of relative electrical permittivity ε_r (Table 1), between the anomalous body (e.g. void or loose zone in the levee) and the surrounding medium (e.g. compacted part of the levee) or between two geological materials (Fig. 4A) decides upon the value of the reflection coefficient R [-], and in consequence upon the amplitude of the reflections recorded on the radargram (Fig. 4B). For SORP technique and for media analysed in the paper reflection coefficient may be simplified to formula (1). Relative electrical permittivity ε_r defines also value of velocity of e-m wave in the examined medium (Table 1), according to simplified formula (2). The value of the electrical conductivity σ (Table 1) decides upon the attenuation α of the e-m wave (Table 1); for media analysed in the paper (Fig. 3), which may be treated as lossless media, attenuation might be defined by simplified formula (3).

$$R = \frac{\sqrt{\varepsilon_{r\_medium}} - \sqrt{\varepsilon_{r\_anomaly}}}{\sqrt{\varepsilon_{r\_medium}} + \sqrt{\varepsilon_{r\_anomaly}}} \qquad v = \frac{30}{\sqrt{\varepsilon_r}} \qquad \alpha \approx 1.7 \cdot 10^3 \frac{\sigma}{\sqrt{\varepsilon_r}} \qquad (1, 2, 3)$$

In SORP technique quality of radargram, so in consequence possibility of proper interpretation, may be increased in four ways, i.e.:

1) by carrying out the surveys with changeable orientation (changeable polarisation) of antennae (Fig. 5),
2) by application of procedure of Diffraction Hyperbola Analysis (DHA) which allows to estimate velocity of examined medium,
3) by application of advanced signal processing and digital images analysis,
4) by application of antennae with different constructions and modern electronic solutions.

https://doi.org/10.1051/e3sconf/20183001008

Table 1. Properties of media analysed in the paper [7].

	Relative electrical permittivity	Velocity of e-m wave	Electrical Conductivity	Attenuation of e-m wave
	ε_r [-]	v [m/ns]	σ [mS/m]	α [dB/m]
Air filling the loose zones	1	0.3	0	0
Fresh water filling the loose zones	80	0.03	0.5	0.2
Flood levee (i.e. mixture of sand and gravel, Fig. 3)	from 4 (dry - d) to 30 (water saturated - ws)	from 0.15 (d) to 0.05 (ws)	from 0.1 (d) to 1 (ws)	from 0.1 (d) to 1 (ws)

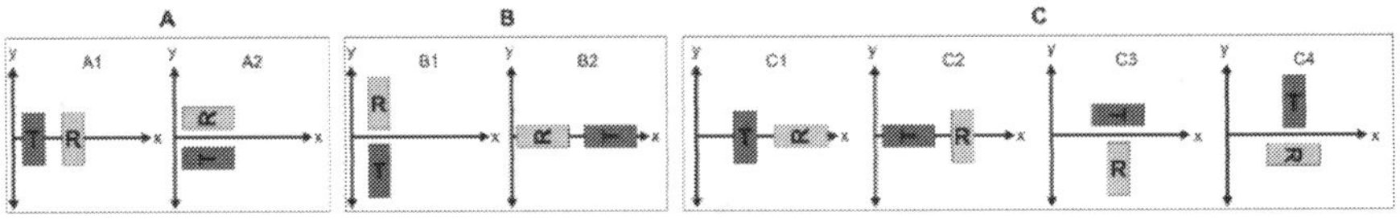

Fig. 5. Different orientations of GPR antennae: A) "co-pole" orientation; B) "end-fire" orientation; C) "cross-pole" orientation.

Results of SORP surveys carried out in 2014 with using of standard orientation of antennae (Fig. 5A – option A1) was shown in Fig. 6 and Fig. 7. Terrain surveys were carried out with the use of ZOND georadar (produced by Radar Systems Inc. – *www.radsys.lv*) and 300 MHz antenna which max. depth penetration is c.a. 10m and mean resolution is c.a. 0.2m was applied for surveys. Traces were recorded with constant distance interval Δx=0.05m. For time-depth conversion mean velocity equal 0.1 m/ns was assumed on the basis of application of Diffraction Hyperbola Analysis procedure (Fig. 6C). All radargrams was subjected to standard signal processing in ReflexW program (*www.sandmeiergeo.de*). In Fig. 6A weathered and loose zone (anomaly no. 1) is easily noticed; leakage in place no. IV was an effect of presence of loose zone located by the basement of the levee (deeper part of anomaly no. 1); inner boundary (anomaly no. 2) is an effect generated by presence of gravel in sand. In Fig. 6B high-amplitude anomaly no. 3 generated by presence of loose zone was recorded; this anomaly was the reason of leakage in place no. III; linear anomaly no. 4 was generated by gravel interbedding located in sand; anomaly no. 5, generated by presence of loose zone, was the reason of leakage of the levee in place no. V. In Fig. 6C weathered and loose zone was recorded by the surface (anomaly no. 6); interesting anomaly no. 7 was recorded in subsoil of the levee and it may indicate the location of highly loose zone or even void which was the reason of leakage in place no. IV. In Fig. 7A two anomalies, i.e. no. 8 (weathered and loose material) and no. 9 (loose zone) were distinguished; leakage observed in 2010 in place no. I was caused by presence of anomalies no. 8 and 9. In Fig. 7B loose and weathered material is clearly visible (anomaly no. 11); gravel interbedding in sand is presented in Fig. 7B as anomaly no. 10. In Fig. 7C one large anomaly no. 12 (i.e. weathered and loose zone) was detected by the surface. Generally body of the levee in the investigation site is disintegrated. Additionally gravel interbeddings located in sand characterise themselves by the higher porosity and water permeability. Mentioned facts cause that during the next flood leakages and seepages will appear and stability of the levee may be threaten.

Unfortunately in many cases standard orientation of GPR antennae delivers worst information than application of non-standard orientation. In Fig. 8 results of surveys carried out in former plant was presented. Remains of tank pad, located underground is only fragmentary visible in Fig. 8A ("co-pole" orientation) and trenched area where leakage of creosote from tank took a placed is visible.

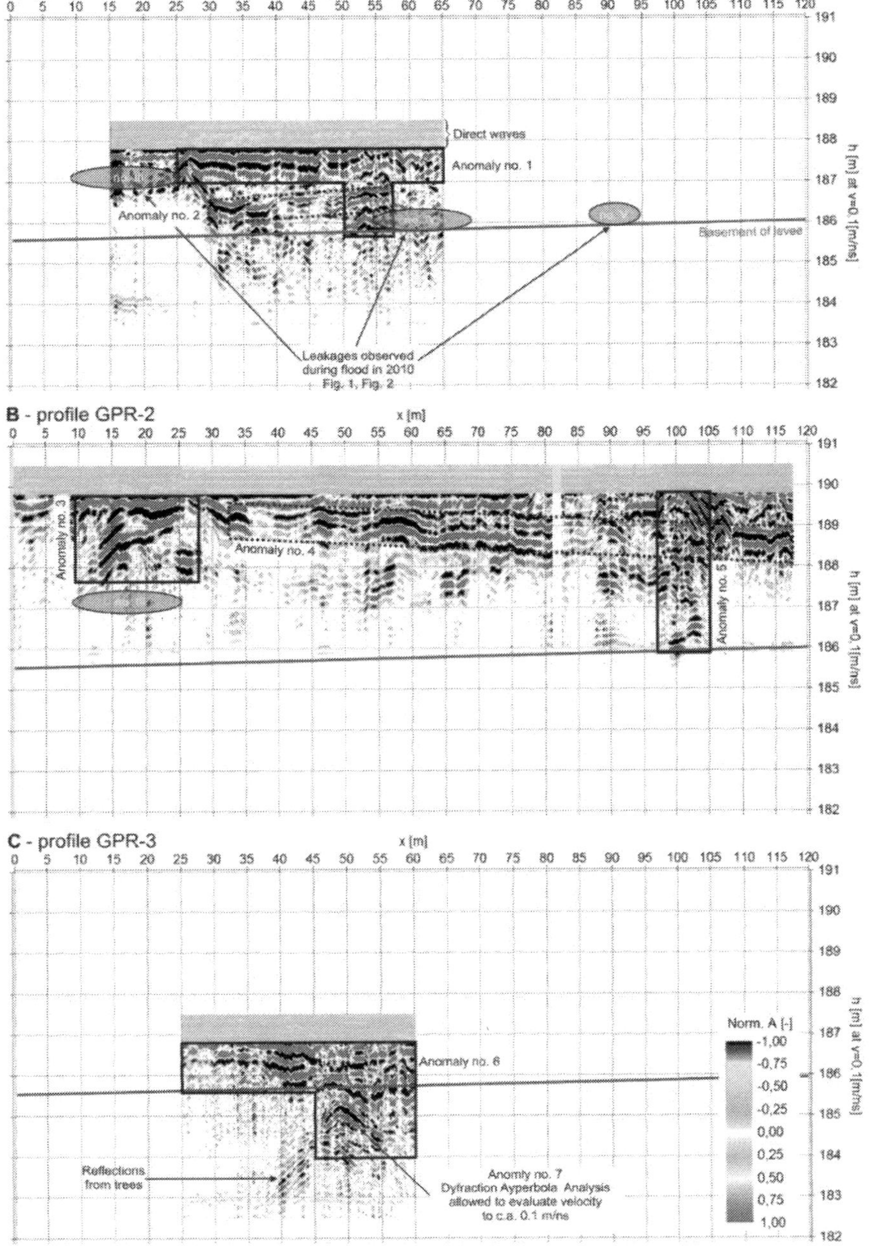

Fig. 6. Radargrams recorded on the levee for profiles: GPR-1 (A), GPR-2 (B), GPR-3 (C)

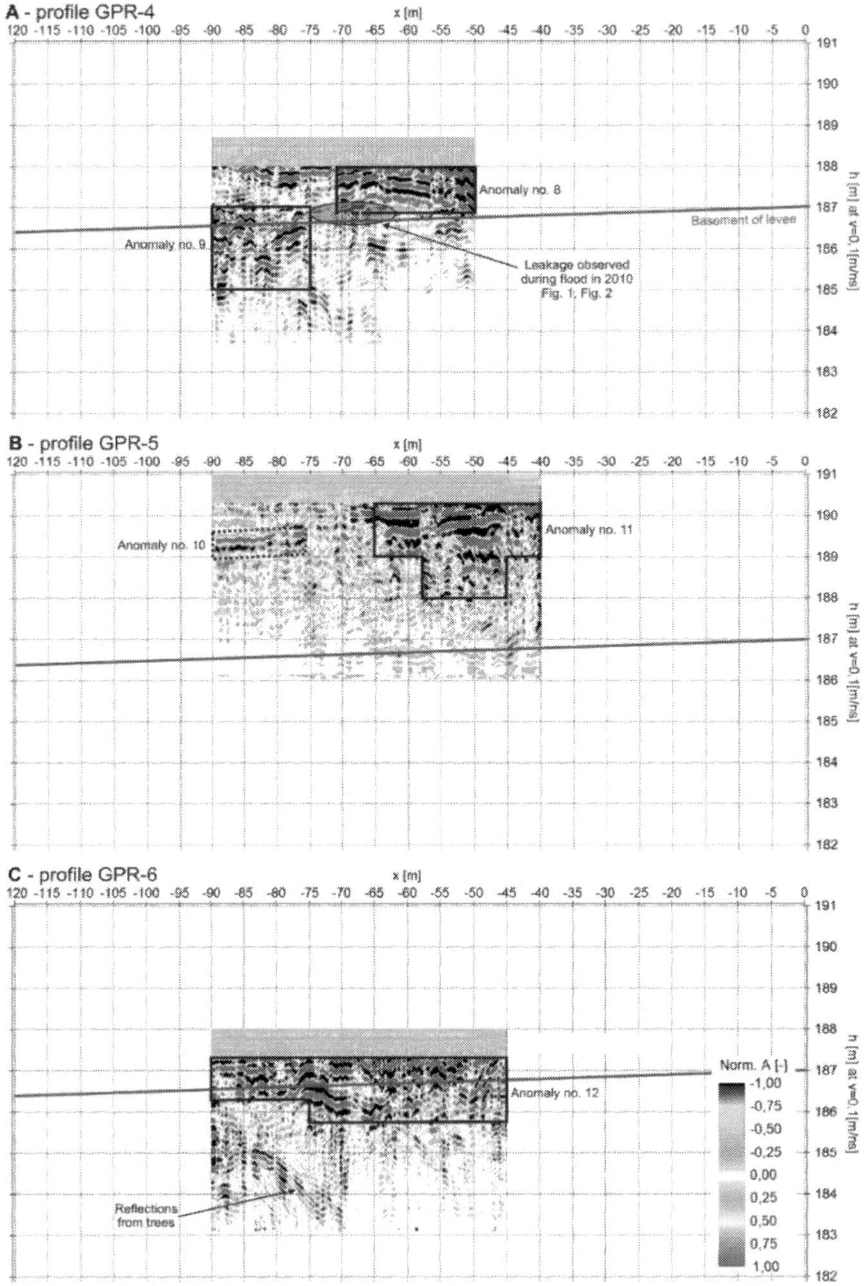

Fig. 7. Radargrams recorded on the levee for profiles: GPR-4 (A), GPR-5 (B), GPR-6 (C)

In Fig. 8B ("cross-pole" orientation) shape of tank pad is much better visible than in Fig. 8A, trench (i.e. brake of continuity of tank pad) is also visible; additionally in Fig. 8B rebars in tank pad may be distinguished. Application of non-standard orientation of antennae (Fig. 8B) allowed also to depict creosote-filled vault which was impossible in Fig. 8A. It is highly probable that application of different orientations of antennae during the surveys [8, 9, 10] on the levee in Wawrzeńczyce site delivered more detailed information. It should be emphasized that the vast majority of the research conducted on the flood embankments is carried out using the SORP technique with standard orientation of

GPR antennae. Not all contractors have bistatic antennas and can not perform tests using different dipole orientations [11, 12].

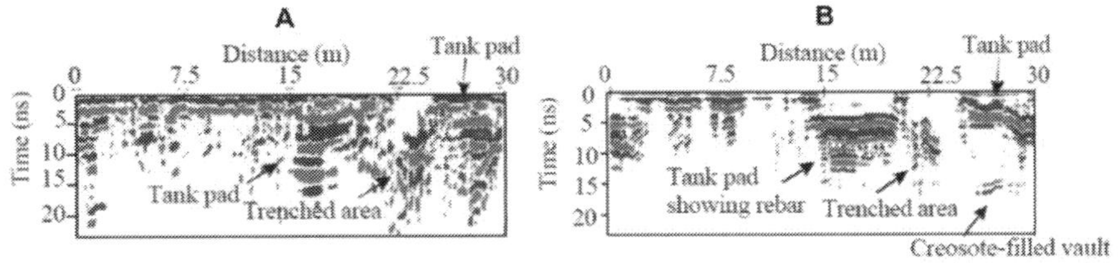

Fig. 8. GPR SORP surveys with the use of standard "co-pole" (A) and non-standard "cross-pole" (B) orientations of antennae [8].

Additional information needed for quantitative interpretation of the results of SORP surveys may deliver application of DHA procedure. This procedure allows to analyse velocity of e-m weave in places of radargram where hyperbolae appeared (like it was shown in Fig. 6C – anomaly no. 7). Information about velocity distribution (Fig. 9A) is very important because it allows to perform proper time-depth conversion of radargram and additionally allows to construct a map of changes of velocity in the examined medium (Fig. 9B); on the basis of such map, distribution of porosity and water saturation of the loose zones may be solved on the basis of formulae (2), Topp formula (4), BHS formula (5) and CRIM formula (6); in consequence quantitative interpretation is possible. Detailed description of velocity analysis and transformation of velocity map into porosity or water saturation maps may be found in works Gołębiowski (2010, 2012). In order to demonstrate DHA procedure, in Fig. 9A shown example radargram with hypothetical diffraction hyperbolae, because real unfortunately was not recorded; in Fig. 9B velocity map solved on the basis of DHA procedure was presented. Unfortunately during SORP surveys of flood levees very seldom few or dozen and so diffraction hyperbolae (like in Fig. 9A) are recorded (see Figs. 6, 7, 8); additionally for construction of proper velocity map, diffraction hyperbolae have to be distributed uniformly on the radargram (Fig. 9A).

$$\theta_w = 4.3 \cdot 10^{-6} \cdot \varepsilon_r^3 - 5.5 \cdot 10^{-4} \cdot \varepsilon_r^2 + 2.92 \cdot 10^{-2} \cdot \varepsilon_r - 5.3 \cdot 10^{-2} \quad (4)$$

$$v = \frac{c}{\sqrt{\left[\theta_w \cdot \varepsilon_{r\_w}^{\alpha} + (1-n) \cdot \varepsilon_{r\_g}^{\alpha} + (n - \theta_w) \cdot \varepsilon_{r\_a}^{\alpha}\right]^{\frac{1}{\alpha}}}} \quad (5)$$

$$n = \left[\frac{\left(\frac{c}{v}\right)^2 - \varepsilon_{r\_w}}{\varepsilon_{r\_w} - \varepsilon_{r\_g}}\right] \cdot \left[\frac{\varepsilon_{r\_w}}{\left(\frac{c}{v}\right)^2}\right]^{\beta} \quad (6)$$

where θ_w is volumetric water content, n is porosity, ε_{r_w} ε_{r_g} ε_{r_a} are respectively relative electrical permittivity of water, ground, air, c is velocity of e-m wave in vacuum, α is anisotropy factor, β is ground structure factor.

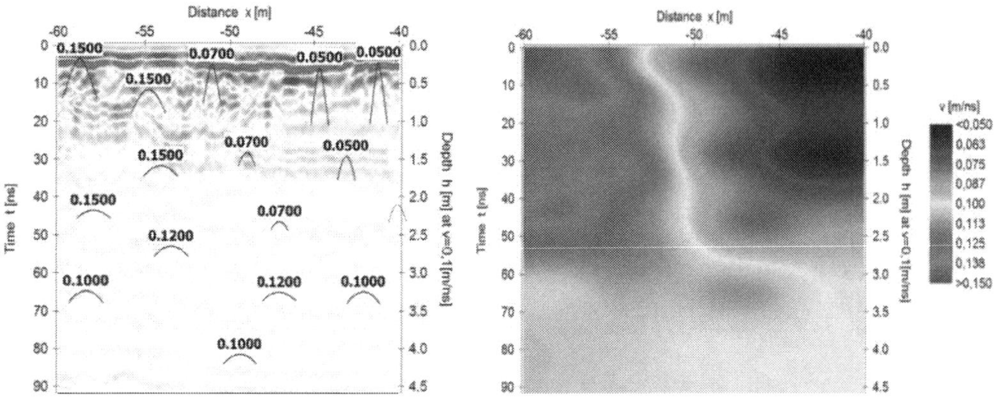

Fig. 9. A) Radargram with hypothetical diffraction hyperbolae; B) Velocity map obtained after application of DHA procedure.

Increasing of information or better resolution of radargram may be obtained by application of advanced procedures of signal processing. As an example, in Fig. 10 shown radargram recorded also in the Vistula river levee in Cracow but in other site than this presented in Fig. 1. In Fig. 10A, in zone A, direct waves were recorded and interpretation of anomalies was impossible. In zone B, stochastically distributed reflections were recorded which originate from loose zones existed in the levee; in this zone, two sub-zones B1 and B2 might be distinguished; the left sub-zone B1 has slightly higher amplitudes which may be interpreted as a more disintegrated medium than the right sub-zone B2; however, it is difficult to depict small regions with a higher leak risk in zone B. As shown in Fig. 10B application of more advanced processing allowed to outline with high accuracy the most dangerous small loose areas (violet colour) located especially in the sub-zone B1.

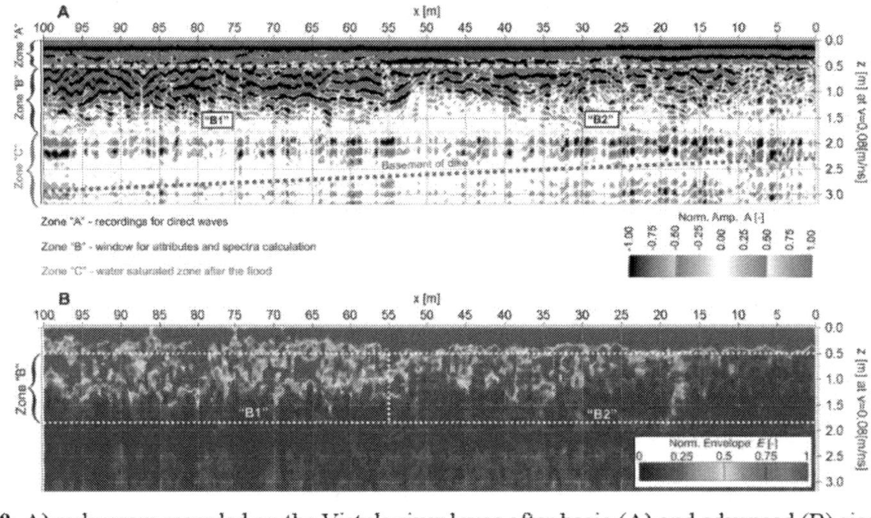

Fig. 10. A) radargram recorded on the Vistula river levee after basic (A) and advanced (B) signal processing [1].

Increasing of quality and resolution of radargrams may be also achieved by application of antennae with different constructions and using of modern electronic solutions of georadar systems. As an example, in Fig. 11 radargrams recorded with standard (classic) system (Fig. 11A) and modern HDR (High Dynamic Range) system (Fig. 11B) were presented.

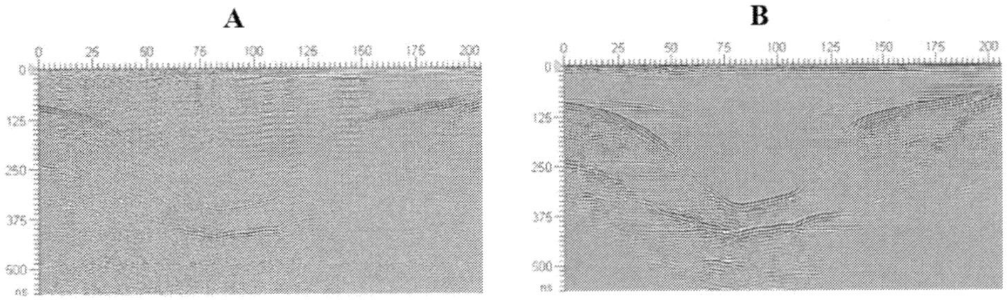

Fig. 11. A) radargrams recorded by using of standard (A) and modern (B) GPR systems (*www.malagpr.com.au*)

4 Conclusions

1. Short-offset reflection profiling (SORP) technique is valuable technique for non-invasive, continuous, fast and cheap examination of flood levees.
2. SORP should be used for interpolation of information delivered by punctual techniques of levee examination like boreholes, geotechnical soundings or measurements with sensors.
3. Disadvantage of SORP technique is fact that only qualitative interpretation may be performed. To improve quality and resolution of radargram, changeable orientation surveys should be carried out and recorded data have to be subjected to more advanced processing.
4. Procedure of Diffraction Hyperbola Analysis (if there is possibility) should be applied which allows to convert from qualitative to quantitative interpretation.
5. It obvious that application of modern solutions of georadar systems delivers better results of surveys than application of classic systems what was presented in the paper.

The works were financed by grant of Cracow University of Technology, no. Ś-2/335/2017/DS.

References

1. H. Marcak, T. Gołębiowski, The Use of GPR Attributes to Map a Weak Zone in a River Dike. *Exploration Geophysics*, **45**(2), 125-133 (2014)
2. T. Gołębiowski, S. Tomecka-Suchoń, J. Farbisz, Application of Complex Geophysical Methods for Non-Invasive Examination of Technical Condition of River Dikes. Proc. of European Symp. on Modern Problems of Flood Protection, Paris-Orlean, , 233-241 (in Polish) (2012)
3. T. Gołębiowski, Application of the GPR Method for Detection and Monitoring of Objects with Stochastical Distribution in the Geological Medium. AGH Ed., p. 251 (in Polish) (2012)
4. T. Gołębiowski, Velocity Analysis in the GPR Method for Loose-Zones Detection in the River Embankments. Proceedings of International Conference "GPR 2010", Lecce, Italy (2010)
5. A. Kot, Expertise of Technical Condition of Flood Gate no. 1 Located in Left Flood Levee of the Vistula River in the Village of Wawrzeńczyce. Firm Aquin, Oświęcim (in Polish) (2013)

6. J. Karczewski,Ł. Ortyl, M. Pasternak, Fundaments of GPR Method. AGH Ed., p. 346 (in Polish) (2012)
7. Mala, Basic and Advance Radar Theory. Documentation of MALA GeoScience, Sweden (2003)
8. E.D. Guy, J.J. Daniels, S. Radzevicius, Demonstration of Using Crossed Dipole GPR Antennae for Site Characterization. *Geophysical Research Letters,* **26**(22), 3421-3424 (1999)
9. S. Radzevicius, Dipole antenna properties and their effects on ground penetrating radar data. *Disertation.* Bell&Howell Information and Learning Company (2001)
10. M. Pasternak, Radar penetration of the ground, WKŁ Ed., (in Polish) (2014)
11. R. Mydlikowski, G. Beziuk, A. Szynkiewicz, Detection of inhomogeneities in structure of flood embankments by means of D.C. resistivity, gpr and frequency electromagnetic method measurements – short note, Acta Geodyn. Geomater., Vol. 4, No. 4 (148), 83-88, (2007)
12. P. Popielski et al., The application of geophysical methods and numerical modeling for the evaluation of embankment technical conditions, Technical Transactions Environment Engineering, 1-Ś/(2013)

Evaluation of the possibility of using the water of the Bystrytsya-Nadvirnyans'ka River in Cherniiv (Ukraine) to supply the population with drinking water

D. Pietrzak*, O. Mandryk**, K. Wątor*, E. Kmiecik* and A. Zelmanowych**

* Faculty of Geology, Geophysics and Environmental Protection, Department of Hydrogeology and Engineering Geology, AGH University of Science and Technology, Adama Mickiewicza 30 Av., 30-059 Cracow (E-mail: *pietrzak@agh.edu.pl; wator@agh.edu.pl; ewa.kmiecik@agh.edu.pl*)
** Ivano-Frankivsk National Technical University of Oil and Gas, ul. Karpacka 15, 76019 Iwano-Frankiwsk, Ukraine (E-mail: *omandryk@mail.ua*)

Abstract
The article presents the results of the research carried out in order to assess the possibility of using surface water of the Bystrytsya-Nadvirnyans'ka River in Cherniiv (western Ukraine), for the public supply of water intended for human consumption. For this purpose an existing database that contains the results of analyses of surface water samples collected in 1999, 2002, 2005, 2008, 2011 and 2014 was used. Each year, from 8 to 13 samples were collected from the Bystrytsya-Nadvirnyans'ka River in Cherniiv. Physicochemical analyses of the samples taken included the determination of pH value, temperature, TDS, alkalinity, hardness, dissolved oxygen, BOD_5, COD, suspended solids and ions: Ca^{2+}, Mg^{2+}, Na^+, K^+, Fe^{2+}, NH_4^+, Cu^{2+}, Cl^-, SO_4^{2-}, PO_4^{3-}, HCO_3^-, NO_2^-, NO_3^-. These chemical analyses were verified by calculation of errors based on the ionic balance.
The results of the analyses were referred to the polish applicable requirements for surface water used for public supply of water intended for human consumption and to the regulation regarding the classification of the surface water status and environmental quality standards for priority substances. The results indicate that water of the Bystrytsya-Nadvirnyans'ka River in the area of Cherniiv was out of the class in the years 1999 and 2002 due to exceeding the limit values for category A3 for Cu^{2+}. On the basis of incomplete assessment of the status of the Bystrytsya-Nadvirnyans'ka River water (due to the tests limitation to the physical and chemical components) determined that the water has a bad status because it exceeded the limits for class II for Cl^-, SO_4^{2-}, NO_3^- and TDS. In the samples collected in 1999 and 2002 it is also observed exceeding the maximum limit concentrations for Cu^{2+}.

Keywords
Surface water, ecological status, chemical status, Cherniiv, Bystrytsya-Nadvirnyans'ka

INTRODUCTION

Through Ukraine flows many great rivers like Dnieper and Dniester. They make up large surface water resources that can be used for public supply. For this purpose appropriate regulations are established to check the water suitability for human consumption and also to define a series of processes through which water have to pass to purify and treat it.

The requirements for surface water which is used to supply the population with water intended for consumption in Poland are set in the Regulation of the Minister of the Environment (RME, 2002). This document defines the requirements that surface water should fulfil, determines the frequency of water sampling, the reference methodology of analysis and the assessment methods, whether the water fulfil the required conditions. The water is classified into three categories A1, A2 and A3 on the base of limit values for water quality indicator included in the regulation.

In Poland the requirements for assessing the status of the surface water bodies (SWB) are specified in the Regulation of the Minister of Environment of the 21 July 2016 (RME, 2016). It defines the criteria and methodology for assessing the status of SWB, including the classification of biological, physicochemical and hydromorphological elements for the 5 surface water quality classes. According to the regulation, the assessment of surface water status is based on the ecological and chemical status of a given SWB.

Assessment of an ecological status is made by assignment one of the five water quality classes to SWB. In turn, the classification of the chemical status is done by comparison the results of analysis with the environmental quality standards for priority substances and other pollutants.

The article presents the assessment of possibility of using the Bystrytsya-Nadvirnyans'ka River water, in Cherniiv, to supply population with drinking water according to the Polish regulations (RME, 2002). This assessment was based on the existing database containing the results of field and laboratory measurements conducted during 6 years (1999, 2002, 2005, 2008, 2011, 2014). In addition, an incomplete assessment of surface water status with the accordance to (RME 2016) was made on the basis of available data.

CHARACTERISTICS OF THE RESEARCH AREA

The study area includes surroundings of Cherniiv. It is a village located in the Ivano-Frankivsk Oblast of western Ukraine, in the Tysmenytsia District, alongside the Bystrytsya-Nadvirnyans'ka River (Fig. 1). Its surface area is 23.35 km^2, and the population is 3.9 thousand (https://pl.wikipedia.org).

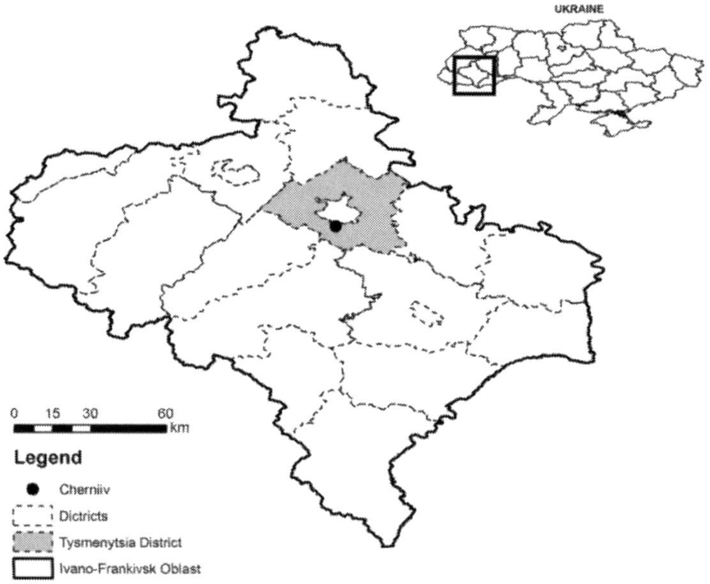

Figure 1. Location of the research area

The Bystrytsya-Nadvirnyans'ka River originates on the northern foothills of Small Syvulia mountain (Gorgany mountain massif). Its total length is 82 km and the area of its catchment is 795 km^2. It joins the Bystrytsia of Solotvyn north of Ivano-Frankivsk creating the Bystrytsia River and finally flows into Dniester as its right-bank tributary. The river has some tributaries: Manyavka, Sadzhavka, Radchanka, and Great Lukavets (https://alchetron.com).

METHODOLOGY

To determine the usefulness of surface water as a water for human consumption, and to assess the status of water in the Cherniiv area, an existing database was used. It contains the results of analyses of surface water samples collected in 1999, 2002, 2005, 2008, 2011 and 2014. Each year, from 8 to 13 samples of surface water were collected from the Bystrytsya-Nadvirnyans'ka River in Cherniiv. Sampling and analysis were done in accordance with the guidelines of Ukrainian Ministry of Ecology and Natural Resources of Ukraine. The analysis included determination of TDS, total hardness, pH, alkalinity, temperature, dissolved oxygen (DO), biochemical oxygen demand (BOD), chemical oxygen demand (COD) and ions: Ca^{2+}, Mg^{2+}, K^{+}, Fe^{2+}, NH_4^{+}, Cu^{2+}, Cl^{-}, HCO_3^{-}, SO_4^{2-}, NO_2^{-}, NO_3^{-}, PO_4^{3-}. In further calculation average results from each year were used.

Due to the lack of detailed data on the methodology, it was possible only to verify the correctness of the analysis by calculation of the errors based on the ionic balance. The values of this error were within the range from 3.76% to 6.68%. For sum of main ions [mval/L] lower than 5 mval/L the maximum allowed error is 10% according to polish standard (PN-C-04638-02). On this basis it was found that all analyses were conducted correctly.

RESULTS OF WATER PHYSICOCHEMICAL ANALYSES

The dominant hydrochemical types of Bystrytsya-Nadvirnyans'ka River in Cherniiv, according to Szczukariew-Prikłoński classification, are two-ionic water: bicarbonate-calcium (HCO_3-Ca) in 2005, 2008 and three-ionic water: bicarbonate-sufate-calcium water (HCO_3-SO_4-Ca) in 2011 and 2014, bicarbonate-calcium-potassium (HCO_3-Ca-K) in 1999 and bicarbonate-calcium-magnesium (HCO_3-Ca-Mg) in 2002.

Average value of mineralization for all samples taken during these years is 222.4 mg/L. The highest mineralisation is observed in 2014 (238.4 mg/L) and the lowest in 2011 (203 mg/L). The mineralization is similar for all years and does not show upward or downward trends.

Table 1 presents minimum, maximum and average concentrations of selected physicochemical elements in the tested points as well as threshold values for good status (RME, 2016, Attachment 1, Type of river: 12 - flysch stream) and limit values of the analysed water quality indicators (RME, 2002, Attachment 1). Bold values indicate that threshold values for good status are exceeded.

ASSESSMENT OF SUITABILITY OF SURFACE WATER AS SUPPLY WITH WATER USED FOR HUMAN CONSUMPTION

The assessment of suitability of surface water as water for human consumption the average values of the measured physicochemical elements from each year were compared with the threshold values of the water quality indicators (Tab. 1). The exceeding of threshold values (RME, 2002) for category A3 (the worst category) were found in 2 samples taken in 1999 and 2002. In both cases it concerns copper concentration. In the next years (2005, 2008, 2011, 2014) the copper concentration was not measured, but it does not mean that it was absent in this water. On the contrary it can be expected that during these years the copper concentration also could be much higher than limit value for category A3. The other indicators for all studied years are in the range of the threshold values for the category A1. According to the regulation (RME, 2002), it is permissible to exceed the limit values for a selected category not marked with an asterisk (*) in 10% of samples if they are not hazardous to human health and in subsequent water samples, no threshold values were exceeded. But in this case high concentrations of copper are observed in both years 1999 and 2002. Even though copper is not particularly hazardous for human health, it belongs to substances that are harmful to the aquatic environment.

https://doi.org/10.1051/e3sconf/20183001009

Table 1. Statement of the selected physicochemical elements of the Bystrytsya-Nadvirnyans'ka River's water

Elements	1999	2002	2005	2008	2011	2014	Mean	Min	Max	Threshold value for good status, II class (RME, 2016)	Threshold values of the water quality indicators (RME, 2002)		
											A1	A2	A3
TDS [mg/L]	**221.6**	**203.1**	**225.2**	161.9	200.2	**236.8**	**208.1**	161.9	**236.8**	203	-	-	-
Total hardness [mg $CaCO_3$/L]	107.0	111.1	114.7	105.9	107.6	118.5	110.8	105.9	118.5	144	-	-	-
pH [-]	8.22	7.91	7.72	7.75	7.43	7.67	7.78	7.43	8.22	7.4-8.4	6.5-8.5	5.5-9	5.5-9
Alkalinity [mg $CaCO_3$/L]	97.6	87.59	99.6	95.60	91.59	101.6	95.60	87.59	101.6	219.0	-	-	-
Temperature [°C]	8.88	11.5	9	10.62	10.08	9.67	9.96	8.88	11.5	24	25*	25*	25*
Suspension [mg/L]	4.125	5.8	3.42	9.62	3.33	9	5.88	3.33	9.62	17.3	25	3	35
DO [mg O_2/L]	9.79	9.96	10.07	10.06	11.05	10.83	10.29	9.79	11.05	>8.9	-	-	-
BOD [mg O_2/L]	2.55	2.24	2.11	1.5	1.38	1.6	1.90	1.38	2.55	2.9	3	5	7
COD [mg O_2/L]	-	13.66	10.08	8.45	5.78	7.03	9.00	5.78	13.66	14	25	30	30
Cations and anions [mg/L]													
Ca^{2+}	33.63	33.4	37.08	34.62	33.17	36.83	34.79	33.17	37.08	51	-	-	-
Mg^{2+}	5.04	6.7	5.38	4.11	5.93	6.82	5.66	4.11	6.82	11.7	-	-	-
NH_4^+	0.27	0.3	0.15	0.4	0.21	0.12	0.24	0.12	0.4	0.54	-	-	-
Cl^-	**18.88**	**19.8**	**18.33**	11.31	9.82	**13.5**	**15.27**	9.82	**19.8**	12.8	250	250	250
SO_4^{2-}	24.5	23.8	24.75	26.54	26.5	**34.25**	26.72	23.8	**34.25**	28.2	250*	250*	250*
NO_2^-	0.04	0.03	0.02	0.04	0.007	0.024	0.027	0.007	0.04	0.082	-	-	-
NO_3^-	**6.98**	5.26	4.33	3.19	1.97	2.38	4.02	1.97	**6.98**	6.64	50*	50*	50*
PO_4^{3-}	-	0.031	0.035	0.02	0.021	0.031	0.028	0.02	0.035	0.21	0.4	0.7	0.7
Substances that are particularly harmful to the aquatic environment [mg/L]													
Cu^{2+}	**3.03**	**2.3**	-	-	-	-	**2.67**	**2.3**	**3.03**	0.05	0.05*	0.05	0.5

Explanations: TDS – total dissolved solids, DO – dissolved oxygen, BOD – biochemical oxygen demand, COD – chemical oxygen demand, * derogations allowed due to exceptional conditions, as defined in the Regulation (RME, 2002).

On this basis surface water of the Bystrytsya-Nadvirnyans'ka River should be classified as out of the category in 1999 and 2002 because of high concentration of copper, what mean that they cannot be used for human consumption. During years 2005, 2008, 2011 and 2014 this surface water belong to category A1 and could be used for public supply. Surface water which belongs to category A1 are water requiring simple physical treatment, in particular filtration and disinfection.

To use surface water of the Bystrytsya-Nadvirnyans'ka River for human consumption, it is necessary to control copper concentration in the river and also to undertake measures to eliminate the causes of this high concentration. High copper concentrations may be associated with electrotechnical, pharmaceutical, rubber and dye industry or anthropogenic pollution (leachate from municipal landfills). Copper is also a part of plant protection products and some mineral fertilizers (Witczak et al., 2013), which easily can reach the surface water along with the precipitation water. This is the most probable cause of its high concentration in surface water, because of an agricultural use of this area.

ASSESSMENT OF THE CHEMICAL STATUS

Relating the values of the physicochemical elements of surface water in the individual monitoring points to the threshold values for good water status (RME, 2016), exceeding of threshold values for class II (Tab. 1) were found for: TDS and Cl^- in samples collected in 1999, 2002, 2005 and 2014, SO_4^{2-} (2014), NO_3^- (1999) and Cu^{2+} in samples collected in 1999 and 2002 (Fig. 2). Not fulfilling the requirements for class II means a status below good for physicochemical elements.

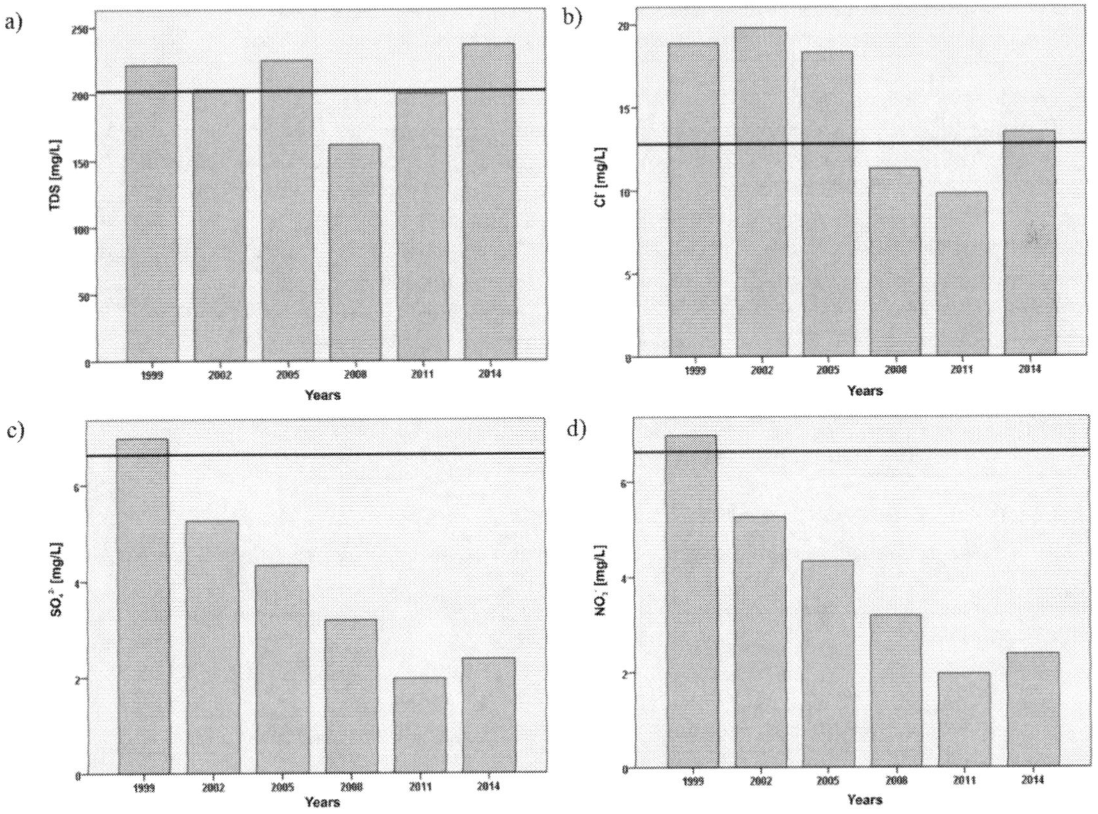

e)

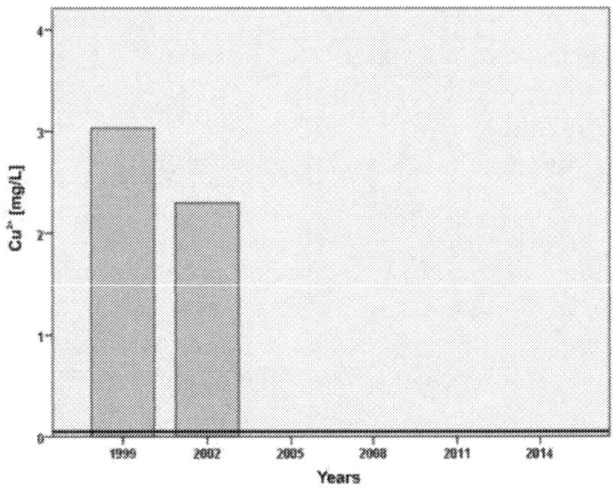

Figure 2. Exceeding threshold values (RME, 2016 – black lines) for: a) TDS, b) Cl^-, c) SO_4^{2-}, d) NO_3^- and e) Cu^{2+}.

According to the regulation, not fulfilling the requirements for class I (very good status) means that the status is below good for hydromorphological elements.

In order to classify ecological status it is necessary to include also biological elements. Due to the absence of biological analysis, in this paper the assessment of ecological status was made only on the basis of the physicochemical analyses results. For the whole investigated area the analysed waters belong to water with poor ecological status in the range of analysed indices.

According to the Regulation (RME, 2016), chemical status is assessed on the basis of not less than 12 results of priority substances and other pollutants measurements. However, only copper was identified in the tested samples, hence the assessment of chemical status is incomplete. Exceeding the maximum allowable concentrations for copper is observed only in samples collected in 1999, 2002. This exceedance determines that the chemical status of the surface water is below good for years 1999 and 2002. For other years the chemical status of this surface water is good most probably because of absence of copper concentration measurement.

The assessment of SWB status is made by comparison of the results of the classification of ecological and chemical status. Classifying studied surface water to poor ecological status and chemical status below good (for 1999 and 2002) or good chemical status (for 1999, 2002, 2005, 2014) results in poor surface water status. Hence, in the area assessment, surface water of the Bystrytsya-Nadvirnyans'ka River in Cherniiv is characterized by poor status.

CONCLUSIONS

The assessment of the usefulness of the surface water of the Bystrytsya-Nadvirnyans'ka River in Cherniiv, in the context of their use as a supply of drinking water for human consumption, was based on the existing database containing the results of field and laboratory analyses conducted during 6 years (1999, 2002, 2005, 2008, 2011, 2014). Each year from 8 to 13 samples were collected from the river and then have been analysed in the laboratory. These results (average values of indicators for each year) have allowed the classification of surface water to the appropriate category of the suitability for consumption and the incomplete assessment of the status of these water.

The dominant hydrochemical type of Bystrytsya-Nadvirnyans'ka River in Cherniiv, according to Szczukariew-Prikłoński classification, are two-ionic water: bicarbonate-calcium (HCO_3-Ca) in

2005, 2008 and three-ionic water: bicarbonate-sufate-calcium water (HCO_3-SO_4-Ca) in 2011 and 2014, bicarbonate-calcium-potassium (HCO_3-Ca-K) in 1999 and bicarbonate-calcium-magnesium (HCO_3-Ca-Mg) in 2002.

Based on the results of analyses, water of Bystrytsya-Nadvirnyans'ka in Cherniiv is out of the category for years 1999 and 2002, which means that the water cannot be used for human consumption, because of high concentration of copper. For other years the analysed surface water is classified to category A1 and could be used for public supply, however the copper concentration was not measured during these years. In addition, incomplete assessment of the status of these water has shown that both in point and area assessment it has a poor status.

Presented evaluation is a simplified assessment. In order to obtain a reliable results, it is necessary to carry out a study covering the full range of indicators listed in the Regulation in an accredited laboratory, in accordance with the sampling collection procedures described in international standards (e.g. ISO 5667-3, 5667-6, 5667-14).

REFERENCES

RME, 2002: Regulation of the Minister of the Environment on the requirements to be met by surface water used to supply the population with drinking water. Dz.U. 2002.204.1728.

RME, 2016: Regulation of the Minister of the Environment on the classification of surface water bodies status and environmental quality standards for priority substances. Dz.U. 2016.1187.

https://alchetron.com/Bystrytsia-of-Nadvirna-1959396-W, access July 2017

https://pl.wikipedia.org/wiki/Czerniej%C3%B3w_(Ukraina), access July 2017

PN-C-04638-02:1989 Water and sewage - Ionic balance of water - How to calculate the balance of ionic water, PKN Warszawa

ISO 5667-3:2012. Water quality -- Sampling -- Part 3: Preservation and handling of water samples. www.iso.org.

ISO 5667-6:2014. Water quality -- Sampling -- Part 6: Guidance on sampling of rivers and streams, www.iso.org.

ISO 5667-14:2014. Water quality -- Sampling -- Part 14: Guidance on quality assurance and quality control of environmental water sampling and handling. www.iso.org.

Witczak S., Kania J. and Kmiecik E. (2013) Katalog wybranych fizycznych i chemicznych wskaźników zanieczyszczeń wód podziemnych i metod ich oznaczania [Guidebook on selected physical and chemical indicators of groundwater contamination and methods of their determination]. Inspekcja Ochrony Środowiska, Warszawa, ISBN: 978-83-61227-13-7, http://www.gios.gov.pl/images/dokumenty/raporty/ebook2_20130422.pdf (in Polish).

Issues of the presence of parasitic protozoa in surface waters

Eliza Hawrylik[1*]

[1] Department of Chemistry, Biology and Biotechnology, Faculty of Civil and Environmental Engineering, Bialystok University of Technology, Wiejska 45E Street, 15-351 Bialystok, Poland

Abstract. Parasitic protozoa are very numerous organisms in the environment that play an important role in the spread of water-borne diseases. Water-borne epidemics caused by parasitic protozoa are noted throughout the world. Within these organisms, intestinal protozoa of the genera *Cryptosporidium* and *Giardia* are ones of the most serious health hazards for humans.
This paper focuses on the problem of the presence of parasitic protozoa in surface waters. Characteristics of the most frequently recognized pathogens responsible for water-borne outbreaks were described, as well as sources of contamination and surface waters contamination due to protozoa of the genus *Cryptosporidium* and *Giardia* were presented. The methods of destroying the cysts and oocysts of parasitic protozoa used nowadays in the world were also presented in a review.

1 Occurrence of parasitic protozoa in surface water

Parasitic protozoa are unicellular animal organisms that are very numerously isolated from natural waters, soils, food and media contaminated with animal manure and diseased humans. They are cosmopolitans, and are found in all countries of the world. They play an important role in the spread of water-borne diseases [1,2]. Water is the most common source of infection with protozoa, such as *Entamoeba histolytica*, *Giardia intestinalis*, *Cryptosporidium* sp., *Cyclospora cayetanensis*, *Toxoplasma gondii*. The first epidemics caused by *G. intestinalis* and *E. histolytica* were already described in 1946, and the etiology of human infection caused by *Cryptosporidium* sp. infection was presented in 1976. Since then, epidemics caused by parasitic protozoa have been recorded all over the world. Most of the infections and deaths caused by parasitic protozoan diseases affect people in developing countries, but they can also cause serious diseases in developed countries [3,4].

The most commonly diagnosed pathogens in water-borne epidemics are *Giardia* sp. and *Cryptosporidium* sp. Until 2010, 524 human lesions attributed to water-borne parasitic protozoa were described, of which 285 were caused by *Cryptosporidium* sp., and 202 cases were caused by protozoa *Giardia* sp. [3]. According to Efstratiou et al. [5], in the period 2011-2016, at least 381 cases of epidemics caused by parasitic protozoa transmitted by water were documented. Almost half of them (49%) occurred in New Zealand, 41% of

* Corresponding author: e.hawrylik@doktoranci.pb.edu.pl

outbreaks were recorded in North America and 9% in Europe. The most common etiologic factor was *Cryptosporidium* spp. reported in 63% (239) outbreaks, while *Giardia* spp. was mentioned in 37% (142) cases.

Increasingly common cryptosporidioses and giardioses result from infecting people with parasitic protozoa and occur as a result of direct or indirect contact through:

- consumption of water contaminated with protozoa (drinking water, surface water, swimming pool water, jacuzzi),
- consumption of undercooked contaminated food,
- contact with objects, surfaces contaminated with feces of humans or animals infected with parasitic protozoa.

It is assumed that the infective dose for healthy adults is low, and amounts to less than 10 oocysts, while in the case of immune-compromised people, a single oocyst may cause cryptosporidiosis [6,7].

Table 1. Parasitic diseases caused by protozoa [8].

Kind of the parasite	Triggered diseases
Entamoeba histolytica	Intestinal amebiasis, amoebic dysentery
Toxoplasma gondi	Toxoplasmosis
Giardia intestinalis	Giardiasis, lambliasis
Cryptosporidium parvum	Cryptosporidiosis
Naegleria fowleri	Naegleriasis

1.1 Protozoa of the genus *Cryptosporidium*

Cryptosporidium protozoa are organisms of *Apicomplexa* type, *Eucoccidiriorida* order, *Cryptospridae* family. They parasitize intracellularly, mainly in epithelial cells of the digestive tract of humans and animals. At least 19 species and several parasitic genotypes are known in many mammals [9]. These protozoa cause a disease called cryptosporidiosis in humans and animals. In immunologically healthy individuals, *Cryptosporidium* causes disease in the form of short-term diarrhea, and in people with immune deficiencies (mainly children and people with immune-compromised organisms), chronic, life-threatening diarrhea [10,11]. The most common parasite in Europe is *C. parvum*, the largest epidemiological significance is *C. parvum* type II [12], while *C. hominis* is most commonly found in the USA and Australia [2].

Invasive forms of *Cryptosporidium* (oocysts) are commonly found in surface waters, in which they survive for a long time [2]. There are two types of *Cryptosporidium* oocysts - thin-walled (there are about 20% of them), which are not excreted outside the host and thick-walled (80%), which escape into the environment, mainly to water and soil [13]. *Cryptosporidium* oocysts show very high resistance to adverse external conditions. They are able to survive freezing in the temperature range from -15 to -20°C, and in the isotonic solution at a temperature of 4°C, they survive even 18 months. In addition, oocysts are particularly resistant to chlorine [14]. Only after 4 hours at a concentration of 30 mg Cl_2/dm^3 of water, the number of oocysts decreased by 99% [6,15]. In addition, they show resistance to iodine and bromine in doses used in the water treatment process [16].

Estimates indicate that 65 million oocysts of *C. parvum* gets into the environment daily from the sewage treatment plant servicing about 50,000 inhabitants, and up to 14,000 oocysts can be detected in a liter of untreated wastewater [17]. Research conducted in the USA (Arizona) showed that in 24 samples of raw wastewater from 2 treatment plants, the average concentration of *Cryptosporidium* oocysts was 74-100 oocyst/dm^3 [18]. According to WHO [6] and Smith [19], *Cryptosporidium* oocysts were detected in the range from 3.3 to 20,000 oocyst/dm^3 in purified wastewater, 0.006-2.5 oocyst/dm^3 in surface waters contaminated with agricultural sewage, 0.66-500 oocyst/dm^3 in recreational waters, and 0.006-4.8 oocyst/dm^3 in drinking water.

1.2 Protozoa of the genus *Giardia*

The protozoa of the genus *Giardia* are flagellates belonging to the type *Sarcomastogophora*, order *Diplomonadida*, family *Hexamitidae* [9]. *G. intestinalis (G. lamblia)* is an invasive species for humans. It is believed that 2-5% of the population in developed countries is infected with this protozoan, while in Poland, a few to a dozen or so percent of adults are infected [20,21]. Other species of this genus, such as *G. muris, G. agilis, G. ardeae* and *G. psittaci*, are non-invasive to humans [22,23].

Protozoa of the genus *Giardia* are located in the digestive tract of humans and animals, and their vegetative forms (trophozoites) occur only in the host organism. In unfavorable conditions, they develop cysts that are excreted into the environment in feces. They are resistant to environmental factors - in chlorinated water at 4-18°C, they can survive up to 3 months, but only 4 days at 37°C [24]. In water from a river or lake, cysts survive up to several months. Among disinfection methods used, UV rays and ozonation are more effective than chlorine [6].

According to studies carried out by Kitajima et al. [18] and Hatam-Nahavandi et al. [25], 1600-4900 cysts from *Giardia* genus (Iranian area) may be present in 1 dm^3 of raw sewage or their number may even reach 4800-6400 cysts (USA, Arizona).

Giardiosis, like cryptosporidiosis, mainly occurs in children and people with reduced system immune. *Giardia* infection is considered a "dirty hands disease". One of the most common transmission routes is the human-environment system. *G. intestinalis* causes inflammation of the small intestine, morphological changes, disturbances in absorption of amino acids, fatty acids, simple sugars and some vitamins (e.g. A, B_{12}), which leads to significant malnutrition. The infection is manifested by the occurrence of increased nausea, vomiting and bloodless diarrhea [26,27].

2 Sources of water infections with protozoa of the genera *Cryptosporidium* and *Giardia*

The basic way to fight diseases caused by parasitic protozoa is to recognize the sources of their origin. Parasitic protozoa get into the water as a result of pollution, which can come from municipal and agricultural sewage, organic fertilizers, as well as manure of livestock and wild animals [16].

The reservoir of *Cryptosporidium* are people, farm animals (mainly cattle, sheep and goats) and free-living animals (beaver, muskrat, roe deer, deer, wild boar, fox and squirrel), as well as laboratory animals (guinea pig, mouse and rat). The occurrence of *Cryptosporidium* spp. in dogs and cats as well as in domestic and wild birds as well as breeding and wild boar rabbits has also been reported [12,20,28]. According to Siński [13], the main source of environmental contamination with *Cryptosporidium* spp. are infected calves and lambs, which can expel about 10 billion oocysts a day, which makes them widely present in the natural environment. Carriers of this protozoon can also be marine fish, crustaceans, molluscs and cephalopods. Research on the waters of the Cheasepeak Bay in the USA has shown that the mussels are contaminated with *C. parvum* oocysts due to anthropogenic and agricultural pollution entering the waters of the Bay. In addition, *Cryptosporidium* vectors can also be flies that carry most oocysts in their digestive tract and then contaminate the food [29,30].

Table 2. Protozoa from the *Cryptosporidium* genera pathogenic to human [13].

Species of the parasite	Main host	Casual host
Cryptosporidium andresoni	cattle	human, sheep, camel
Cryptosporidium baileyi	hen, turkey	duck
Cryptosporidium canis	dog	human
Cryptosporidium hominis	human	sheep, pigs, cattle
Cryptosporidium muris	rodents, other mammals	human
Cryptosporidium parvum	ruminants, human	rodents
Cryptosporidium suis	pigs	human

The main source of surface water pollution with *Giardia* protozoa cysts are animals infected with *G. intestinalis*. In Poland, infection by *Giardia* was found, among others, in cattle (2.2%), sheep (1.3%), dogs (6.1%) and cats (1.3%). The highest percentage of *Giardia* spp. infections was found in wild rodents, i.e. bank vole and common vole - 93.9% and 96.3%, respectively [31,32]. Earlier reports also mention the responsibility of beavers for water-borne giardioses - it was found that the intensity of water contamination in streams populated by beavers is from 1 to 65 cysts per 100 dm^3 of water [33].

3 Contamination of surface waters with parasitic forms of protozoa

Numerous analyses carried out in the USA, Great Britain and Canada clearly indicate that contamination with parasitic protozoa can occur in various types of waters, i.e. surface, ground, underground, treated, lakes, ponds and swimming pools [34]. The greatest biological pollution concerns flowing and standing surface waters used by people and animals (farmed and wild). Quality of water in a given water reservoir depends on many factors. Environmental studies have shown that water contamination by parasitic protozoa depends primarily on:

- degree of pollution and topography of the area, where the water reservoir is located,
- climate - rainfall can only slightly influence the number of oocysts in surface water, where their survival depends on temperature,
- social factors - infrastructure of water departments and their management.

In addition, the quality of water in the reservoirs is affected by the activity of recreational centers, changes in the direct reservoir catchment and transformation of the lakes' coastline [35,36,37].

Cryptosporidium spp. oocysts are quite common all over the world. Research carried out on approximately 6 thousand surface water samples from the USA showed that 20% of them were contaminated with oocytes, and their average density was 2 oocysts/100 dm^3 of water [38]. Surveys of water sources in Great Britain revealed the presence of *Giardia* cysts in as many as 46% of samples of raw drinking water, 57% of samples containing cysts were found in surface waters. At the same time, a higher percentage of positive tests were recorded always in summer and autumn [2].

Studies on the occurrence of *Cryptosporidium* spp. and *Giardia* spp. in the environment have been conducted in Poland for over 30 years. These are not routine tests, and they mostly cover analyzes of various types of waters. Although raw and purified sewage can be the most serious source of contamination with cysts and oocysts of parasitic protozoa, not much Polish research deals with determining the number of *Cryptosporidium* spp. and *Giardia* spp. in wastewater. Drinking water, as a source of transmission of pathogenic intestinal protozoa, has been studied by numerous authors. There are also frequent studies upon the presence of parasitic protozoa in recreational waters - lakes or swimming pools.

According to Bojar and Kłapeć [2], high contamination with *Cryptosporidium* oocysts is found in surface waters (83%), a lower percentage of positive samples is found in pre-treated (13%) and tap water (22%). With regard to protozoa of the genus *Giardia*, the surface water pollution with cysts is 57%, whereas in the case of pre-treated and tap water, it is identical to that of *Cryptosporidium*. Research on the waters of Lake Malta and the Warta river in Poznań showed that the largest amount of intestinal protozoa occurs between August and November, which indicates the seasonal nature of contamination [39]. Based on the analyzes carried out, Matuszewska et al. [40] found the presence of parasitic protozoa in the environmental samples of water originating from surface intakes. The *Giardia* cysts were detected in 100% of water samples in the number of 1.0-4.5 cysts per 10 dm^3 of water from the intake at Zalew Zegrzyński and from the intake on the Vistula river in the number of 1.0-38.9 cysts per 10 dm^3 of water. *Cryptosporidium* oocysts were present in 50% of surface water samples from the Zalew Zegrzyński intakes and 47.6% samples from the Vistula river, and their number was similar in both cases and ranged from 0.5 to 2.5 oocysts per 10 dm^3 of water. In other studies, Matuszewska et al. point to the presence of parasitic protozoa cysts in surface waters, from which water is subsequently captured by waterworks, the purpose is to supply the population with drinking water [41,42]. Research conducted on outdoor water reservoirs performing the bathing function in Cracow indicated the regular occurrence of live *Cryptosporidium* sp. Oocysts [43]. This allows to suppose that the problem of water contamination with parasitic protozoa of the genus *Cryptosporidum* and *Giardia* may apply to the entire territory of Poland.

4 Prevention of water-borne infection caused by protozoa of the genus *Cryptosporidium* and *Giardia*

Water disinfection, as an element of the treatment process, has various effects on the disposal of parasitic protozoa present in this medium. The action of chlorine is usually a sufficient barrier to protect water against pathogenic microorganisms. In the case of parasitic protozoa, it is necessary to use much higher doses of disinfectants than in the case of bacteria. With the chlorine dose of 0.08 mg/dm^3 at 1-2°C and pH 7, 99% reduction of most vegetative forms of bacteria can be achieved, while in the case of protozoa, similar effect can be achieved only when the dose of chlorine is increased to 3.3 mg/dm^3. At the same time, it is effective only against *Giardia*, but it does not kill *Cryptosporidium*. Also, when using monochloramine, a similar relationship is noted [1,44,45]. Among the disinfection methods used, UV and ozonation appear to be more effective than chlorine compounds in relation to parasitic protozoa. In order to inactivate 99% *Giardia*, the effective dose is 59 mJ/cm^2 and for *Cryptosporidium* 10 mJ/cm^2. In turn, effective doses of ozone are 1.9 mg min/dm^3 and 40 mg min/dm^3, respectively, at 1°C and pH 6-9. It should be emphasized that not only the use of chemicals, but also conventional methods of water treatment such as coagulation, flocculation, sedimentation, filtration cannot always ensure that water intended for consumption will be free from parasitic protozoa of the genus *Cryptosporidium* and *Giardia* [6].

Table 3. Disinfectants used to inactivate *Cryptosporidium parvum* and *Giardia intestinalis* [46]

Disinfectant C*t	***Cryptosporidium parvum***	***Giardia intestinalis***
UV ($mJcm^{-2}$)	19	80
chlorine	7.200	03-121
chloramines	7.200	1.470
chlorine dioxide	78	17
ozone	5-10	1.3

C*t – Factor that is the product of the concentration of disinfectant after time t from the introduction of the disinfectant to the water at its first recipient, expressed in mg/dm^3, providing 99% inactivation at 5°C and pH 7.0.

As mentioned before, the most serious source of parasitic protozoa reaching the environment (surface water) is wastewater, both raw and purified. This is due to the fact that the conducted wastewater treatment processes do not change the lifespan of the protozoan dispersal stages [16]. Studies on the effectiveness of the parasitic protozoa removal in the process of wastewater treatment conducted throughout the world illustrate the variation in the degree of reduction of cysts and oocysts reaching the aquatic environment. According to the literature, in the case of protozoa of *G. intestinalis* in the USA, cyst removal rate reaches 97% and 99% [18], while in Germany, the estimated average removal rate for all wastewater treatment plants is 92% [47], in Spain, the yield is 82% [48], and in Sweden, it is 67-87% [49]. These differences may result mainly from the dissimilarity of applied waste water treatment technologies and the amount of organisms contained in the raw sewage. One should be aware that detection, estimation of abundance and species identification of parasitic protozoa in sewage is an important issue from the point of view of engineering and environmental protection as well as health protection. No less important is the desire to develop more effective methods of destroying the cysts and oocysts in treated wastewater discharged into the waters of receivers.

This work was carried out as part of the own work No. MB/WBiIŚ/3/2017, funded by the Ministry of Science and Higher Education.

References

1. R. Matuszewska, Rocz Panstw Zakl Hig **58(3)**, 569-577 (2007)
2. H. Bojar, T. Kłapeć, MONZ **17(1)**, 45-51 (2011)
3. J. Plutzer, P. Karanis, WATER RES., **101**, 318-332 (2016)
4. F.M. Rosado-Garcíaa, M.Guerrero-Flórez, G. Karanis, M.C. Hinojosa, P. Karanis, INT J HYG ENVIR HEAL., **220**, 783-798 (2017)
5. A. Efstratiou, J.E. Ongerth, P. Karanis. Water Res., 114, 14-22 (2017)
6. WHO, Guidelines for Drinking - water Quality, 4th ed. (2011)

7. J. Li, H. Dong, R. Wang, F. Yu, Y. Wu, Y. Chang, Ch. Wang, M. Qi, L. Zhang, IJP-PAW, **6**, 8-15 (2017)
8. E. Hadaś, M. Derda, Ł. Skrzypczak, M. Cholewiński, PHiE, **95(4)**, 795-802 (2014)
9. L. Xiao, R. Fayer, Int J Parasitol, **38**, 1219-1237 (2008)
10. W.Q. Betancourt, D.C. Duarte, Mar Pollut Bull, **85(1)**, 268-273 (2014)
11. H. Leclerc, L. Schwartzbrod, E. Dei-Cas, Crit Rev Microbiol, **28(4)**, 371-409 (2002)
12. J. Sroka, J. Karamon, T. Cencek, Życie Wet., **84(6)**, 476-478 (2009)
13. E. Siński, Mikroorganizmy chorobotwórcze i potencjalnie chorobotówcze w ekosystemach wodnych i wodociągowych, Wyd. "Seidel-Przywecki" Sp. z o.o. (2010)
14. N. J. Ashbolt, Curr Envir Health Rpt, **2**, 95–106 (2015)
15. C.M. Carey, H. Lee, J.T. Trevors, Water Res, **38**, 818-862 (2004)
16. M. Połus, R. Kocwa-Haluch, Technical transactions Environmental Engineering, **6(108)**, 183-193 (2009)
17. A. Butarewicz, Organizmy patogenne w osadach ściekowych – ich wykrywanie i unieszkodliwianie. Oficyna Wydawnicza Politechniki Białostockiej (2013)
18. M . Kitajima, E. Haramoto , B.C. Iker, C.P. Gerba,. Sci Total Environ., **484**, 129–136 (2014)
19. H.V. Smith, JRSM, **83(10)**, 629-631 (1990)
20. J. L. Gundłach, A.B. Szadzikowski, Parazytologia i parazytozy zwierząt. Państwowe Wydawnictwo Rolnicze i Leśne (2004)
21. A. Bajer, M. Bendarska, E. Siński, Wiad Parazytol, **55(4)**, 301-304 (2009)
22. M.M. Marshall, D. Naumowitz, Y. Ortega, Ch.R. Sterling, Clin Microbiol Rev, **10**, 67-85 (1997)
23. W. Magdzik, D. Naruszewicz-Lesiuk, A. Zieliński, Choroby Zakaźne i pasożytnicze – epidemiologia i profilaktyka. α- medica Press (2004)
24. K. Niewiadomska, T. Pojmanska, B. Machnicka, A. Czubaj, Zarys parazytologii ogólnej, Wydawnictwo PWN (2001)
25. K. Hatam-Nahavandi , A. Hossein Mahvi , M. Mohebali, H. Keshavarz, I. Mobedi, M. Rezaeian, J ENVIRON HEALTH SCI., **13(1)**, 1-13 (2015)
26. K. Ekdahl, Y. Andersson, Am J Trop Med Hyg, **72**, 825-830 (2005)
27. E. Li, S. L. Stanley Jr., Yamada' s Textbook of Gastroenterology, **121**, 2322 – 2336 (2016)
28. G. Robinson, R.M. Chalmers, Zoonoses Public Hlth, **57(7-8)**, 1-13 (2010)
29. T.K. Graczyk, Wiad Parazytol, **50**, 40 (2004)
30. T.K. Graczyk, B.H. Grimes, B. Szostakowska, R. Knight, W. Kruminis-Łozowska, M. Racewicz, L. Tamang, A. J. Da Silva, P. Myjak, Wiad Parazytol, **50**, 41 (2004)
31. A.C. Majewska, W. Kasprzak, A. Werner, B. Kozakiewicz, Acta Parasitol, **43**, 1-3 (1998)
32. W. Zygner, H. Wędrychowicz, Postep Microbiol, **47(3)**, 287-291 (2008)
33. E.W. Akin, W. Jakubowski, Water Sci Tech, **18**, 219-226 (1986)
34. P. Karanis, C. Kourenti, H. Smith, J Water Health, 5, 1-38 (2007)
35. J.B. Rose, T.R. Silfco, A. Gennaccoro, Fems Microbiol Rev, **26**, 113-123 (2002)

36. N. Kuczyńska-Kippen, B. Messyasz, B. Nagengast, Abstrakt: 22nd International Symposium. The Phytological Section of the Polish Botanical Society – „Algae and biological state of waters – a threat or a support?, 61 (2003)
37. I.C. Vidal Siqueira-Castro, J.A. Greinert-Goulart, T.R.Bonatti, S. Yamashiro, R.M.B. Franco, ENVIRON SCI POLLUT R., **23(11)**, 11357-11362 (2016)
38. M.J. Messner, R.L. Walpert, In Water Quality Technology Conference Proceedings. Denver (2000)
39. A.C. Majewska, Przegląd Eureka, **8**, 32 (2003)
40. R. Matuszewska, M. Szczotko, B. Krogulska, Rocz Panstw Zakl Hig **63**, 499 – 506 (2012)
41. R. Matuszewska, M. Szczotko, M. Bartosik, B. Krogulska, Ochr Sr, **3(31)**, 67-69 (2011)
42. R. Matuszewska, M. Szczotko, R. Giziński, B. Krogulska, Ochr Sr, **3(35)**, 49-52 (2013)
43. M. Polus, Przemysł Chemiczny, **T. 95, nr 1**, 107-109 (2016)
44. S. Morita, A. Namikoshi, T. Hirata, K. Ogum, H. Katayama, S. Ohgaki, N. Motoyama, M. Fujiwara, App Environ Microb, **68**, 5387-5393(2002)
45. A. Horman, R. Rimhanen-Finne, L. Maunula, C.H. Bonsdorff, N. Torvela, A. Heikinheimo, M. L. Hanninen, App Environ Microb**, 70(1)**, 87-95 (2004)
46. J. B. Rose, T.R. Silfco, J Food Protect, **62**, 1059-1070 (1999)
47. C. Gallas-Lindemann, I. Sotiriadou, J. Plutzer, P. Karanis, Epidemiol. Infect., **14**, 9–21 (2013)
48. J.A. Castro-Hermida, I. García-Presedo, M. González-Warleta, M. Mezo, Water Res., **44**, 5887–5896 (2010)
49. B. Berglund, O. Dienus, E. Sokolova, E. Berglind, A. Matussek, T. Pettersson, P. Lindgren, Sci. Total. Environ., **598**, 821-827 (2017)

Evaluation on the Quality of Bangkok Tap Water with Other Drinking Purpose Water

A. Kordach*, C. Chardwattananon*, K. Wongin*, B. Chayaput* and N. Wongpat**

*Water Supply Section, Water Surveillance Division, Water Quality Department, Metropolitan Waterworks Authority, 400 Prachacheun Road, Tungsonghong, Laksi, Bangkok 10210, Thailand
(Email: Auttapol.k@mwa.co.th)
**Director of Water Quality Department, Metropolitan Waterworks Authority, 400 Prachacheun Road, Tungsonghong, Laksi, Bangkok 10210, Thailand
(Email: Nisapas@yahoo.com)

Abstract
The concern of drinking purposed water quality in Bangkok, Nonthaburi, and Samutprakarn provinces has been a problem for over fifteen years. Metropolitan Water Works Authority (MWA) of Thailand is fully responsible for providing water supply to the mentioned areas. The objective of Drinkable Tap Water Project is to make people realize in quality of tap water. Communities, school, government agencies, hotels, hospitals, department stores, and other organizations are participating in this project. MWA have collected at least 3 samples of water from the corresponding places and the samples have to meet the World Health Organization (WHO) guidelines level. This study is to evaluate water quality of tap water, storage water, filtered water, and filtered water dispenser. The water samples from 2,354 attending places are collected and analyzed. From October 2011 to September 2016, MWA analyzed 32,711 samples. The analyzed water parameters are free residual chlorine, appearance color, turbidity, pH, conductivity, total dissolved solids (TDS), and pathogenic bacteria; *E.coli*. The results indicated that a number of tap water samples had the highest number compliance with WHO guidelines levels at 98.40%. The filtered water, filtered water dispenser, and storage water were received 96.71%, 95.63%, and 90.88%, respectively. However, the several samples fail to pass WHO guideline level because they were contaminated by *E.coli*. The result is that tap water has the highest score among other sources probably because tap water has chlorine for disinfection and always is monitored by professional team round-the-clock services compared to the other water sources with less maintenance or cleaning. Also, water quality reports are continuously sent to customers by mail addresses. Tap water quality data are shown on MWA websites and Facebook. All these steps of work should enhance the confidence of tap water quality.

Keywords
Tap water; water quality; water supply; Metropolitan Water Works Authority (MWA); drinking water quality

INTRODUCTION

Metropolitan Waterworks Authority (MWA) is a State Enterprise, which has a major responsibility for tap water production in Bangkok, Nonthaburi, and Samut Prakarn Provinces in Thailand. Service areas cover 2,358-kilometer square and they are divided into 18 branches. MWA has four Water Treatment Plants and ten pumping stations. The production capacity is about 5.9 million cubic meters per day. The Chao Phraya River and Mae Klong River are water resources. Integrated water quality monitoring is managed for water resources and tap water quality. Bio monitoring and online raw water monitoring are used for raw water sources. Moreover, generally, the water samples are analyzed with Laboratories certified by ISO/IEC 17025 every month. In the crisis, the water is closely monitored with water treatment process by certified ISO 9001:2015, ISO 14001:2015 and HACCP. Furthermore, tap water is monitored, analyzed, and quality controlled by our scientists round-the-clock. (MWA Annual Report 2016) During the flooding crisis of Bangkok, although, MWA tap water had confidently complied with WHO guidelines level 2011 (Kitkaew, et.al, 2013), customers have lost trust on drinking water supply so one of the strategies is building trust by conducting a project "Drinkable Tap Water Project." Building up confidence for the customers is the key factor, the activities composed of 1) in-house water quality check, 2) maintaining water quality in storage tanks consultancy, 3) filter container or install correct pumps on communities, government agencies, educational institutions, hotels, hospitals, and private enterprises. Water quality for participating organizations which comply to WHO guidelines for drinking will be certified, and water quality will be continually mornitored at least once a year.

Table 1. Recommended minimum sample numbers for faecal indicator testing in distribution systems* of MWA referenced by WHO guidelines.

Type of water supply and population	Total number of samples per year
Point sources	Progressive sampling of all sources over 3-to 5-years cycles (maximum)
Piped supplies	
< 5000	12
5000-100 000	12 per 5000 population
> 100 000-500 000	12 per 10 000 population plus an additional 120 samples
> 500 000	12 per 50 000 population plus an additional 600 samples

* Parameters such as chlorine, turbidity and pH should be tested more frequently as part of operational and verification monitoring.

In distribution systems, we also have plans and goals in water quality monitoring comply with WHO guidelines. As shown in Table 1, recommended minimum sample numbers for faecal indicator testing is 12 per 50,000 plus an additional 600 samples in the case of population is more than 500,000. (WHO 2011) MWA has clients about 10 million population in our services area (Annual Report 2016) so we have sampling plan in distribution systems about 3,000 samples per year. The number of samples collected from tap water pipes in distribution systems at customers' places cover all our services area; 18 branches, are shown in Fig.1. The distribution of collected sample is determined by the density of population in each service areas to analyze physical, chemical, and biological parameters.

Table 2. Annual list of analyzed MWA Tap water quality parameters

Order	Water Quality	Parameters	Frequency
1	Physical	Color, odor, turbidity	Everyday
2	Chemical	Free residual chlorine, iron, conductivity, total dissolve solid	everyday
		Insight chemical 40 parameters	Once/month
3	Bacteria	*E.coli*	Everyday
4	Heavy metal	Lead, chromium, cadmium, mercury, copper, zinc, arsenic	Once/month
5	Carcinogen	Total organic carbon (TOC), Trihalomethane (THMs)	Once/month
6	Agricultural toxin	Organochlorine, organophosphate	Twice/year
7	Pathogenic bacteria	*Vibrio Cholerae* serotype o:1 , *Vibrio Cholerae* serotype non o:1, *Samonella* sp., *Shigella* sp., *Staphylococcus aureus*, *Clostridium perfringens*	Three times /year
8	Protozoa	*Cryptosporidium* spp., *Giardia* spp., *Cyclospora* spp.	Once/year
9	Virus	Rotavirus, poliovirus, Hepatitis virus , Norovirus	Once/year
10	Radioactive	Beta and alpha	Twice/year
11	Microcystin	MC-LR, MC-YR, MC-RR	4 times/years
12	Phenol	Phenol	4 times/years
13	Cyanine	Cyanine	4 times/years

Moreover, we plan to monitor water quality monitoring in toxic parameters, which cause deleterious effects to health; such as heavy metal, pesticide, agricultural toxin, carcinogen (trihalomethane), radioactive, and pathogenic bacteria, as shown in Table 2.

This study is to assure that tap water quality produced under responsibility of MWA, conforms to WHO guidelines for drinking water quality and does not differ from the other drinking water purpose. Many researchers found that tap water and bottled water had quite the same water quality. (Kornprabha et al. 2005, Marie Eliza Z. S. et al 2008, and Rosa et al 2011) In addition, this research disseminate tap water quality data to users in order to build their trust on MWA tap water quality and realize the importance of frequently cleaning water storage as well as related equipment in their places. The finding can resume the users' confidence in tap water consumption.

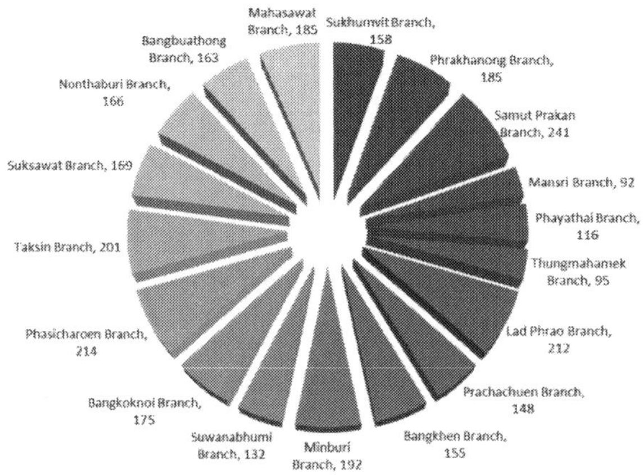

Fig. 1 Scheme of total water samples shown by number of sample per branches per years

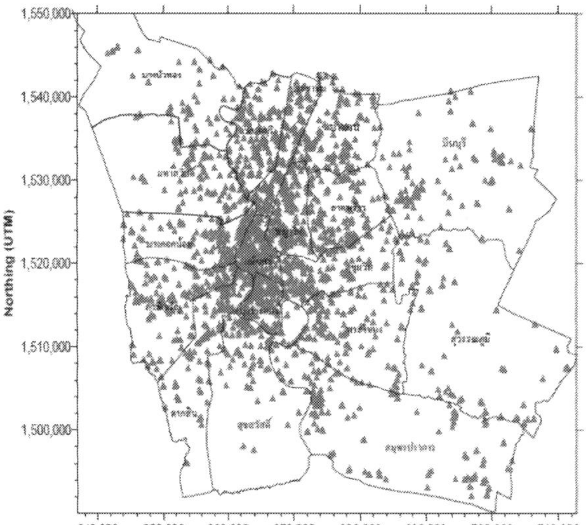

Fig.2 Observation data map of annual water sampling points

MATERIAL AND METHODS

Water samples from participating in Drinkable Tap Water Project amount 2,354 places were collected during October 2011 to September 2016. The samples were collected from tap water pipes in distribution systems, storage tanks, filter tanks, filter dispensers, for at least three samples each places. The samples were analyzed in seven parameters; free residual chlorine (analyzed in the sampling points), appearance color, turbidity, pH, conductivity, total dissolve solid (TDS), and *E.coli*. The analytical methods are shown in Table 3. The samples were collected in appropriate containers. Polyethylene plastic bottles (250 ml) were used for physical and chemical parameters. For bacteria parameter, we used volume 120 ml glasses, which had Sodiumthiosulfate ($Na_2S_2O_3$) 3% amount of 0.1 ml, maintained in metal box to prevent being broken during transportation. The bacteria containers must be sterilized at 170± 10 °C in 2 hours. (APHA 2012)

Table 3. Analytical methods and references of seven parameters for water quality evaluation covered physical, chemical, and biological aspects

Parameters	Analytical methods	References
Free residual Chlorine	DPD Colorimetric Method	APHA,AWWA,WEF 4500-CL G.
Physical characteristic		
Appearance color	Visual Comparison Method	APHA,AWWA,WEF 2120 B.
Turbidity	Nephelometric Method	APHA,AWWA,WEF 2130 B.
Conductivity	Electrical Conductivity Method	APHA,AWWA,WEF 2510 B.
pH	Electrometric Method	APHA,AWWA,WEF 4500-H^+ B.
Chemical characteristic		
Total Dissolved Solids (TDS)	Total Dissolved Solids Dried at 180 °C	In house method based on APHA,AWWA,WEF 2540 C.
Biological characteristic		
Escherichia coli (*E.coli*)	Fluorogenic Substrate Test	APHA,AWWA,WEF 9221 F.

Note; APHA = American Public Health Association, WEF = Water Environment Federation, AWWA = American Water Works Association

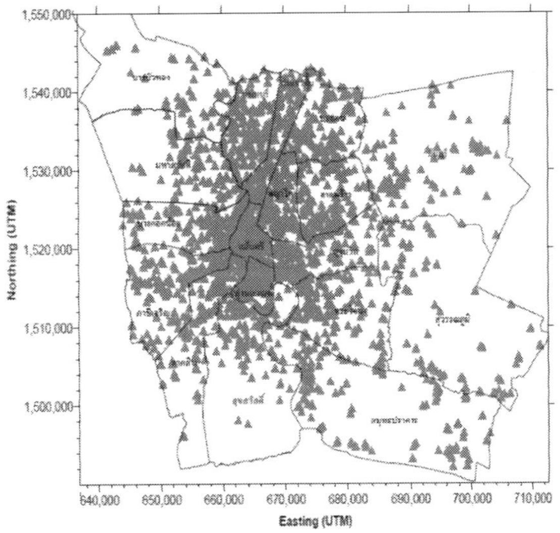

Fig.3 Observation data map of water sampling points during October 2011 to September 2016

FINDINGS AND DISSCUSSTION

The samples from participating were collected during October 2011 to September 2016, MWA analyzed 32,711 samples. Seven parameters according to MWA water quality standard (MWA Standard 2013) were implemented for analysis, some of which were reported in this paper as shown in Fig.4. The results showed that tap water, filtered water, filtered dispenser water, and storage water were clear appearance and low turbidity. Their appearance colors were less than 15 TCU. Average turbidity was in the range of 0.17 to 23.4 NTU, the lowest turbidity of which was from filtered water because it had filter for removing particles. Average conductivity ranged from 2 to 3700 µS/cm. The high conductivity was from contaminated tap water and filter water. In case of tap water, that is probably because the effect of seawater intruded during drought season, resulting from conventional water technology. In case of filter water, that it feasible because filters did not clan wash. Average total dissolved solids (TDS) tended to be same way of conductivity in the range of 1 to 2220 mg/l possibly due to the mineral in the water. Average pH values in the range of 5.80 to 10.55. Their free residual chlorine values were in the range of 0.00 to 2.00 mg/l. The highest free residual chlorine values were from tap water. However, the others were lower free residual chlorine values than the one probably resulting from loss of chlorine during stored in storage tanks or trapped in the filters. Moreover, the results indicate that four types of water were not difference in statistical significance at 0.95-confidence level, as shown in Table 4.

Table 4. Water quality of water tap water, filtered water, filtered water dispenser, and storage water

Parameters	Tap water			filtered water			Filtered water dispenser			Storage water		
	max	min	P 95	max	min	P 95	max	min	P 95	max	min	P 95
Free residual chlorine	2.00	0.00	1.06	1.13	0.00	0.00	1.03	0.00	0.00	1.71	0.00	0.14
Turbidity	15.5	0.23	1.13	14.0	0.17	0.77	17.8	0.20	0.67	23.4	0.24	1.15
pH	10.55	6.48	7.70	9.89	5.80	7.76	9.53	5.93	7.85	10.12	6.29	7.76
Conductivity	2374	120	501	2264	3	502	3240	2	488	3700	8	503
Total Dissolved Solid (TDS)	1424	72	301	1358	2	301	1944	1	293	2220	5	302

Besides, the results revealed that a number of MWA tap water samples from distribution system had the highest number in compliance with WHO guidelines values for drinking water quality 2011 at 98.40%. The filtered water, filtered water dispenser, and storage water were registered at 96.71%, 95.63%, and 90.88%, respectively as shown in Fig. 5. Interestingly, the consequence of this study showed that a number of water samples classified by water type annually met WHO guidelines levels had a tendency to be same results as those classified by water type during fiscal year 2011 to 2016 (October, 2011 – September, 2016). Results were illustrated in Fig. 6 as following. Furthermore, the water quality of all water types tended to increase compliance with WHO guidelines levels.

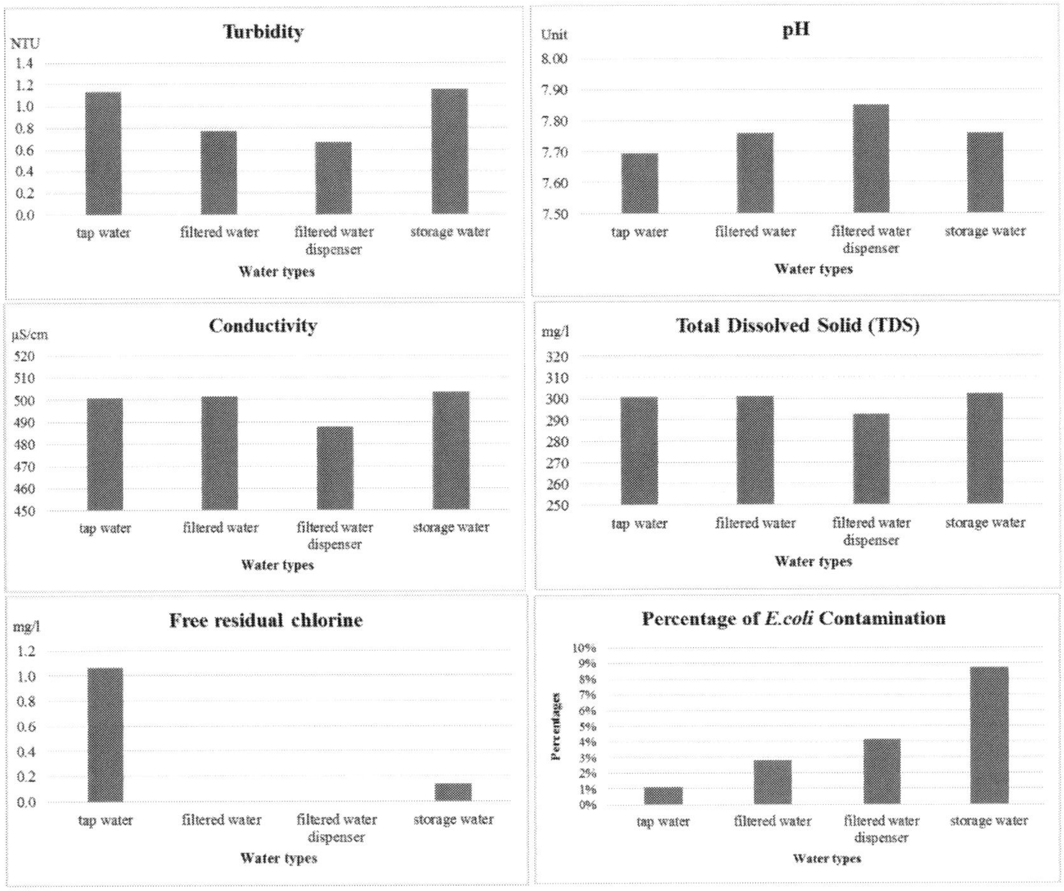

Fig. 4 Average water quality from tap water, filtered water, filtered water dispenser, and storage water classified by parameters

As shown in Fig. 4, the results indicated that the parameters, which were not compliance with WHO guidelines levels, were presence of bacteria *E.coli*, at 3.78% of all samples. The pH, turbidity, and TDS gotten above WHO guidelines level were recorded 0.43%, 0.28%, and 0.07%, of all samples respectively.

The results showed that the storage water had a highest number of samples, which did not meet WHO guidelines values, at 8.73%. The percentage of 4.13, 2.82, and 1.10 were not compliance with WHO guidelines levels, which were from filtered water dispenser, filtered water, and tap water, respectively. According to free residual chlorine chart (fig. 4), the tap water had higher chlorine values than the others. Therefore, it had a lowest chance to contaminate with bacteria *E.coli* from various causes such as broken of a storage tank, lack of water equipment maintenance or stain accumulation resulting in *E.coli* contaminated. WHO recommended in distribution system should be free residual chlorine above 0.2 mg/l to prevent post contamination. (WHO 2011) Moreover, tap water from distribution systems of MWA were compliance with WHO guidelines values probably because water quality monitoring was frequently monitored as WHO guidelines values both quality and quantity. Besides, MWA has 50 online monitoring stations for surveillance tap water quality. The stations are scattering settled cover our services area. Free residual chlorine, turbidity,

https://doi.org/10.1051/e3sconf/20183001011

conductivity, and pH are shown in online monitoring so that this information is used for immediately water quality management. In addition, chlorination at nine pumping stations is one of the MWA methods to improve tap water quality.

Table 5. WHO guideline values for drinking water quality 2011

Parameters	Units	WHO guidelines values for drinking water quality 2011
Free residual Chlorine	mg/l	-
Appearance color	True Color Unit (TCU)	Less than 15
Turbidity	Nephelo Turbidity Unit (NTU)	4.0
Conductivity	µS/cm	-
pH	-	6.5 - 8.5
Total Dissolved Solid (TDS)	mg/l	1000
Escherichia coli (*E.coli*)	P-A/100 ml	Absence

However, MWA tap water from distribution system compliance with WHO guidelines values is safety used and drunk. At present, many houses have storages tank for reserving water and solving low water pressure problem. People do not direct use water supply from distribution pipes so they water used is storage water. Filtered water and filtered water dispenser are used for drink because they need remove chlorine odor and reduce particles in the storage tanks. Nevertheless, the storage tanks and filter are a lot benefits. If they do not take care of equipment, they will make problems. For example, sediment in a bottom tank is collected when it does not clean or expiration filter results in biological contaminated and odor occurred. Therefore, taking care tanks and involved equipment have an effect on water quality.

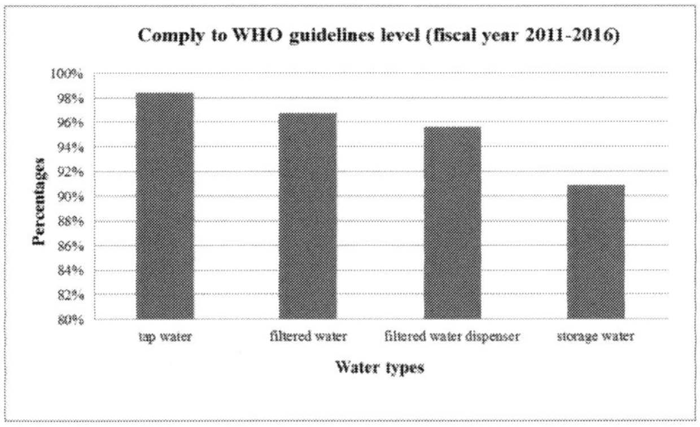

Fig. 5 The percentage of numbers of water sample compliances with WHO guideline values for drinking water quality 2011 classified by water type during fiscal year 2011 to 2016 (October, 2011 – September, 2016)

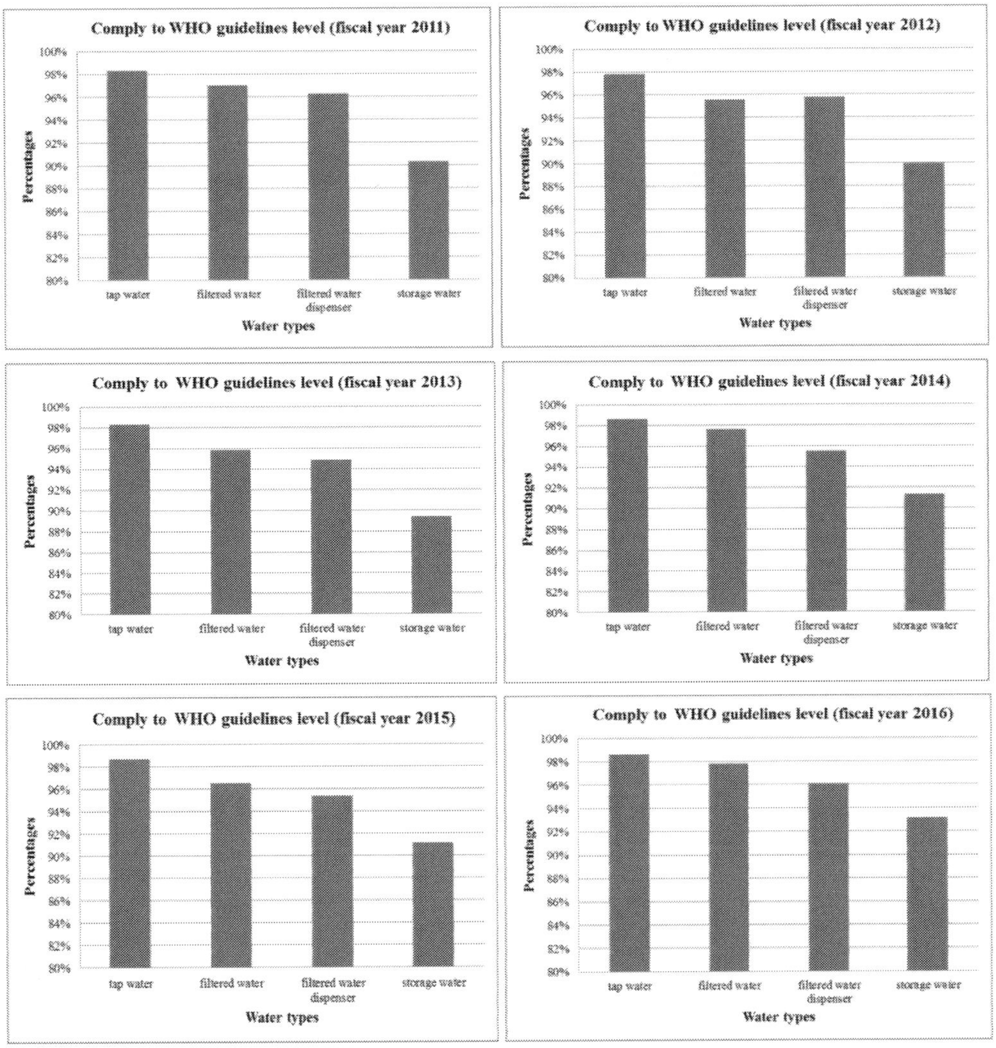

Fig.6 The percentage of numbers of water sample compliances with WHO guidelines values for drinking water quality 2011 classified by water type annually

CONCLUSIONS AND RECOMMENDATIONS

Consequence of this study, several samples were determined and complied with WHO guidelines for drinking water quality 2011. Tap water quality from pipelines in distribution system were not difference water quality from filter water, filtered water dispenser, and storage water. A number of tap water samples from those were a little many number samples complied with WHO level than any other samples. This is because MWA realized the importance of water quality and determined to improve water quality in the future. In order to build the trust of water supply as drinkable tap water, MWA has created Water Quality Integrated Center. The Water Quality Integrated Center was a hub of water quality management and control of remote automatic chlorination at pumping station. This systems will also increase an adequate amount of free residual chlorine in pipelines to prevent post contamination in faraway from water treatment plants. Moreover, MWA had sent water quality report of attended places to our customers by mail addresses. Customers could access water quality information with themselves on the website http://twqonline.mwa.co.th, Facebook clean water clinic, or MWA on mobile application. The dissemination water quality information of tap water to customers would increase trustworthy of water supply and realize that the tap water gained international standard. Customers reached safe use and drink. However, using storage tanks, filter tanks, or filter dispensers should attend to always clean tanks and related equipment or change expired filter to eliminate bacteria contamination risk for safety water.

ACKNOWLEDGEMENT

The authors are grateful to Metropolitan Waterworks Authority of Thailand for financially supporting for the project and permission to publish these results. We also thank Water Quality Analysis Division for analyzing water samples. Helpful proof reading from Asst. Prof. Dr. Bhakaporn Kuljirundhorn and Ms. Suchayada Mahavana are gratefully acknowledged. Also, we would like to acknowledge the Director of Water Quality Department for their useful suggestion and supporting.

REFERENCES

Duangta K., Chatchai C., Sopa C. (2013) Confidence of Tap Water for Drinking Purpose in Bangkok Metropolitan and Its Vicinity after Flooding Crisis. Proceeding *International Conference on Environmental and Hazardous Substance Management (EHSM 2013)*. Bangkok. Thailand. **O-J-08**, 1-8

Kornprabha K., Frank S., Andreas W., Jutta M., and Thomas P. K. (2005). Chemical water quality in Thailand and its impacts on the drinking water production in Thailand. *Science of the Total Environment*. **340**, 57-70

Marie Elize Z.S., Rosangela G.S., Marcio G., Iven C.F., Eliana H.E., Tania U.N., Celso V.N., and Benedito Prado P.D.F. (2008). *Int. J. Hyg. Environ. Health*. **211**, 504-509

Metropolitan Waterworks Authority (MWA): *Annual Report 2016 Metropolitan Waterworks Authority*. https://www.mwa.co.th/download/pln0201/annual2016/Annual_Report_2016_Eng.pdf (accessed 15 May 2017)

Metropolitan Waterworks Authority (MWA): *Metropolitan Waterworks Authority Standard*. (In Thai). https://www.mwa.co.th/download/prd01/tws/who2011.pdf (accessed April 2017)

Rosa C., Franco F., and Paolo T. (2011). Drinking water quality: Comparing inorganic components in bottled water and Italian tap water. *Journal of Food Composition and Analysis*. **24**, 184-193

Standard Methods for the Examination of Water and Wastewater (2012). 22nd edn, American Public Health Association, Washington DC, USA. Part 9060 A.2

Standard Methods for the Examination of Water and Wastewater (2012). 22nd edn, American Public Health Association, Washington DC, USA. Part 9060 B.

World Health Organization (WHO) 2011. Guidelines for Drinking Water Quality, 4rd ed, Geneva

The occurrence of antibiotic resistance genes in tap water – a review

Agata Siedlecka[1,*]

[1]Wrocław University of Science and Technology, Faculty of Environmental Engineering, Wyb. Wyspiańskiego 27, 50-370 Wrocław, Poland

Abstract. The study presents a review of the occurrence of genetic determinants of antibiotic resistance in tap water. The aim of this study was also to compare the applied methods for antibiotic resistance genes (ARGs) investigations in tap water. As the concentration of ARGs in treated, drinking water is expected to be very low and may cause problems in a standard isolation procedure, the special emphasis is placed on the applied procedures of DNA extraction and their efficiency. The study presents the first attempts to obtain DNA directly from tap water. Further efforts must be put to determine the final amount of obtained DNA and the presence of chosen ARGs among the molecules.

1 Introduction

The phenomenon of antibiotic resistance of bacteria is an emerging threat all over the world [1-4]. Nowadays, due to the high availability of antibiotics, the overuse of drugs is inevitable both in developed and developing countries. The increasing prevalence of antibiotic-resistant bacteria (ARB) is the obvious result of the situation [5-10]. As the prokaryotic genomes could be regarded as the molecules of high plasticity, their rapid adaptation to the new environmental conditions, i.e. presence of antibiotic agents, is expected [6, 9]. This phenomenon is explained, among others, by the association of class 1 integrons to gene cassettes containing antibiotic resistance genes (ARGs). Therefore, investigations of ARGs are frequently supplemented by the *intI* gene research [11-13]. As ARB could cause incurable infections, the constant monitoring of their prevalence should be implemented in all environmental components, including wastewaters, soil, air, and waters [6]. Resistant bacteria, if present in wastewaters, could reach the catchment of a drinking water treatment plant with purified, but poorly disinfected wastewaters and then enter the water supply systems, in the case of insufficient water treatment processes [13, 14].

Although the major part of bacteria, including resistant strains, is retained in water treatment plants, the ARGs can enter a water supply system, because the dimensions of genes (which are indeed small DNA fragments) allow them to omit most of the treatment processes, including membrane units [15-17]. Moreover, disinfection processes could cause the lysis of bacterial cells in some circumstances, leading to the release of free DNA

* Corresponding author: agata.siedlecka@pwr.edu.pl

molecules, which could contain the ARGs [9]. Drinking water treatment plants with biological activated carbon [18] or biological treatment tanks [19] may even increase normalized copy numbers of ARGs. To deactivate genes, the disinfectant must reach cell DNA in quantity not altered by previous contacts with other cell structures [9].

There are two major ways of spreading resistance among the bacteria: vertical gene transfer (in the process of proliferation) and horizontal gene transfer, i.e. conjugation, transformation, and transduction, promoted by mobile genetic elements (MGEs) [20, 21]. All of them contribute to the prevalence of the antibiotic resistance. However, the transformation is the most important, as in consequence of contact with ARGs, initially pathogenic, but susceptible (curable) strains may become pathogenic and resistant [6, 16]. From this point of view, these genes could be regarded as the new contaminants [9, 22] and should be mandatorily monitored and suppressed.

In view of the fact, that drinking water is one of the main sources of potable water, its highest quality should be provided regardless of other factors. This means that the occurrence of ARGs should be investigated in water supply systems, as the intake of potable water containing these genes could lead to the transfer of resistance to human pathogenic strains [23-27]. Figueira et al. claimed that *Enterobacteriaceae* are 'vehicles' of resistance in drinking water [28]. The ability to transfer ARGs via conjugation by *E. coli* isolated from tap water was confirmed also by other studies [29, 30].

Moreover, most of the bacteria dwelling in tap water cannot be cultivated in laboratory conditions and remains undetectable. That may lead to underestimation of the problem [31]. Furthermore, the routinely applied method for detection of ARB, i.e. Kirby-Bauer disc-diffusion method, is dedicated to human or animal pathogens, which present diverse optimum condition (temperature 35°C and rich nutrient medium like Mueller-Hinton agar) [32] than autochthonous tap water microorganism (which are psychrophilic and accustomed to habitats poor in nutrients) [33, 34]. Therefore, the other way to investigate the antibiotic resistance phenomenon is needed and the detection of ARGs seems to be an appropriate solution.

2 Investigations of ARGs in tap water

In investigations of ARGs in tap water distribution systems, two main, different approaches may be distinguished:

•the isolation of ARB, identification of pure strains and determining the ARGs presence in isolates

•the metagenomic approach, which demands to obtain all DNA present in a water sample, including microorganisms' genomes and free DNA, to amplify or sequence the DNA and to apply bioinformatics analyses.

Regarding the human pathogenic or opportunistic pathogenic ARB in tap water, the first approach may be considered as sufficient, if appropriate laboratory conditions would be provided during the cultivation of bacteria. Nevertheless, the second approach may include the identification of all microorganisms to species, genes in the request and their amounts and, most importantly, does not demand the cultivation of bacteria in laboratory conditions, what in fact almost always leads to omitting many unculturable or non-viable species [35]. Moreover, the presence of free DNA, released from bacterial cells due to various processes, could be regarded as a dangerous contaminant itself [22], serving as a transferable element in the phenomenon of transformation (i.e. uptake and incorporation of DNA in a cell). From that point of view, not ARB, but simply ARGs present in tap water are the real threat factors.

In fact, the first approach is more common in literature. To date, there is a plethora of data concerning the presence of ARB in tap water [36, 37], but due to the development of molecular biology, insight into ARGs, omitting the cultivation of ARB, becomes a trend [27, 38].

3 ARGs occurring in ARB and DNA isolated from tap water – a review

Techniques commonly used for ARGs investigations include conventional PCR and its modifications, which allow to detect desired sets of genes and quantitative PCR (qPCR), i.e. real-time PCR, which allows to determine the copy number of selected genes. The PCR conditions, sequences of primers and additional information necessary to perform the research are often presented in literature (or in supplementary data). Newly discovered ARGs are constantly deposited in databases, as GenBank, for instance (https://www.ncbi.nlm.nih.gov/genbank/).

Another method for specific genes screening is a microarray with oligonucleotide probes. It could be used in search for virulence genes [39] as well as resistance genes [40, 41]. Although 99% correlation between PCR and microarray approaches was provided [40], the latter seems to be less common in the literature concerning ARGs. Furthermore, Henriques et al. applied also DGGE analysis to determine the molecular diversities of obtained *bla* PCR amplicons [42].

Together with ARGs, MGEs searching is often implemented, as this approach may indicate the prevalence and distribution of genes among the microorganisms [4, 27]. Predominantly, the MGEs investigations are focused on the *intI* gene encoding the class 1 integronase [11, 13].

The most frequently investigated ARGs are: *bla* genes encoding resistance to β-lactams (including the ESBL phenomenon), *ampC* genes encoding resistance to ampicillin, *tet* genes encoding resistance to tetracyclines, *sul* genes encoding resistance to sulphonamides, *erm* genes encoding resistance to erythromycin, *cat* and *cmr* genes encoding resistance to chloramphenicol, *qnr* genes encoding resistance to quinolones, among others [14, 16, 43].

The search for ARGs began with screening the hospital, virulent, pathogenic isolates. For instance, the presence of various *bla* genes in *Enterobacteriaceae* strains isolated from hospital patients, including members of the *blaCTX-M*, *blaSHV*, and *blaTEM* families, was confirmed in many studies [7, 43-45]. The findings of Adesoji et al. indicated, that not only enteric, but also environmental strains carry *bla* genes [46], but this phenomenon is often neglected, as the environmental strains are perceived as harmless. Genes encoding β-lactamases and integrons were detected also in Gram-negative bacteria isolated from estuary [47]. Nevertheless, the occurrence of ARGs in tap water isolates and tap water itself became a fact.

Among 164 multidrug-resistant isolates from raw, treated, and municipal tap water in Nigeria, 61 were carrying *bla* genes. The most common gene family was *blaTEM*, present in 50 out of 61 PCR-positive isolates, which consisted mainly of *Klebsiella spp.* and *Proteus spp.* strains (18.03 % and 14.75 % of the investigated isolates, respectively). However, the authors admitted that some phenotypically ESBL strains did not exhibit *bla* genes in PCR assays [46]. This fact may lead to the conclusion that there are still undiscovered genes conferring resistance to antibiotics or that the mechanisms of resistance are related with different mechanisms, encoded by other genes, as, for example, efflux pumps.

Khan et al. detected *sul1* and *sul2* genes in isolates from tap water samples, but these genes were never found together in the same strain. Moreover, the *sul1* gene and the *intI* gene were not always present in the same strains, indicating that the *sul1* gene could be associated with other MGEs or with bacterial chromosome. The authors were not able to confirm the presence of *qac* genes (encoding disinfectant resistance) in any of investigated strains [21].

In quantitative studies of ARGs, their abundance is often normalized to eubacterial 16S rRNA present in the sample [38, 48]. Samples of DNA extracted from filtrated, disinfected and tap water were investigated in terms of ARGs presence and the effect of chlorination on their occurrence by Shi et al. The *blaTEM-1*, *tetA*, *tetG*, *sulI*, *ermA* and *ermB* genes were present in every sample, while *ampC* was found only in disinfected and tap water samples, indicating the promotion phenomenon of water treatment and distribution processes on these ARGs. The other gene, *aphA2* (encoding resistance to aminoglycoside), seemed to disappear in tap water and three genes: *tetB*, *tetC* and *drfA17* (encoding resistance to trimethoprim) remained undetected in each sample [38]. Results of qPCR showed that from eight ARGs taken into consideration in the study of Shi et al., the abundance of seven was increased after the chlorination unit. However, distribution of treated water decreased abundances of all eight ARGs. Interestingly, ARGs present in filtrated, disinfected and tap water could encode 9, 11 and 5 types of antibiotic-resistant proteins, respectively. Moreover, the class 1 integronase gene dominated the MGEs in each water sample, and the trend of the presence of integrons was similar to that of ARGs [38].

Studies of Shi et al. and Xi et al. demonstrated that the chlorination process reduced the abundance of the *sul1* gene, but in the same time increases the richness of other genes [38, 48]. Other treatment units also may influence ARGs quantity and diversity in drinking water [18, 19].

In the study of Xi et al., the copy number of majority of tested ARGs (*cat*, *cmr*, *blaTEM*, *sulI*, *sulII*) normalized to the copy number of 16S rRNA genes were higher in tap water samples than in treated water; the exception was the *blaSHV* gene (no significant difference observed) and the *tetO* and *tetW* genes, detected only in source water. Some samples demonstrated correlations between ARB and ARGs occurrences, i.e. phenotype resistance profiles corresponded with detected genes encoding resistance to insufficient antibiotics, though there was no significant correlation between ARGs and the total number of bacteria, nor 16S rRNA quantity [48].

Vaz-Moreira et al. detected the *intI1*, *blaTEM* and *sul1* genes at copy numbers normalized to water volume. From 11 tested taps, *intI1* was quantified in 2 taps at all sampling dates, while *blaTEM* and *sul1* were detected in all, but quantified in 9 and 2 taps, respectively, without temporal recurrence in both cases. The *qnrS* gene was not detected in any sample. Low quantities of *intI*1 and *sul1* were attributed to the phenomenon of hiding these genes carriers by complex microbiota of tap water [49]. Nevertheless, the quinolone resistance genes were detected in *Aeromonas spp.* from tap water samples [50].

The increase of ARGs in drinking water distribution systems was confirmed in other studies. The amount of ARGs in tap water was from 6.4 to 109.2-fold higher than in treated water, produced in the water treatment plant in Hangzhou (China). The positive correlation between ARGs and MGEs in water distribution system was also confirmed. As high-throughput qPCR was applied in the study, the total number of 285 genes were searched. As a result, from 112 to 122 genes were found in tap water samples, depicting the diversity of ARGs in this medium [17]. The elevated level of ARGs in tap water, in comparison to treated water, was confirmed also by Zhang et al. [51].

4 Extraction of DNA from ARB and tap water samples

In every research approach, the most crucial step is the collection of samples. The applied procedures may affect all the subsequently obtained results. In the case of molecular biology investigations, the most essential stage is the extraction of DNA [52, 53].

DNA may be extracted from microorganisms in many ways, including the employment of commercial kits and other physical-chemical methods. Khan et al. extracted DNA by mixing strains with PBS, freezing and thawing the suspension at -80°C and 70°C, respectively, and centrifugation [21]. The simplest, but efficient approach was made by Osińska et al., as loopfuls of bacterial colonies were suspended in sterile water, heated at 95°C for 10 min and centrifuged (5000 rpm, 5 min, 4°C) [54].

Hwang et al. compared five methods of DNA extraction, including two commercial kits (dedicated for soil samples) and three classical phenol/chloroform procedures on monocultures of bacteria belonging to different genera (representatives of strains characterized by various cell wall properties). The authors used three approaches to determine DNA extract quality, quantity and purity: spectrophotometric NanoDrop technology, fluorescent-based quantitation Q-bit Quantitation Platform and agarose (0.8 %) gel electrophoresis using a DNA/HindIII fragment ladder. The applied methods revealed significant differences in obtained DNA amounts, reflected in bacterial diversity, but the DNA purity was sufficient in every approach. In general, the DNA obtained by commercial kits was pure enough to perform PCR assay (the minimum A260/A280 ratios was 1.40) and solutions obtained by phenol/chloroform methods required further purification steps in some cases, what could hinder PCR investigations [55]. The alternative solution was implemented by Zhang et al., who combined the application of the commercial kit with further phenol-ethanol methods [51].

As mentioned above, the extraction of DNA from bacterial colonies obtained from tap water and cultivated in laboratory conditions, even if rich in ARGs, may lead to the neglecting of the majority of microorganisms dwelling in tap water pipelines, thereby the metagenomic approach is likely to be an alternative. Nevertheless, DNA obtained from environmental samples may be contaminated with many PCR inhibitors. In the case of tap water, these inhibitors may be humic substances and corrosion, for example, which interfere with the DNA extraction process, altering the PCR results in consequence [55, 56]. Nevertheless, Henriques et al. successfully purified freshwater samples on 5 μm membranes, then concentrated microorganisms on 0.2 μm pore-size filters, washed them in TE buffer, centrifuged (15 200 g, 10 min) and further processed by two commercial kits [42]. In a need for adequate DNA extraction from environmental samples, Bag et al. proposed THSTI method using enzymes, buffers, glass and zirconia beads, among others [52].

4.1 Extraction of DNA directly from tap water – a preliminary research

The aim of this preliminary research was to determine the minimal volume of tap water required to obtain a sufficient amount of DNA using commercial kits. The assumption was made, that the established volume needs to be useful in further comparative research, i.e. should not exceed 5 L of tap water, as larger volumes would be inappropriate in laboratory processing.

DNA was extracted simultaneously from 1 L, 3 L and 5 L of tap water. The output volume of tap water (collected from the same tap, according to PN-EN ISO 19458:2007 procedure) was filtrated in sterile conditions on 0.2 μm membranes (Whatman). Afterwards, the membranes were shaken with 1xPBS buffer for 3 h at room temperature (200 rpm), then centrifuged respectively (9 000 rpm, 15 min). Obtained pellet was subjected to DNA isolation by kits, according to with manufacturers' instructions. Unfortunately, the amount of obtained pellet in this approach was evidently too small to obtain sufficient concentration of DNA, even in 5 L sample. The obtained concentrations were insufficient to perform further research. Therefore, another approach and altered procedure must be applied. The small amount of pellet may be explained by the chosen sampling procedure, as in this study tap water was taken after flushing [PN-EN ISO 19458:2007].

As the issue in question was simply the output volume, not the exact diversity of nucleic acids, the extraction tool of choice was the commercial kit dedicated to Gram-negative bacteria, Genomic Mini (A&A, Poland). According to the literature review, the volumes of fresh water and tap water used in DNA extraction procedures varied widely: from 15 L [49], through 10÷15 L depending on the sample contamination and possible clogging of membrane [51], 4 L [57], 2÷1 L [58-60], 1.5 L [61] to 500 ml [14] or 450 mL [42], depending on the sample origin. However, drinking water samples obtained by other researchers could have been affected by the biofilm dwelling in the taps. The volume collected in the study of Shi et al. was even 2 000 L, but using a different approach for the other objective [38]. In studies applying qPCR analyses, higher amounts of extracted DNA are required to provide the number of copy above the limit of quantification. In this study, the assumption was made, that the required DNA concentration is 20 ng/μL and the purity A260/280 = 1.8÷2.0, what should be the sufficient parameters for further PCR investigations.

5 CONSIDERED SOLUTIONS

As it is known, the quality of tap water is inseparably correlated with water treatment processes. Technologies presently applied in water treatment plants do not provide the sufficient ARGs removing levels [11, 16, 51, 57]. The proposed improvements may be the increase in dosages of chlorine and the intensity of UV irradiation or sequential UV/chlorination [9], but these solutions generate high costs. Among proposed technologies are also: Fenton and ozone oxidation, photocatalytic processes, biodegradation and nanoparticles (the last, though applied in medical research, seems to enhance the conjugative transfer of ARGs among bacteria in terms of water treatment) [9]. Besides the monitoring of ARB and ARGs, the analyses of antibiotics and drugs concentrations in wastewaters, freshwater, and tap water should be implemented, as these agents are probably responsible for resistance promotion [62-64]. According to Schwartz et al., the exposure of susceptible and resistant *Pseudomonas aeruginosa* strains to wastewater and tap water did not affect the expression of the intrinsic ARGs and MGEs [65], so the medium itself is not the main factor of the prevalence of resistance.

The significant issue in ARB and ARGs research is the discrepancy between phenotype and genotype [38] of bacteria dwelling in tap water. This hindrance may be overcome with the application of newly designed technologies for high-throughput sequencing and bioinformatics investigations of obtained sequences. The increasingly used in environmental microbiology studies Next Generation Sequencing (NGS) approach becomes more available and allows to gain seemingly more information of tested sample [66-68]. However, as indicated in this study, the successful implementation of NGS in tap water samples research requires proven, reliable and reproductive procedures of high-efficiency DNA extraction from this medium. This step needs to be broken down and adapted to routine laboratory practice in the case of ARGs monitoring in tap water samples.

References

1. A. Rodríguez-Rojas, J. Rodríguez-Beltrán, A. Couce and J. Blázquez, Int. J. Med. Microbiol., **303**, 293-297 (2013)
2. C.L. Ventola, P.T., **40**(4), 277-283 (2015)

3. M.Frieri, K. Kumar and A.Boutinc, Journal of Infection and Public Health, **10**, 369-378 (2017)
4. A. MacGowan and E. Macnaughton, MEDICINE, **45**(10), 622-628
5. D.M. Fernando, H.M. Tun, J. Poole, R. Patidar, R. Li, R. Mi, G.E.A. Amarawansha, W.G.D. Fernando, E. Khafipour, A. Farenhorst and A. Kumar, Appl. Environ. Microbiol., **82**(15), 4767-4775 (2016)
6. J.L. Martinez, Science, **321**(5887), 365-367 (2008)
7. A. Ostholm-Balkhed, M. Tarnberg, M. Nilsson, L.E. Nilsson, H. Hakan, A. Hallgren, J. Antimicrob. Chemother., **68**, 2144–2153 (2013)
8. WHO, *Global Report on Surveillance* (2014)
9. V.K. Sharma, N. Johnson, L. Cizmas, T.J. McDonald, K. Hyunook, Chemosphere, **150**, 702-714 (2016)
10. L.Price, J. MacDonald, L. Melone, T. Howe, P. Flowers, K. Currie, E. Curran, V. Ness, D. Waddell, S. Manoukian, A. McFarland, C. Kilpatrick, J. Storr, A. Twyman, B. Allegranzi and J. Reilly, Lancet Inf. Dis., doi: **10.1016/S1473-3099(17)30479-6** (2017)
11. A. Farkas., A. Butiuc-Keul, D. Ciatarâş, C. Neamţu, C. Crăciunaş, D. Podar and M. Drăgan-Bularda, Sci. Total Environ., **443**, 932-938 (2013)
12. S. Jia, P. Shi, Q. Hu, B. Li, T. Zhang and X.-X. Zhang, Environ. Sci. Technol., **49**, 12271-12279 (2015)
13. A. Di Cesare, E.M. Eckert, S. D'Urso, R. Bertoni, D.C. Gillan, R. Wattiez, G. Corno, Water Res., **94**, 208-214 (2016)
14. H.-C. Su, Y.-S. Liu, C.-G. Pan, J. Chen, L.-Y. He and G.-G. Ying, Sci. Total Environ., **616–617**, 453-461 (2018)
15. M. Kabsch-Korbutowicz, K. Majweska-Nowak, *Membrane separation processes in environmental protection* (Wrocław University of Science and Technology, Wrocław, 2011)
16. S. Bergeron, R. Boopathy, R. Nathaniel, A. Corbin, G. LaFleur, Int. Biodeterior. Biodegradation, **102**, 370-374, (2015)
17. L. Xu, W. Ouyang, Y. Qian, C. Su, J. Su., H. Chen, Environ. Pollut., **213**, 119-126
18. J. Zheng, T. Chen and H. Chen, Sci. Total Environ., **612**, 1-8 (2018)
19. X.Guo, J. Li, F. Yang, J. Yang and D.Yin, Sci. Total Environ., **493**, 626-631 (2014)
20. Z. Markiewicz, Z.A. Kwiatkowski, *Bakterie, antybiotyki, lekooporność* (Wydawnictwo Naukowe PWN, Warszawa, 2008)
21. S. Khan, C.W. Knapp, T.K. Beattie, Environmental Processes, **3**, 541–552 (2016)
22. J.M. Philip, U.K. Aravind and C.T. Aravindakumar, Chemosphere, **190**, 307-326 (2018)
23. F. Baquero, J.L. Martinez and R. Canto, Curr. Opin. Biotechnol., 19, 260-265 (2008)
24. R. Tao., G.-G. Ying, H.-C. Su, H.-W. Zhou, J.P.S. Sidhu, Environ. Pollut., 158, 2101-2109 (2010)
25. A. Pruden, Environ. Sci. Technol., **48**(1), 5-14 (2014)
26. S. Vincenti, G. Quaranta, C. De Meo, S. Bruno, M.G. Ficarra, S. Carovillano, W. Ricciardi and P. Laurenti, Sci. Total Environ., **499**, 47–54 (2014)
27. J.L. Balcázar1, J. Subirats and C.M. Borrego, Front. Microbiol, **6**(1216), 1-10 (2015)
28. V. Figueira, I. Vaz-Moreira, M. Silva, M.M. Celia, Water Res., **45**, 5599-5611 (2011)
29. S.K. Walia, A. Kaiser, M. Parkash and G.R. Chaudhry, J. Environ. Sci. Health, **39**(3), 6501-662 (2004)
30. B.L. Coleman, M.I. Salvadori, A.J. McGeer, K.A. Sibley, N.F. Neumann, S.J. Bondy, I.A. Gutmanis, S.A. McEwen, M. Lavoie, D. Strong, I. Johnson, F.B. Jamieson, M. Louie, ARO Water Study Group, Epidemiol. Infect., **140**, 633-642 (2012)
31. I. Vaz-Moreira, C. Egas, O.C. Nunes, C.M. Manaia, FEMS Microbiol. Ecol. **83**, 361-374 (2013)

32. R. Schwalbe, L. Steele-Moore, A.C. Goodwin (ed.), *Antimicrobial Susceptibility Testing Protocols* (CRC Press, Boca Raton, 2007)
33. T. Traczewska, M. Sitarska, I. Biedroń, *Ekologiczne i techniczne aspekty powstawanie biofilmu w wodzie* (Wrocław University of Science and Technology, Wrocław, 2014)
34. A. Siedlecka and K. Piekarska, *9th Conference on Interdisciplinary Problems in Environmental Protection and Engineering EKO-DOK 2017*, E3S Web of Conferences, **17** (2017)
35. S.A. Waller, A.I. Packman and M. Hausnerc, J. Microbiol. Methods, **144**, 8-21 (2018)
36. C. Narciso-da-Rocha, I. Vaz-Moreira, L. Svensson-Stadler E.R.B. Moore and C.M. Manaia, Appl. Microbiol. Biotechnol., **97**, 329-340 (2013)
37. C. Narciso-da-Rocha, I. Vaz-Moreira, C.M. Manaia, Sci. Total Environ. **466–467**, 127-135 (2014)
38. P. Shi, S. Jia, X.-X. Zhang, T. Zhang, S. Cheng, A. Li, Water Res., **47**, 111-120 (2013)
39. M.F. Anjum, M. Mafura, P. Slickers, K. Ballmer, P. Kuhnert, M.J. Woodward, R. Ehricht, Appl. Environ. Microbiol., **73**(17), 5692–5697 (2007)
40. M. Batchelor, K.L. Hopkins, E. Liebana, P. Slickers, R. Ehricht, M. Mafura, F. Aarestrup, D. Mevius, F.A. Clifton-Hadley, M.J. Woodward, R.H. Davies, E.J. Threlfall and M.F. Anjum, Int. J. Antimicrob. Agents, **31**, 440–451 (2008)
41. J.G. Frye, R.L. Lindsey, G. Rondeau, S. Porwollik, F. Long, M. McClelland, C.R. Jackson, M.D. Englen, R.J. Meinersmann, M.E. Berrang, J.A. Davis, J.B. Barrett, J.B. Turpin, S.N. Thitaram and P.J. Fedorka-Cray, Microb. Drug Resist., **16**(1), 9-19 (2010)
42. I.S. Henriques, A. Moura, A. Alves, M.J. Saavedra, A. Correia, FEMS Microbiol. Ecol., **56**, 418-429 (2006)
43. G. Wu, M.J. Day, M.T. Mafura, J. Nunez-Garcia, J.J. Fenner, M. Sharma, A. van Essen-Zandbergen, I. Rodríguez, C. Dierik, K. Kadlec, A.-K. Schink, J. Wain, R. Helmuth, B. Guerra, S. Schwarz, J. Threlfall, M.J. Woodward, N. Woodford, N. Coldham and D. Mevius, PLoS One, **8**(9), 1-10 (2013)
44. F. Randrianirina, S. Vedy, D. Rakotovao, C.-E. Ramarokoto, H. Ratsitohaina, J.F. Carod, E. Ratsima, M. Morillon, A. Talarmin, J. Hosp. Infect., **72**, 23-29 (2009)
45. D. Ojdana, P. Sacha, P. Wieczorek, S. Czaban, A. Michalska, J. Jaworowska, A. Jurczak, B. Poniatowski, E. Tryniszewska, International Journal of Antibiotics, 1-8 (2014)
46. A.T. Adesoji and A.A. Ogunjobi, Biomed. Res. Int., **http://dx.doi.org/10.1155/2016/7149295**, 1-9 (2016)
47. I.S. Henriques, F. Fonesca, A. Alves, M.J. Saavedra, A. Correia, Res. Microbiol., **157**, 938–947 (2016)
48. C. Xi, Y. Zhang, C.F. Marrs, W. Ye, C. Simon, B. Foxman, J. Nriagu, Appl. Environ. Microbiol., **75**(17), 5714–5718 (2009)
49. I. Vaz-Moreira, O.C. Nunes, C.M. Manaia, Sci. Total Environ., **586**, 1141-1149 (2017)
50. V. Figueira, E.A. Serra, I. Vaz-Moreira, T.R.S. Brandâo, C.M. Manaia, J. Water Health, **10.1**, 1-10 (2012)
51. S. Zhang, W. Lin, X. Yu, FEMS Microbiol. Ecol., **92**(5), 1-9 (2016)
52. S. Bag, B.Saha, O. Mehta, D.Anbumani, Naveen Kumar, M. Dayal, A. Pant, P. Kumar, S. Saxena, K.H. Allin, T. Hansen, M. Arumugam, H. Vestergaard, O. Pedersen, V. Pereira, P. Abraham, R. Tripathi, N. Wadhwa, S. Bhatnagar, V.G. Prakash, V. Radha, R.M. Anjana, V. Mohan, K. Takeda, T. Kurakawa, G.B. Nair and B. Das, Sci Rep, **6**(26775), 1-9 (2016)
53. A. Felczykowska, A. Krajewska, S. Zielińska and J.M. Łoś, Acta Biochim. Pol., **62**(1), 151-160 (2015)

54. A. Osińska, E. Korzeniewska, M. Harnisz, S. Niestępski, Sci. Total Environ., **577**, 367–375 (2017)
55. C. Hwang, F. Ling, G.L. Andersen, M.W. LeChevallier, W.-T. Liu, Microbes and Environ., **27**(1), 9-18 (2012)
56. I.G. Wilson, Appl. Environ. Microbiol., **63**(10), 3741-3751 (1997)
57. X. Bai, X. Ma, F. Xu, J. Li, H. Zhang, X. Xiao, Sci. Total Environ. **533**, 24–31 (2015)
58. K. Lautenschlager, N. Boon, Y. Wang, T. Egli and F. Hammes, Water Res., **44**, 4868-4877 (2010)
59. K. Lautenschlager, C. Hwang, W.-T. Liu, N. Boon, O. Koster, H. Vrouwenvelder, T. Egli and F. Hammes, Water Res., **47**, 3015-3025 (2013)
60. E.I. Prest, J. El-Chakhtoura, F. Hammes, P.E. Saikaly, M.C.M. van Loosdrecht and J.S. Vrouwenvelder, Water Res., **63**, 179-189 (2014)
61. E.P. Holinger, K.A. Ross, C.E. Robertson, M.J. Stevens, J. Kirk Harris and N.R. Pace, Water Res., **49**, 225-235 (2014)
62. J. Fick, H. Soderstrom, R.H. Lindberg, C. Phan, M. Tysklind and D.G.J. Larsson, Environ. Toxicol. Chem., **28**(12), 2522-2527 (2009)
63. Yiruhan, Q.-J. Wang, C.-H. Mo, Y.-W. Li, P. Gao, Y.-P. Tai, Y. Zhang, Z.-L. Ruan, J.-W. Xu, Environ. Pollut., **158**, 2350-2358 (2010)
64. D.I. Andersson, D. Hughes, Drug Resist Update, **15**, 162-172 (2012)
65. T. Schwartz, O. Armant, N. Bretschneider, A. Hahn, S. Kirchen, M. Seifert, A. Dötsch, Microb. Biotechnol., **8**(1), 116–130 (2015)
66. B.F. Tan, C. Ng, J.P. Nshimyimana, L.L. Loh, K.Y.-H. Gin and J.R. Thompson, Front. Microbiol., **6**(1027), 1-10 (2015)
67. C.R. Proctor and F. Hammes, Curr. Opin. Biotechnol., **33**, 87-94 (2015)
68. R.P.A. Pereira, J. Peplies, I. Brettar and M.G. Höfle, BMC Microbiol., **17**(79), 1-14 (2017)

Impact of Catchment Area Activities on Water Quality in Small Retention Reservoirs

Katarzyna Oszczapińska[1,*], *Iwona* Skoczko[1], and *Joanna* Szczykowska[1]

[1]Department of Technology in Engineering and Environmental Protection, Bialystok University of Technology, Wiejska St. 45E, 15-351 Bialystok

Abstract. The aim of the study was to evaluate catchment area impact on small water reservoirs condition in Podlasie. The researches were conducted in two different catchment areas. Topiło reservoir, located in Podlasie area in the south-east of Białowieża Forest, has typical sylvan catchment. Second reservoir, Dojlidy, is located also in Podlasie, in the south-east of Białystok as a part of Dojlidy Ponds. In contrast to Topiło, Dojlidy has agricultural catchment. Water samples collected from five sites along each reservoir were analysed for the presence of total nitrogen and phosphorus, chlorophyll "a", reaction, turbidity and conductivity. Researches took place in spring, summer and autumn 2013 (Topiło Lake) and 2014/2015 (Dojlidy). The lowest trophic state was observed in autumn and the highest in summer. Because of the high loads of phosphorus received by the reservoirs, this element did not limit primary production. Calculated TSI values based on total phosphorus were always markedly higher than calculated on chlorophyll-a and total nitrogen. Both reservoirs demonstrated TSI indexes specific to hypertrophic lakes due to large amount of total phosphorus.

1 Introduction

There is close relationship between surface water quality and its catchment area character. Chemical substances scouring, conditioned on the type of land use, geological substratum structure, kind of soil and climate conditions, has significant impact on chemical water components changes [1].

Right functioning of small retention reservoir depends on water quality. Objects located in agricultural catchment are particularly exposed to fertilizers, pesticides and organic matter eluted from the soil. Small and medium reservoirs depth aids eutrophication. Often it comes to supply watercourse pollution [2].

The most common biogenic sources in surface waters are, mentioned earlier, fertilizers used in agricultural catchments. Their influent on soils in catchment basin was tested by Czyżyk and Rajmund (2013), who after six-year studies proved that more nitrogen has permeated to the soil from mineral than natural fertilizers [3].

On the other hand, sylvan catchments exhibit smaller biogenic runoff thanks to forest cover absorption abilities [4].

* Corresponding author: k.oszczapinska@doktoranci.pb.edu.pl

2 Methods

Studies were conducted in two different catchment areas. Topiło reservoir, located in Podlasie area in the south-east of Białowieża Forest, has typical sylvan catchment. It was built in 1932-1933 as a wooden logs store, but in 1975 tank was excluded from this type of use. Topiło Lake catchment area with Perebel River involves Białowieża Forest and Hajnówka urban area. Topiło location bears on industrial and municipal pollution character. Reservoir has 21.1 ha water surface and only 1 m of average depth [5].

Fig. 1. Location of control points within Topiło Lake [6].

Second reservoir, Dojlidy, is located also in Podlasie area, in the south-east of Białystok as a part of Dojlidy Ponds. In contrast to Topiło Dojlidy, with 34 ha and 1.75 m average depth, has agricultural catchment. It was built in 1962 by damming Biała River. The research object is being used as a swimming pool, but in the period of the study it was excluded from this type of use because of modification works made by the Water Sports Centre "Dojlidy". It is also a place for extensive, non-commercial fish farming. In addition, the reservoir has the task of regulating water surface runoff [7].

Fig. 2. Location of control points within Dojlidy [6].

Along each reservoir five sites were appointed. The sampling points were distributed in specific areas within Topiło and Dojlidy to allow the observation of changes dynamics in the ecological status of the objects. Within Topiło Lake points were located consecutively: first near the Perebel River flow into the reservoir, second next to the Topiło-Łozice bridge, third beside water dam with railroad, fourth by the side of wooden dock and fifth on the Perebel River, behind the reservoir (Fig. 1). The first point in Dojlidy was located near the Biala River flow into the reservoir. The second point included central, coastal part of the object. The third point was located nearby the bridge on the beach. The fourth point was the place of water outflow over the front dam, while the fifth point was on the Biala River, behind the reservoir (Fig. 2). Researches took place in spring, summer and autumn 2013 (Topiło Lake) and 2014/2015 (Dojlidy). Each time of the year the samples were collected twice, from a depth of about 0.4 m. They were analysed for the presence of the following indicators: total nitrogen, total phosphorus, chlorophyll "a", reaction, turbidity, and conductivity. Different parameters were determined by a UV-VIS spectrometry according to the following methods [8-10]:
- total nitrogen – Total Kjeldahl Nessler Method, HACH® Company,
- total phosphorus – PhosVer® 3 with Acid Persulfate Digestion Method, HACH® Company,
- chlorophyll "a" – spectrophotometric, monochromatic method with feopigments alfa correction,
- reaction – electrometric method, CP-401 Elmetron pH-meter,
- turbidity – nephelometry method, HACH® Company,
- conductivity – electrometric method, Slandi CM 204 conductometer.

3 Results and discussion

Concentrations of total nitrogen and total phosphorus in water were characterized by remarkable variability. Seasonal changes in these indicators of water quality during research period were recorded. The lowest concentrations of total nitrogen were observed in summer and autumn, while the highest in spring (Table 1). According to Environment Minister Regulation from 21 July 2016 [11] total nitrogen concentration for I and II water quality class in reservoir should not be higher than 2 mg/dm^3. Topiło and Dojlidy reservoirs, with relatively large total nitrogen amount, cannot be classified to first and second water quality class.

Table 1. Average, minimum-maximum total nitrogen amount in Topiło and Dojlidy.

Total nitrogen [mg/dm^3]	Topiło					Dojlidy				
Control point	1.	2.	3.	4.	5.	1.	2.	3.	4.	5.
Spring	6.40 5,96-7,27	4.29 3,84-4,78	5.08 4,59-5,27	3.94 3,52-4,42	3.58 2,65-4,20	6.13 4,09-6,89	1.25 0,58-2,02	2.63 2,45-2,89	0.63 0,21-0,98	2.13 0,87-2,54
Summer	2.09 0,65-3,10	2.08 0,83-2,95	2.29 1,25-3,00	1.67 0,68-2,92	1.88 0,87-3,21	3.25 3,11-3,85	1.38 1,01-1,55	1.88 1,12-2,05	3.88 2,22-4,10	2.88 2,55-3,06
Autumn	1.04 0,50-1,16	1.21 0,74-1,35	1.79 0,85-2,02	0.67 0,42-0,89	0.50 0,38-0,74	0.38 0,11-0,86	1.50 1,01-1,87	0.88 0,52-1,02	0.75 0,33-1,16	0.63 0,21-0,95

Highest total phosphorus concentrations in both reservoirs were notified in spring and autumn. Total phosphorus amount was greater in Topiło Lake, like total nitrogen concentration (Table 2). The two studied objects were marked by large total phosphorus amount, which was much higher than pollution limit for first and second water quality class (accordingly ≤ 0.045 mgP/dm^3 and ≤ 0.060 mgP/dm^3) [11].

Table 2. Average, minimum-maximum total phosphorus amount in Topiło and Dojlidy.

Total phosphorus [mg/dm^3]	Topiło					Dojlidy				
Control point	1.	2.	3.	4.	5.	1.	2.	3.	4.	5.
Spring	2.24 1,76- 2,84	1.95 1,62- 2,35	1.85 1,56- 2,19	1.33 0,74- 2,02	1.47 0,93- 2,10	2.17 1,52- 2,29	2.30 1,68- 2,57	1.95 1,85- 2,32	1.59 1,02- 1,70	1.88 1,60- 2,04
Summer	1.09 0,90- 1,24	0.90 0,75- 1,10	1.53 0,62- 2,20	1.22 0,54- 1,95	0.98 0,31- 1,33	0.58 0,23- 0,66	0.55 0,39- 0,98	0.42 0,30- 0,49	0.63 0,58- 0,71	0.48 0,36- 0,52
Autumn	0.68 0,55- 0,92	1.31 1,06- 1,62	2.44 2,21- 2,96	2.66 2,07- 2,81	0.30 0,12- 1,04	2.30 2,16- 2,39	1.14 0,98- 1,22	0.93 0,86- 1,08	1.80 1,23- 2,14	1.00 0,68- 1,19

Seasonal differences in nutrients concentrations in water of the reservoirs relate to plant's vegetation [12]. Factor that caused high concentrations of both nitrogen and phosphorus could be probably sediments covering bottom of the reservoirs, which in Topiło came from wooden logs stored in the past. Because of lack of regular sediment removal thick layer in Dojlidy have agglomerated [13].

The content of chlorophyll "a" in waters body of both studied reservoirs was also subject to seasonal changes. Largely higher concentrations were noticed in Topiło with maximum 34.94 μg/dm^3 in summer, while the highest average chlorophyll "a" amount in Dojlidy reached 16.02 μg/dm^3. Chlorophyll "a" concentration increase was observed while the intensive autotrophic microbes growth. It slightly decreased in autumn, as the vegetation season was ending (Table 3).

A beneficial condition for phytoplankton development is low average depth below 2 m and connected with that biogenic compound dilution in smaller water volume [14].

Table 3. Average, minimum-maximum chlorophyll "a" amount in Topiło and Dojlidy.

Chlorophyll "a" [g/dm^3]	Topiło					Dojlidy				
Control point	1.	2.	3.	4.	5.	1.	2.	3.	4.	5.
Spring	13.36 4,01 - 19,04	19.22 5,30 - 33,37	24.90 6,68 - 41,04	17.86 5,32 - 29,70	15.69 4,26 - 27,37	14.69 5,85 - 16,04	11.35 3,70 - 15,37	16.02 4,69 - 18,22	2.67 2,01 - 3,34	4.01 3,94 - 4,69
Summer	34.12 3,75 - 36,04	34.94 34,13 - 35,50	32.58 32,52 - 33,04	31.47 30,63 - 32,24	24.52 17,80 - 25,97	4.01 2,01 - 5,67	12.28 4,52 - 13,6	7.12 6,94 - 8,34	6.23 4,01 - 7,58	2.00 0,67 - 3,56
Autumn	14.04 5,68 - 23,04	12.03 5,52 - 19,37	15.34 6,04 - 21,13	14.03 5,97 - 23,37	6.01 4,01 - 7,68	1.78 0,37 - 2,67	5.34 2,01 - 7,90	3.56 1,67 - 5,52	2.67 0,97 - 5,01	0.89 0,30 - 1,12

Summary statistics for turbidity, pH, and conductivity (Table 2) did not indicate impaired water quality in any of the reservoirs. Measured water turbidity was never greater than 37 NTU in Topiło, and 29 in Dolidy, measured pH values were never less than 6,88 in Topiło, and 7.25 in Dojlidy. Water conductivity was never greater than 321 μS/cm in Topiło, and 518 in Dojlidy.

Much higher water turbidity was measured in Topiło, especially during summer and autumn researches. Dojlidy in turn was characterized by lower turbidity (Table 4).

Table 4. Topiło and Dojlidy average, minimum-maximum water turbidity.

Turbidity [NTU]	Topiło					Dojlidy				
Control point	1.	2.	3.	4.	5.	1.	2.	3.	4.	5.
Spring	5 3 - 8	6 5 - 6,6	3,5 2 - 5,7	3,5 1,5 - 5	2 1,4 - 3	10 6 - 12	1 0,5 - 2	9 7 - 10	29 23 - 32	13 11 - 14
Summer	31 25 - 39	24 20 - 26	12 9 - 21	7 5 - 11	20 15 - 24	9 8 - 11	3 1 - 4	1 0,5 - 2	5 3 - 6	1 0,5 - 3
Autumn	23 17 - 24	37 25 - 41	29 22 - 33	21 17 - 22	20 7 - 24	13 10 - 14	4 3 - 7	3 1 - 4	22 17 - 23	25 21 - 28

Summer researches demonstrated similar level in both objects. On the other hand, Topiło reservoir was marked by slightly lower water reaction in spring and autumn than Dojlidy (Table 5).

Table 5. Topiło and Dojlidy water reaction (median, minimum-maximum).

Reaction [pH]	Topiło					Dojlidy				
Control point	1.	2.	3.	4.	5.	1.	2.	3.	4.	5.
Spring	7.02 6,94- 7,60	6.88 6,62- 7,30	6.97 6,85- 7,33	7.03 6,94- 7,28	6.95 6,87- 7,40	7.25 7,02- 7,41	7.64 7,34- 7,89	7.80 7,54- 8,01	7.90 7,71- 8,12	7.41 7,22- 7,85
Summer	7.37 7,21- 7,95	7.76 7,57- 7,96	7.97 7,65- 8,01	8.01 7,89- 8,11	7.32 7,11- 8,00	8.03 7,79- 8,15	8.09 7,85- 8,21	8.25 8,12- 8,31	7.57 7,49- 7,62	7.84 7,62- 7,96
Autumn	7.35 7,19- 7,56	7.53 7,32- 7,69	7.46 7,28- 7,70	7.10 6,95- 7,25	7.40 7,29- 7,49	7.94 7,80- 8,11	7.87 7,76- 8,09	8.03 7,91- 8,07	8.12 8,01- 8,19	7.47 7,25- 7,90

Water conductivity in both reservoirs was on the level characteristic for natural surface waters. Autumnal results in Topiło were within range 62 – 69 μS/cm (Table 6). Dojlidy, in contrast to Topiło, stood out with lower conductivity variety during the period of researches.

Table 6. Topiło and Dojlidy average, minimum-maximum water conductivity.

Conductivity [µS/cm]	Topiło					Dojlidy				
Control point	1.	2.	3.	4.	5.	1.	2.	3.	4.	5.
Spring	191 185-204	178 162-180	186 180-191	174 171-179	178 169-190	518 501-523	330 296-360	336 333-345	162 156-180	303 224-354
Summer	273 252-289	282 267-289	272 263-278	291 272-299	321 290-332	248 202-300	164 125-185	153 122-160	252 236-268	169 125-205
Autumn	69 62-74,5	64 58-102	63 56-69	62 59,5-66	62 58-64,5	268 222-296	160 145-172	157 150-168	159 148-162	173 121-200

During the study period, the values of analysed indicators exhibited high variability and showed characteristic patterns of seasonal changes. Carlson (1977) and Kratzer-Brezonik (1981) Trophic State Indexes were used to compare two examined reservoirs [15, 16]. Because of low average deepness Secchi depth hasn't been measured. TSI results are presented in Figure 3.

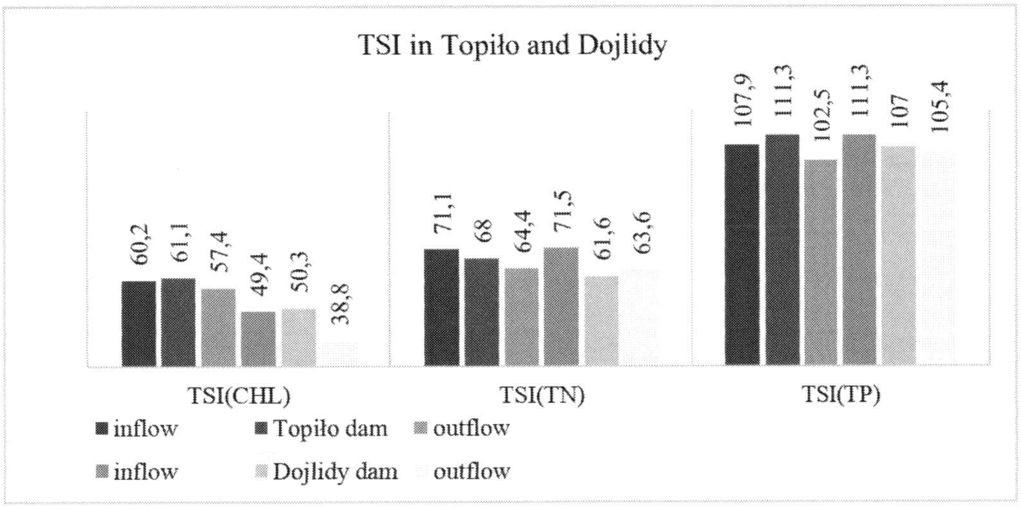

Fig. 3. Caption of the Figure 1. Below the figure.

The lowest trophic state was observed in autumn and the highest in summer. Because of the high loads of phosphorus received by the reservoirs, this element did not limit primary production. TSI values calculated based on total phosphorus were always markedly higher than calculated on chlorophyll "a" and total nitrogen. Both reservoirs demonstrated TSI indexes specific to hypertrophic lakes due to total phosphorus large amount.

Numerous studies have documented declines in water quality, habitat, and biological assemblages as the extent of agricultural land increases within catchments [17-20]. Researchers commonly report that streams draining agricultural lands support fewer species of sensitive insect and fish taxa than streams draining forested catchments [20-22]. Higher inputs of sediments, nutrients, and pesticides accompany increased agricultural land use [23, 24].

Conducted researches proved that type of catchment activities has an impact on examined reservoirs water quality. Due to elevated degradation vulnerability Dojlidy reservoir was characterized by relatively high nutrients amount [25]. Moreover, researches led in 2014 by „ROLEX" Sp. z o. o. company pointed out that sediments samples from Dojlidy reservoir responded to the cultivated soil chemical composition [13].

Topiło reservoir and its tributary named Perebel River pollution level has typical natural character, which was proved by microbiological tests. They were led by i.a. Professor Edward Pierzgalski from Forest Research Institute within the expertise to determine the reasons for the Topiło water resources reduction. Researches results excepted anthropogenic impact on water quality, despite Topiło village nearness, by Topiło and Perebel high microbiological purity [26]. It should be noted that trophic state in both reservoirs was on similar, elevated level. Organic matter from wooden logs stored in the past caused lasting many years pollution.

4 Conclusions

1. Nutrients amount in both reservoirs was relatively high.
2. The average annual values of the TSI indexes qualify the studied reservoirs as hypertrophic.
3. Trophic state of studied reservoirs strictly dependent on the processes taking place in their vicinity.
4. Despite woody catchment area Topiło Lake water pollution was on similar level as Dojlidy with agricultural catchment area.

References

1. M. Sidoruk, A. Skwierawski, Ecol. Chem. Eng, **13(52)**, 337-343 (2006)
2. A. Górniak, M. Grabowska, International Rev. Hydrobiol., **83**, 307-310, (1998)
3. F. Czyżyk, A. Rajmund, Ecological Engineering, **33**, 29-34 (2013) (in Polish)
4. General Directorate of The State Forests, *Principles of silviculture, State forests*, 11-14, (2012)
5. Hajnówka Forest District unpublished data
6. Google Maps (2017)
7. A. Kluczko, *Elaboration for Białystok Sport and Recreation Center*, (2014) (in Polish)
8. HACH Company Loveland, *Water Analysis Handbook*, Colorado, (1992)
9. W. Hermanowicz, J. Dojlido, W. Dożańska, B. Koziorowski, J. Zerze, *Water and wastewater physicochemical testing,* (1999) (in Polish)
10. PN-C-05560-02:1986. Water and wastewater. Chlorophyll content in surface waters tests. Chlorophyll alfa measurement in planktonic algae with spectrophotometric, monochromatic method with feopigments alfa correction (in Polish)
11. Environment Minister Regulation from 21 July 2016 (Dz. U. 2016 poz. 1187) (in Polish)
12. J. Koc, J. Koc-Jurczyk, K. Solarski, Zeszyty Naukowe Polskie Towarzystwo Gleboznawcze, 121–128 (2009) (in Polish)
13. S. Rola, *Bathymetric tests of Dojlidy reservoir in Białystok*, (2014) (in Polish)

14. Z. Kajak, *HYDROBIOLOGY - LYMNOLOGY. Inland waters ecosystems,* (1998) (in Polish)
15. R. E. Carlson, Limnol. Oceanogr., **22**(2), 361-369, (1977)
16. C. R. Kratzer, P. L. Brezonic, Water Resour. Bull., **17**, (1982)
17. C. Richards, L. B. Johnson, G. E. Host, Can. J. Fish. Aquat. Sci., **53**, 295-311, (1996)
18. N. E. Roth, J. D. Allan, D. L. Erickson, Landsc. Ecol., **11**, 141-156, (1996)
19. R. A. Sponseller, E. F. Benfield, H. M. Valett, Freshwater Biol., **46**, 1409-1424, (2001)
20. L. Wang, J. Lyons, P. Kanehl, R. Gatti, Fisheries, **22**, 6-12 (1997)
21. D. Genito, W. J. Gburek, A. N. Sharpley, Freshw. Biol., **17**, 109-119, (2002)
22. D. R. Lenat, J. K. Crawford, Hydrobiol., **294**, 185-199, (1994)
23. C. M. Cooper, J. Environ. Qual., **22**, 402-408, (1993)
24. L. B. Johnson, C. Richards, G. E. Host, J. W. Arthur, Freshw. Biol., **37**, 193-208, (1997)
25. K. Oszczapińska, J. Szczykowska, *Wybrane zagadnienia z zakresu ochrony środowiska i energii odnawialnej*, 269-283 (2016) (in Polish).
26. E. Pierzgalski, J. Tyszka, A. Boczoń, M. Janek, M. Wróbel, A. Stolarek, K. Pachuta, P. Oględzki, M. Frąk, P. Sikorski, L. Komecka, S. Czachorowski, L. Pietrzak, M. Ksepko, *Operat zarządzania wodami i ochrony ekosystemów wodnych (do Planu Ochrony Białowieskiego Parku Narodowego),* Forest Research Institute, 1–162, (2010)

Wise use of water in smart cities - possibilities and limitations

Joanna Bąk[1,*]

[1]CUT, Institute of Water Supply and Environmental Protection, 31-155 Cracow, 24 Warszawska Street, Poland

Abstract. The need to save water is due, inter alia, to the paradigm of sustainable development. There are many ways to minimize the consumption of high quality water supplied by the water supply network. These include the simplest way and those complex, requiring additional installation. The lack of water is a big problem, but not only water deficit are dangerous. There is a possibility of secondary water pollution in the water supply network due to changes in network parameters. Changes in these parameters may occur due to reduced demand for water by residents and, as a result, reduced water flow - at the same pipe diameter. The article includes a review with comparative analysis of various classification systems for the tap fittings and other sanitary equipment, such as the Water Efficiency Label (WELL) in Europe or the Water Efficiency Labelling and Standards (WELS) in Australia. Several types of perlators and flow regulators were compared in the research section. This equipment was tested in the household. The possibilities of minimizing water consumption by using them was collated. In addition, the work also analyses the evolution of water consumption in Poland in recent years and their possible relationship with the threats quality of drinking water supplied to consumers.

1 Introduction

Water is essential for human life and other living organisms. Access to high quality water is a key issue for human life. Therefore, it is very important to protect water resources and minimize water consumption in households, public buildings, businesses and factories. The need to save water also arises from sustainable development goals and the paradigm of sustainable development. This task is of particular importance in modern and constantly developing cities, where the population is constantly growing. Remember that even in such a situation, minimizing water consumption is most dependent on the water demand of individual human beings (bypassing the industry). In implementing the directions for sustainable progress and, in particular, sustainable consumption, it is extremely important to draw attention at the same time to the need to adapt existing infrastructure to the changing needs of society.

[*] Corresponding author: jbak@pk.edu.pl

The aim of this paper is to try to answer the question of the current water saving potential in modern smart cities and whether there are any drawbacks or limitations for wise use of water. The purpose of the research was to find out if drinking water in household can still be saved in cheap ways. If so, how much?

1.1 Possibilities of wise use of water

There are many ways to minimize the consumption of high quality water supplied by the water supply network. These include the simplest and cheapest ones such as flow regulators, as well as those requiring complex systems and additional installations such as the use of grey and rain water. Note that their effectiveness varies. Sometimes a simple solution that is a small investment may prove to be more effective than a complex installation requiring complete refurbishment for existing buildings. Ways to reduce water use in households can be classified into several groups. These include the following:
- consumer behaviour
- kitchen and bathroom equipment
- use water from other source than the water supply network (greywater and rainwater).

The consumer behaviour group includes eliminating consumer misbehaviour (not turning off water while brushing or shaving), and attempting to replace user habits (washing dishes under running water, drinking cold tap water) with greener activities. In the group of wise ways to use water for kitchen and bathroom equipment can be separated several subgroups. Belong to them:
- armature
- utensils and sanitary appliances
- white goods (dishwasher, washing machine)
- accessories (flow regulators, aerators, restrictors).

The capabilities of this group of solutions are constantly evolving. New technologies help. Among the more modern fittings solutions are the thermostatic faucets, the electronic sensors, the two-stage armature opening system and change the position of faucet lever in cold water mode. There are also innovative projects and solutions. Among them there are integrated sanitary utensils - a urinal combined with a washbasin and a sink combined with a toilet bowl through a flush. The second of these solutions can be found in Japan.

The dissemination of information and education of the general public is a very important factor in both the user-habits and kitchen and bathroom solutions. Also important are ways to convince people to save water, and here in turn can help smart city.

1.1.1 Wise use of water in smart city

The smart city vision is the idea of urban development using Information and Communication Technologies (ICT) and Internet of Things (IOT) technology. Through the use of ICT and IOT, it is also possible to support the wise use of water in cities and metropolises. By using appropriate monitoring sensor systems - data can be collected. Collections of such data may serve different purposes. Among them can be mentioned their analysis. Another goal may be an attempt to convince and encourage water users to save it. Examples of such practices may include the MK Smart project in the southern part of Great Britain in the Broughton area of Milton Kenyes [1]. From May 2016 to May this year, there was an attempt called Water Monitor [2]. Project participants (more than 100 households) through the website could check the weekly water consumption in their households and compare it with others, and also use advice on further minimization.

New technologies used within the broadly understood concept of smart city can help to indicate areas for action and upgrading. They can help highlight the problem and try to convince users to use water wisely and be their coach. It is important to note that the basics principle of water saving in households is man and his choice and the decision.

1.2 Systems of evaluate tap fittings and its equipment

Another way to support the decision-making process of water consumers in terms of saving it is the special evaluate systems of sanitary tapware and other equipment using water. Elements such as inter alia taps, showers, flow controllers, lavatories, urinals, clothing washing machines or dishwashers are subjects of evaluation.

There are several systems in the world for the classification of sanitary fittings and its fittings. Belong to them: the European Water Label scheme (formerly WEPLS - Water Efficient Product Labelling Scheme), the European scheme WELL (Water Efficiency Label) and the Australian WELS standards (Water Efficiency Labelling and Standards). Table 1 provides a comparison of basic information on the above three assessment systems. One can list both the example of the obligatory system covered by all the products sold within its jurisdiction and the non-mandatory system. For the latter, manufacturers voluntarily submit their goods for evaluation to gain added value to the product in the form of a mark of a prestige rating system. There are similarities and differences, but the goal of all systems is the same - saving water and energy.

Table 1. Comparison of three different systems of evaluate of tapware fittings and other sanitary equipment based on [3-10]

Feature/Name of system	WELL	WELS	Water Label formerly WEPLS
full name of system	Water Efficiency Label	Water Efficiency Labelling and Standards	Water Efficient Product Labelling Scheme (till 2009) Water Label The European Water Label (launched 2012)
country/countries	Europe	Australia New Zealand	Europe
subject to evaluation	voluntary	mandatory	voluntary
more important documents	WELL Classification Scheme [3]	Water Efficiency Labelling and Standards Act 2005 (WELS Act)[4], AS/NZS 6400:2016 Water Efficient Products – Rating and Labelling [5]	Scheme Roadmap & Vision (Updated 2017) [6]
the rating scale	rating using letter (A-F) and star rating	star rating	efficiency rating from >13 l/minute up to max.6 l/minute
the highest mark	A 4 stars for home or 6 stars for public	6 stars	max.6 l/minute

Feature/Name of system	WELL	WELS	Water Label formerly WEPLS
the scope that covers	wash basin bidet valves shower valves shower heads shower hoses urinal flush systems WC flush systems accessories	tap equipment urinal equipment lavatory equipment showers flow controllers clothes washing machines dishwashers	baths WC suites cisterns basin taps shower controls shower handsets grey water recycling units kitchen taps urinal controllers electric showers replacement WC flushing devices supply line flow regulators independent WC pans
more important sources	[3, 7]	[4, 5, 8, 9]	[6,10]

1.3 Water consumption in Poland versus basic water needs

In Poland water consumption in the last 30 years has changed. On the basis of data from Statistical Yearbooks [11] Figure 1 was prepared. It presents the changes in water consumption in Poland per capita between 1986 and 2015. Figure 1 also indicates the upper limit of basic water needs according to [12] and basic water requirements [13]. The highest water consumption in the analysed 30 years in Poland was recorded in 1987 and amounted to 196.2 dm^3/capita/d. From this year to the year 2014, there was a decrease in water consumption (except for 2007 and 2008, where water consumption was unchanged). An interesting situation took place in 2015, as water consumption this year increased again. In the following years it is expected to stabilize water consumption at around 100 dm^3/capita/d or slightly less.

The basic water needs are determined by [12] as falling within the range of 7.5 - 15 dm^3/person/d. The minimum amount of water required depends on, among other things, the climate, available sanitary facilities, religious and cultural customs and also the food that is cooked. In turn [13] suggests to assume a value of 50 dm^3/person/d as the basic water requirement.

The difference between the cited minimum values and the current water consumption level in Poland shows that there is potential and opportunity to further reduce the consumption of tap water. The water consumption of 2015 in Poland (94 dm^3/capita/d) in comparison with the value of [13] shows twice the difference. It should be borne in mind that the current technological development allows for much greater water savings with adequate living comfort than a dozen or so years ago. For this reason, more recent sources point to much lower minimum water needs. For this reason, the current literature sources indicate much lower minimum water requirements.

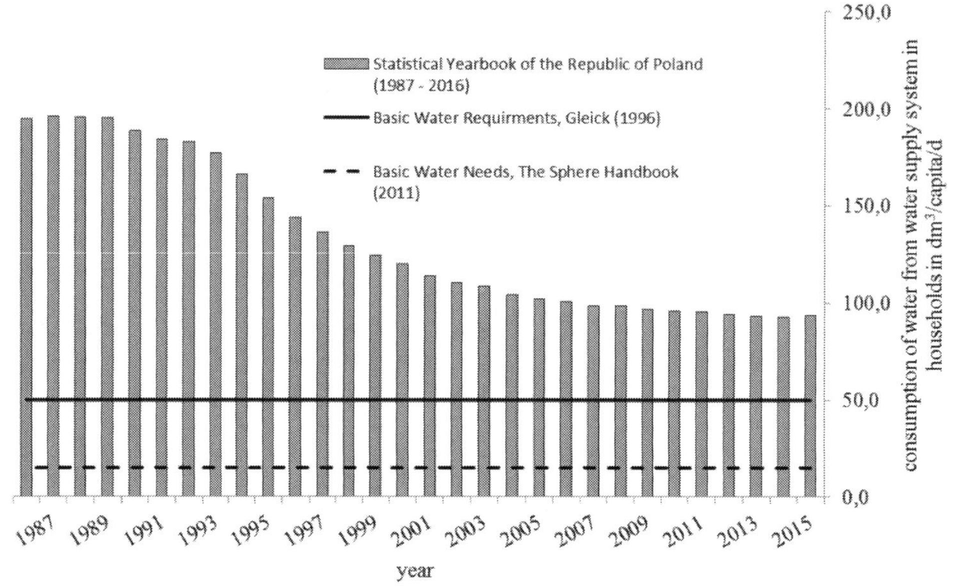

Fig. 1. Consumption of water from water supply system in households in Poland 1986 – 2015, based on data from [11] and basic water needs/requirements [12, 13]

2 Material and methods

The outflow of cold water from the plumbing through the single lever basin mixers feature with different tap equipment (flow controllers and aerators) was investigated. The research was conducted in a household on the first floor in a large city in Poland. For comparison, the outflow from fittings without any tap equipment was also tested. Experiment was conducted as field test in the household in the private building under actual use conditions. For laboratory tests of flow regulators, the conditions set out in (PN - EN 246: 2005) [14] must be complied with.

Five new flow regulators which were marked in numbers 1 to 5 were tested (flow regulator 1 ÷ flow regulator 5). These water saving products have integrated technology which allows for compensating pressure. They are almost pressure independent. They produce different types of stream: spray jet, laminar jet without air intake and most popular aerated stream. They have also different structures, inter alia - honeycomb. In addition, one water saving aerator without special technology for compensating pressure was checked. Figure 2 shows all water consumption minimization products used in the study.

The outflow from the faucet was tested by the substituted vessel method. For this purpose, the water draining from the fitting spout was collected for 20 seconds and then the volume was gauged with measuring cylinder of 500 ml capacity. Time is measured with an electronic stopwatch. The discharge from the faucet equipped with each flow regulator was tested several times in each series and then the result was averaged. Due to the possibility of changing the pressure in the installation, two series of measurements were made during various times of the day for inspection. The first series of measurements was performed during the day between 10.30 and 12.40, and the second at night between 00.30 and 02.15. Both series of research were conducted during the weekend.

Fig. 2. All water-saving products used in the tests performed

At the present time very rarely occurrence of faucet without an aerator. In addition, the outflow from the tapware without any aerator and flow regulator showed some changes in subsequent tests. For these reasons, the reference level in the calculation was the outflow of the fittings provided with the old used water saving product. This has minimized the effect of pressure changes on the results obtained, since all flow regulators (including old product) were almost pressure independent.

The water used for the study has been reused for watering the backyard garden.

3 Results

The results of the study on the outflow from fittings supplied with different type of flow controllers are shown in Figure 3. In three out of seven tested outflows, there was no difference between the daily series and the night series. In other cases, the difference was small and was 0.1 dm^3 / min.

Before the comparison of the different tap equipment, a statistical evaluation of the obtained volume results for the used flow regulator, five new flow regulators and one aerator called tap equipment (daily series and night series) were carried out. The calculations were carried out in the Statistica 10 program.

It was checked whether the averages are the same for individual equipment of the faucet. Before starting a one-way analysis of variance, the assumptions of this parametric test were checked. The normality of the distribution was confirmed for all equipment by Shapiro – Wilk test, with the exception of aerator. Therefore, the homogeneity test and analysis of variance were performed for all measurements in each series and excluding the results for aerator. On the basis of the Brown - Forsythe homogeneity test, there is no basis for rejecting the null hypothesis with the equality of variances for both the daily series ($p = 0.7961$ for all results and $p = 0.7725$ excluding measurements for aerator) as well as for the night series ($p = 0.5962$ and $p = 0.6260$, respectively). Analysis of variance showed the significance of the differences between the considered averages for each series. In addition, post-hoc tests (multiple comparisons) were carried out. The results for the aerator were excluded due to the lack of confirmation of the normal distribution. The Tukey and Scheffe tests confirmed the significance of the differences between the average volume values in each series.

In addition, Kruskal - Wallis non-parametric test was carried out. Daily and night series of results were tested. The obtained test probability level $p = 0.0032$ allowed the rejection of the null hypothesis. The results of the analysis allow to conclude that the type of tap equipment used has a statistically significant impact on the volume of the outflow from the single lever basin mixers feature.

The percentage reduction of the outflow of faucet with new tap equipment compared to the old used flow regulator (but good quality with the possibility of easy lime cleaning) was calculated. Calculations were made for both series of tests. The results are presented in Table 2. Between the results of the various series there were no differences or differences of 1 -2%. By adopting less favourable percentages of reduction, it can be observed that the water flow has been reduced from 34% to 79%. By adopting a hands wash duration of 15 seconds, using newer products it is possible to save from 0.7 to 1.7 litres with a single hands wash. The amount of water consumed for this activity can be minimized up to 0.4 litres (although there are commercially available products that allow even greater reduction the outflow). The outflow from a tap fitted with a new product accounts for even only 20% of the current outflow. By contrast, compared to the outflow from the fittings without any regulating or aerating product, the new products allow to reduce the flow even to 12-14% of the current outflow. In the comparison of all used water saved products with no flow controller state, the stream reduction value ranged from 34% to 88%.

Studies show that it is possible to further minimize water consumption in households (even if good quality flow regulators have been used for some years). It is possible to achieve a reduction in the flow rate of up to about 80%. However, it should be borne in mind that the savings introduced should not cause a loss of comfort of consumers. Extreme flow reduction can be used in households with young children learning to use sanitary utensils.

Research on the use of high efficiency water saving devices in households was conducted i.a. in the United States. A study [15] published in 2004 by the Tampa Water Department and United States Environmental Protection Agency (EPA) showed that modernization of houses into water saving devices (including aerators) allowed to reduce daily water consumption by an average 46% for household. The use of aerators enabled to save of almost 14 m^3 per year.

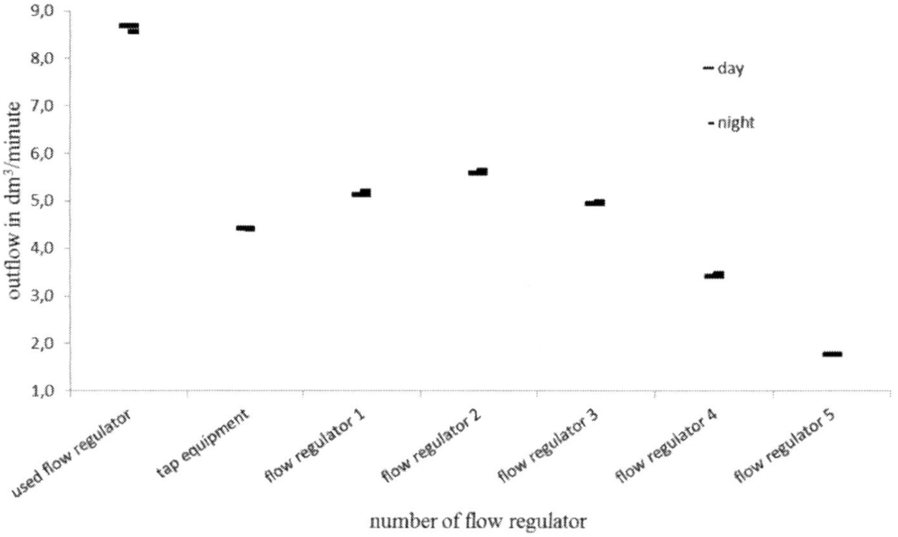

Fig. 3. The outflow from different flow regulators during daylight hours and during the night hours

Table 2. The percentage of reduction in the outflow

Number of flow regulator	reduction in the outflow (%)	
	day series	night series
Tap equipment	49	49
Flow regulator 1*	41	39
Flow regulator 2*	36	34
Flow regulator 3*	43	42
Flow regulator 4*	61	59
Flow regulator 5*	80	79

* statistically significant difference

4 Summation and conclusions

The basic constraint in saving water in households is not money at all, but man. State-of-the-art technologies will not help much without a human being's decision to use them. Already very small investments, like the order of 5 - 10 euros, in modern flow controllers, can allow for significant savings. It is important that the benefits of minimizing water use are multifaceted. Water, energy (in case of reducing hot water consumption) and the environment are saved.

On the basis of the conducted research it can be concluded that replacing the current flow regulator with newer products, the consumption of water in the household can be reduced by up to 79%. The other question is the efficiency of aerators limiting the flow after a certain period of time. This problem sets out a further direction of research in which the water quality (general water hardness) will also be important.

There is another side of this issue. It is less likely, but it should also be considered. It is worth bearing in mind that the vast majority of large and modern towns and cities are already equipped with a water supply network. Thus, pipelines of a certain diameter projected and selected by the designers already exist. If the water consumers will follow the guidelines for minimizing water consumption, they will implement mentioned solutions that will reduce usage of water and, in addition, the process will be large scale and take place in short term, there is a risk of secondary water contamination in the water supply network due to changes in network performance (e.g. velocity of water). Changes in these parameters may occur due to reduced demand for water by the inhabitants, and thus inseparable - reduced flow rate of water - at the same diameter of the pipeline. At the same time, the population is constantly moving into large and modern cities, new housing estates are being built there, so new buyers are emerging. It can therefore be assumed that possible changes in water demand resulting from consumption limitation are balanced by new consumers in the city and there is no significant risk of water contamination for this reason. It is important to monitor and, if necessary, adapt existing infrastructure to the evolving needs of society when adopting new directions for action and implementing sustainable development as well as, in particular, sustainable consumption (water). In addition to modern technology, the effective and substantive education (persuading to implementing changes) of the public in this area is still a key task.

References

1. The MK Smart project http://www.mksmart.org/water/#1 access (2017)
2. MK: Smart Water Monitor watermonitor.mksmart.org access (2017)
3. WELL Water Efficiency Labelling Classification Scheme European Valve Manufacturer Association, http://www.well-online.eu (2016)
4. Water Efficiency Labelling and Standards Act 2005 as amended No.4 (2005)
5. AS/NZS 6400:2016 Australian/New Zealand Standard. Water Efficient Products – Rating and Labelling (2016)
6. Y. Orgil, Scheme Roadmap & Vision, http://www.europeanwaterlabel.eu/scheme_roadmap.asp, access (2017)
7. The WELL Water Efficiency Label website, www.well-online.eu access (2017)
8. WELS product database https://wels.agriculture.gov.au/wels-public, access (2017)
9. The Water Rating website Water Efficiency Labelling and Standards (WELS) scheme, A joint government and industry program, www.waterrating.gov.au access (2017)
10. The European Water Label website www.europeanwaterlabel.eu access (2017)
11. *Statistical Yearbooks of the Republic of Poland*, Central Statistical Office, Warsaw (1987 – 2016)
12. *The Sphere Project The Sphere Handbook. Humanitarian Charter and Minimum Standards in Humanitarian Response*, The Sphere Project, http://www.spherehandbook.org/en (2011)
13. P.H. Gleick, M. IWRA, Water Int, **21**, No.2, 83-92 (1996)
14. PN – EN 246 : 2005, European Standard. Sanitary tapware – General specifications for flow rate regulators – polish version (2005)
15. P.W. Mayer, W.B. DeOreo, E. Towler, L. Martien, D.M. Lewis, *Tampa Water Department Residential Water Conservation Study. The impact of high efficiency plumbing fixture retrofits in single – family homes,* Aquacraft, Inc. Water Engineering and Management (2004)

Measurements of water temperature in fountains as an indicator of potential secondary water pollution caused by Legionella bacteria

Joanna Bąk[1,*]

[1]CUT, Institute of Water Supply and Environmental Protection, 31-155 Cracow, 24 Warszawska Street, Poland

Abstract. At high air temperatures persisting for a long time, water temperature in the fountains may also increase significantly. This can cause a sudden and significant increase in Legionella bacteria, which results in secondary water contamination. This phenomenon with water – air aerosol generated by fountains can be very dangerous for people. During the test, water temperature measurements in fountains in Poland were made. These research tests was conducted in the spring and summer. The research was conducted in order to determine whether there is a possibility of growth of Legionella bacteria. One of the aims of the study was to determine what temperature range occurs in the fountains and how the temperature changes in the basin of the fountain and when the highest temperature occurs. Single temperature measurements were made and also the temperature distribution was measured during daylight hours. The water temperature in most cases was greater than 20^0C, but in no case exceed 26^0C. The paper presents also the review about the effect of water temperature on the presence and bacterial growth. The study confirmed the existence of the risk of increasing the number of bacteria of the genus Legionella in the water in the fountains.

1 Introduction

Not only the excessive quantity or the water deficit are dangerous. The presence in water of certain elements and compounds and selected microorganisms can also have a negative impact on the health and life of humans. Contaminated water is also harmful to the environment and living organisms. Public fountains can be potential sources of Legionella bacteria. The quality of water circulating in urban fountains is extremely important. Due to the formation of water-air aerosols in their vicinity, particular attention should be paid to the presence of Legionella in the water, as the inhalation of the bacteria is by inhalation. Infected aerosol remains unsafe up to 1 km [1].

In drinking water used by consumers at tap water supply points may be small amounts of Legionella bacteria. Circles of public fountains are usually supplied with or supplemented with drinking water. In Polish regulations, these bacteria appear several

* Corresponding author: jbak@pk.edu.pl

times. The Ordinance of the Minister of Health of 13 November 2015 on the quality of water intended for human consumption [2] reiterates the establishment of requirements for their quantity in warm water that was already present in the previous regulation (The Ordinance of the Minister of Health of 29 March 2007 on the quality of water intended for human consumption, [3]). The requirements for these microorganisms also appear in the Ordinance of the Minister of Health concerning the quality of water in swimming pools [4]. This issue is particularly important because of the lack of requirements for the quality control of water in fountains' basins in terms of presence Legionella bacteria.

In cases where there are no favourable conditions for the development of Legionella bacteria in the building, there are no significant risks. Problems begin to appear when certain conditions are present, including increased cold water temperatures or reduced hot water temperatures. Issues related to the development of Legionella bacteria in hot and cold water systems are already described in the technical literature.

Another potential source of Legionella bacteria, which may be fountains, has been observed. It was due to the very hot summer of 2015 in Poland. During sunny days with a high temperature for a considerable part of the day, the water temperature in the fountains may also be significantly elevated. This can cause a sudden and significant increase in the number of Legionella bacteria. These bacteria in a fountain that produces water – air aerosols can be very dangerous to humans.

The aim of the study was to assess whether there is a risk of potential secondary contamination of water in fountains caused by the sudden growth of Legionella bacteria. The factor that made the assessment was the water temperature which is one of the most important factors determining the development of these bacteria.

1.1 Water quality in fountains in Poland

Water quality in fountains in Poland is not a priority issue. Although voivodship sanitary-epidemiological stations issue warnings to avoid bathing in fountains and contact with water coming from them and explaining what might be a threat, quality of water in fountains is not constantly monitored. Water supply is subject to constant control, water quality in swimming pools is also checked, but there is no obligation in Poland to maintain constant monitoring of water in urban fountains.

The quality of water in fountains is sporadically discussed in national scientific articles[5-8]. The research conducted by [7] analysed not only water quality, but also air around the fountains in Warsaw. Some of the selected fountains contained mesophilic bacteria, coliform bacteria, Pseudomonas sp., macroscopic fungi and mannitol – positive staphylococcus. Microorganisms whose presence was examined were found in at least one of the fountains. In another work [6], water from Poznan city fountains was inspected at different times of the year. The presence of mesophilic, psychrophilic and coliform bacteria was detected there. In the remaining articles mentioned, inner fountains or fountains in another country were checked. It should be noted that in the cited studies of fountain waters in two large cities of Poland, the presence of mesophilic bacteria was found in which Legionella could be present.

1.2 Air temperatures in the city in Poland

The biological stability of water is defined as the lack of tendency for secondary microbial growth [9]. Water in fountains should also be biologically stable water. Considering only the Legionella bacteria - there is a danger that on hot days in the fountains there may be conditions favourable to their development due to raising the water temperature. In addition, it should be noted that long-lasting high air temperatures affect the temperature of the surface water (source of drinking water) and the temperature of the ground. As a result, the risk of rising water temperatures in the fountain itself is increasing.

Numerous climate changes are observed. Different sources [i.a. 10] indicate that some extreme phenomena (including heat waves) will increase in strength and occur more frequently. Climate models predict global warming and projections for Poland estimate a temperature rise of up to 3,5 degrees Celsius for the summer [10]. For these reasons, the scope of research has been extended to the analysis of air temperature in Krakow in recent decades.The summer of 2015 in Poland was a period of significantly high air temperatures. The end of spring of 2017 was also very hot. Figure 1 shows a graph of average monthly temperatures for a meteorological station located in Krakow for 2015 and a multi-year average (1971-2000 and 2001-2010) based on data of Central Statistical Office and Institute of Meteorology and Water Management - National Research Institute [11]. By analysing the graph, it can be seen that the average monthly air temperatures recorded in Cracow in 2015 are higher than the multi-year average in 9 months. In April, May and October were lower temperatures.

Analysis of air temperature data confirms that the warmest months in Poland are from May to September each year. For this reason, the study was conducted in July, August and September 2015 and May and June 2017.

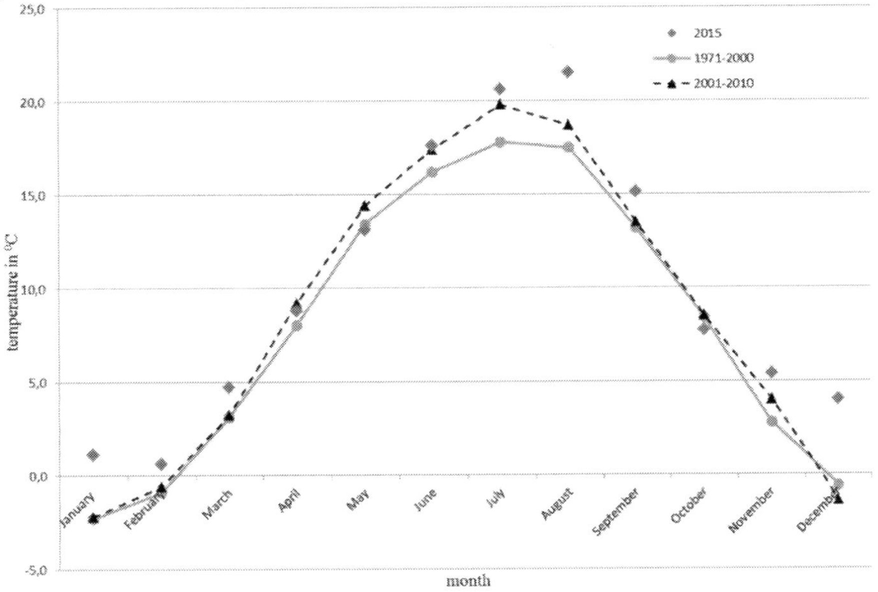

Fig. 1. Average monthly air temperatures for individual months from meteorological station in Krakow in the years 1971 - 2000, 2001 - 2010 and 2015, data of Central Statistical Office and Institute of Meteorology and Water Management - National Research Institute [11] - own elaboration

1.3 Influence of temperature on the growth of Legionella bacteria

Dependence the survival of Legionella bacteria and the water temperature was repeatedly investigated (among others [12- 13]). The literature provides narrower or wider temperature ranges suitable for the growth of these bacteria. This depends, inter alia, on the conditions they concern (natural or laboratory).

Under laboratory conditions, Legionella bacteria grow in the range of 15 ÷ 43°C, with optimum 36°C [14]. In developed recommendations [15] as the optimal growth temperature, but in natural conditions, the range of 32 ÷ 42°C is mentioned. Research [12] suggests a range of temperatures 32 ÷ 42°C with an optimum of 37°C for the multiplication of naturally occurring L. pneumophila in tap water. In the publication [1] it appears that these bacteria settle in the network of utility water at temperatures of 20 ÷ 50^{0}C. According to [14] Legionella bacteria colonize often water distribution systems at temperatures between 20 ÷ 50^{0}C and thus within the same range as in the previous work. In recommendations [15] it is stated that the temperature range of 20 ÷ 45^{0}C is one of the factors contributing to the occurrence and proliferation of artificial reservoirs of water.

According to [12], little or no growth of L. pneumophila was observed at 25^{0}C. In turn, according to [16], little, if any, increase in number of Legionella bacteria is noted below 20^{0}C. Other authors [15] indicate that at temperature less than 20^{0}C Legionella bacteria do not multiply, but survive for a long time. This fact has been demonstrated in studies [17], where it was found that at 5^{0}C in tap water these bacteria can survive for more than 299 days. For these reasons, it is suggested that temperatures below 25^{0}C be recommended for cold water storage and distribution, stating that it would be ideal to be below 20^{0}C [16]. However, recent studies on mutant strains of Legionella bacteria [18] have shown that these bacteria may also grow at temperatures below 20^{0}C under certain conditions.

2 Materials and methods

The study was conducted in two phases: summer 2015 and late spring 2017. In the first stage, water temperatures in selected fountains were measured in one or two series at different times of the day. The results were then compared with the preferred temperature range for the growth of Legionella bacteria. In the second stage the temperature was measured in one fountain, but increasing the frequency of measurements during the day. The research was conducted in different years in search of period of possibly high air temperatures.

Both the selection of the public fountains under study and the months in which the research was conducted were targeted. The research was conducted in the hottest months of the year, i.e. in May, June, July, August and September.

Public fountains were surveyed in Cracow. The work was carried out by the Cracow University of Technology as part of a research project. The research has deliberately chosen fountains located in the very centre of the city which is a popular destination for many tourists. On hot days the fountains located most often in sunny places are an oasis for travellers seeking refreshing.

The number of measuring points was adjusted to the size of the fountain basin and its geometry, so that the average temperature could be representative of the fountain, and the individual measurements provided complete information about the water temperature. Measurement points were evenly located in the basin of the fountain. Fountains with simple geometric shapes of the basin (rectangle, circle, oval) were studied. Sometimes, if temperature of water were aligned in previous tests, the number of measurement points was reduced due to difficulty in accessing them.

Fountains number 1 and 2 flow in closed loop. Fountain number 3 has an open circuit, but the stop time in the fountain's basin is long enough to allow the water to warm up in it. Studies in fountain number 3 have been added for comparative purposes.

In the first stage of the study, one or two series were performed at intervals of approximately 60 minutes. In the second stage, the number of series of tests performed in one day was increased. During each series of tests were recorded: water temperature in the fountain basin at several measurement points, air temperature and air humidity.

Water temperature tests were carried out using a KT-300 digital thermometer. The thermometer resolution was 0.1°C and its accuracy was ± 1°C. Air temperature and humidity were recorded on the WS-9400 weather station.

3 Results and discussion

Results of water temperature tests in the fountains of the first stage of the tests are presented in Tables 1, 2 and 3. In each tabulation prepared for each of the examined fountains, there are the date of measurements, the measurement of the basic parameter, i.e. the water temperature expressed in degrees Celsius and name of measurement points.

Table 1. Results of the water temperature test in the fountain number 1 in Krakow in 2015

Date / Measurement point	Temperature of water [^{0}C]			
	22/07	7/08	20/08	30/09
1	24,0	24,1	17,9	11,1
2	24,1	24,2	18,0	11,1
3	24,1	24,1	18,0	11,1
4	24,0	24,1	18,1	11,0
1	24,8	24,7	-*	-*
2	24,8	24,7		
3	24,8	24,8		
4	24,8	24,8		

* no data

Table 2. Results of the water temperature test in the fountain number 2 in Krakow in 2015

Date / Measurement point	Temperature of water [^{0}C]			
	22/07	7/08	20/08	30/09
1	23,8	24,5	21,1	11,7
2	23,8	24,5	21,1	11,7
3	23,8	24,5	21,3	11,7
4	23,9	24,5	21,3	11,7
5	23,9	24,5	21,3	11,7
6	23,8	24,5	-*	11,7
7	23,9	24,5	21,2	11,7
1	24,1	24,7	-*	-*
2	24,1	24,7		
3	24,1	24,7		
4	24,2	24,8		
5	24,2	24,8		
6	24,2	24,7		
7	24,2	24,7		

* no data

If the name of the measurement point in the statement occurs twice, it means that two series of tests were carried out at a spacing of approximately 60 minutes on a given day.

The air temperature during research tests ranged from 13 to 32,2 degree Celsius. The air humidity was from 34% up to 52%.

Based on the research (first stage) it can be stated that the water temperatures in each fountain were aligned. Usually this difference was 0.1 ÷ 0.2°C, although there were measurement series in which there were no differences in the measurements. Such results may indicate, among other things, the good mixing of water in fountains. In the case of fountain number 3 there is a greater temperature difference, but these are the temperatures measured in two different basins - the lower and the upper.

Table 3. Results of the water temperature test in the fountain number 3 in Krakow in 2015

Date / Measurement point	Temperature of water [°C]	
	7/08	20/08
1	25,1	18,4
2	25,1	18,4
3	25,2	18,4
4	-*	20,6
5	-*	20,7

* no data

In the second stage, the daily temperature distribution was measured in warm days for fountain 2. The number of measurement series varied within days, with an interval of 30 minutes to 3 hours. The water temperature in each of the measurement series was also aligned. The differences were up to 0.2°C, or none at all. The only exception was the measurement series, where differences were up to 0.4°C (Figure 2 – series: 14,00). Probably the cause of this phenomenon was turning off the nozzle for two hours. Figure 2 shows the results of several measurement series from one day of measurements.

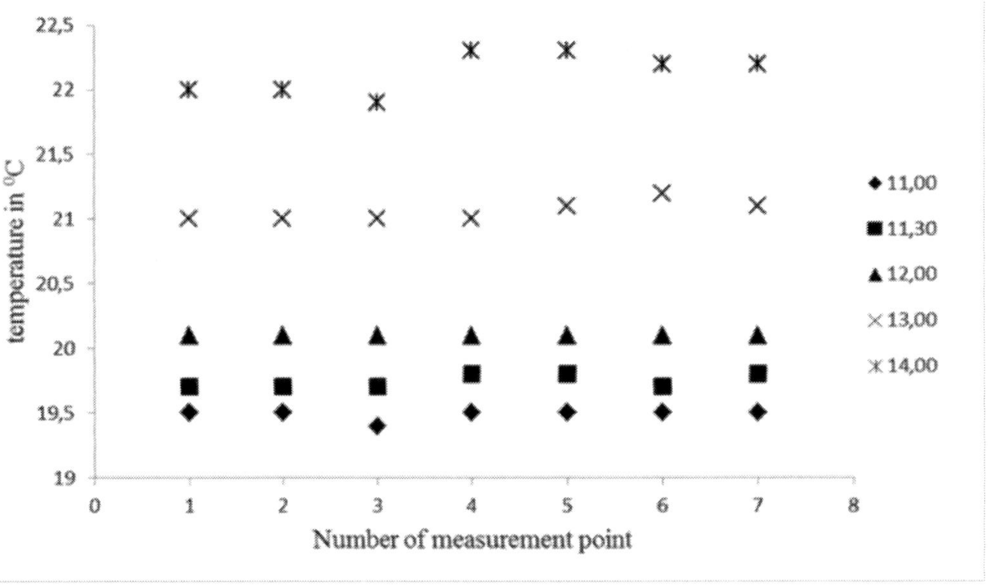

Fig. 2. Temperature of water in fountain's basin for different measurement points in five series of test during spring 2017 (30.05.2017)

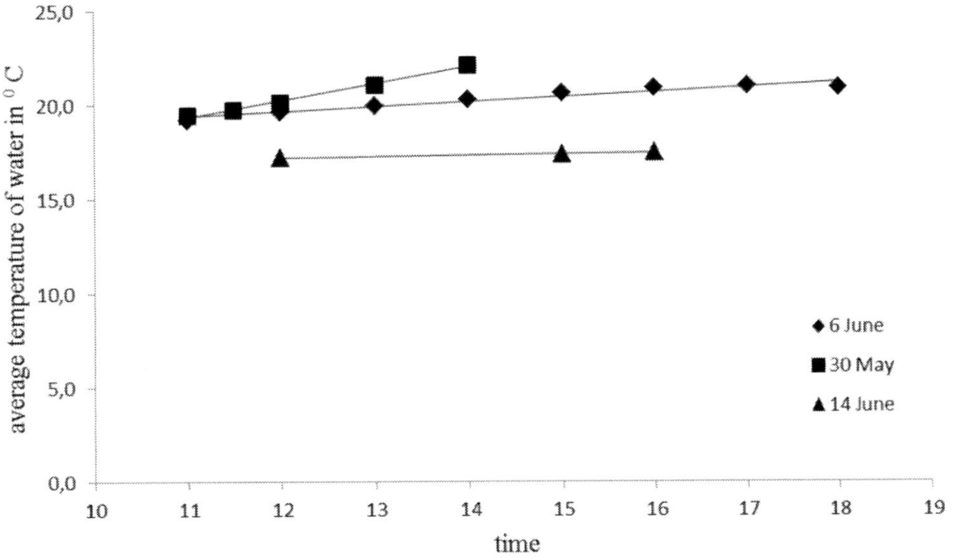

Fig. 3. Average temperature of water in fountain's basin in different days of measurement series during spring 2017

Figure 3 shows the distribution of the average water temperature in the fountain basin in daily hours on different days. It can be observed that there is a linear increase in water temperature.

In most of the first stage measurements, the water temperature in the fountains was above 20^{0}C, so above the temperature at which the Legionella bacteria are in hibernation state [19]. In the second stage the temperature fluctuated from 17.2^{0}C to 22.3^{0}C. In none of the stages, however, there was no temperature higher than 26^{0}C. The highest temperature recorded was 25.2^{0}C.

It can therefore be stated that, despite the high temperature of the air, which has persisted for a long time, there was no significant increase in the water temperature in the fountains (above 26^{0}C) during the measurements. However, it may be disturbing that the water temperature was in many cases above 20^{0}C. According to [16] below 37^{0}C the reproduction rate of these bacteria decreases and below 20^{0}C there is no growth or little. At low temperatures Legionella bacteria can survive for a very long time (according to [17] up to almost 10 months at 5^{0}C). Recorded temperatures indicate the possibility of presence and slow growth of these bacteria.

While studies on the presence and development of Legionella bacteria in tap water or in hot water systems are common, water quality tests for fountains due to these bacteria are rare. Water temperature measurements in fountains in Poland were also conducted by [6] in Poznan and their results are consistent with those presented in the article in Krakow. The highest recorded water temperature in Poznan in spring was 17.9 degrees Celsius (June) and in summer was 22.2 degrees Celsius (July). These measurements were made in the morning, so the temperature during the day could increase significantly. In the presented results in fountain number 2 in Krakow the average water temperature from 19.5 degrees Celsius at 11.00 rose to over 22.0 degrees Celsius at 14.00. It can therefore be concluded that studies [6] also confirm the existence of a potential threat from the Legionella bacteria due to the occurring water temperature in the fountains. It should be noted that in studies [6, 7] of fountain water in two large cities of Poland (Poznan and Warsaw), the presence of mesophilic bacteria was found in which Legionella could be present.

In another studies [20] among the samples of water with a temperature in the range of over 20 degrees and below 50 degrees Celsius, taken from fountains, swimming pools, cooling and heating systems, in 12% Legionella bacteria were detected.

4 Conclusion

Water in fountains is usually disinfected and can be considered as safe. However, there are various breakdowns in the case of urban fountains, and there may be a danger of excessive development of Legionella bacteria due to high temperature of water. As a consequence, secondary water pollution may even lead to outbreaks of legionellosis in the spring and summer. Therefore, the temperature and quality of water surrounding us in urban space is also very important.

Every fountain is different. Differences include the shape of the basin, its dimensions or the volume of the underground tank. It seems reasonable to require at least one single test of water temperature in fountain basin during period of high air temperatures to determine whether there is a risk of inordinate growth of Legionella bacteria. If the study confirms such a threat, it is extremely important to introduce additional technical safeguards which will protect water in fountains.

References

1. W. Magdzik, D. Naruszewicz – Lesiuk, A. Zieliński, *Diseases of infectious and parasitic diseases - epidemiology and prevention*, α – Medica Press Bielsko – Biała, 154 – 159 (in polish) (2007)
2. Rozporządzenie Ministra Zdrowia z dnia 13 listopada 2015 r. w sprawie jakości wody przeznaczonej do spożycia przez ludzi, Dz.U.2015, 1989 (2015)
3. Rozporządzenie Ministra Zdrowia z dnia 29 marca 2007 r. w sprawie jakości wody przeznaczonej do spożycia przez ludzi, Dz. U. 2007 No 61, 417 (2007)
4. Rozporządzenie Ministra Zdrowia z dnia 9 listopada 2015 r. w sprawie wymagań, jakim powinna odpowiadać woda na pływalniach, Dz. U. 2015, 2016 (2015)
5. J. Sperzyńska, M. Michałkiewicz in Dymaczewski Z., Jeż-Walkowiak J., Urbaniak A. *Water supply and water quality*, Poznań, 385 – 394 (in polish) (2016)
6. M. Michałkiewicz, A. Szumigała, Technology of water, **38**, number 6, 83 – 91 (in polish) (2014)
7. E. Miaśkiewicz – Pęska, D. Dąbrowska, Installation market, no. 4, 106 – 109 (in polish) (2016)
8. J. Bąk in Z. Dymaczewski, J. Jeż-Walkowiak, A. Urbaniak *Water supply and water quality*, Poznań, 235 – 246 (in polish) (2016)
9. J. Nawrocki, S. Biłozor – editors, *Water treatment. Chemical and biological processes*, Scientific Publishing House, PWN (in polish) (2000)
10. Z.W. Kundziewicz, P. Kowalczak, *Climate change and their effects*, KURPISZ S.A. (in polish) (2008)
11. Central Statistical Office and Institute of Meteorology and Water Management - National Research Institute, *Environment* (2016)
12. R.B. Yee, R.M. Wadowsky, R.M., Appl. Environ. Microbiol., **43**, No.6, 1330 – 1334 (1982)
13. A. Ohno, N. Kato, K. Yamada, K. Yamaguchi, Appl. Environ. Microbiol., **69**, No. 5, 2540 – 2547 (2003)

14. K. Botzenhart, Legionella in *Guidelines for drinking – water quality. Second edition. Addendum. Microbiological agents in drinking water*, WHO Geneva, 40 – 69 (2002)
15. B. Krogulska, R. Matuszewska, D. Maziarka, A. Krogulski, M. Szczotko, M. Bartosik, Zalecenia dotyczące ograniczenia występowania zanieczyszczeń mikrobiologicznych, w tym bakterii z rodzaju Legionella, w systemach wody technologicznej/chłodniczej i w sanitarnych instalacjach wody ciepłej w zakładach przemysłowych, National Institutes of Public Health, The National Institute of Hygiene, Warsaw (2013)
16. J. Bartram, Y. Chartier, J.V. Lee, K. Pond, S. Surman – Lee, (edition), *Legionella and the prevention of legionellosis*, WHO (2007)
17. S.C. Hsu, R. Martin, B.B. Wentworth, Appl. Environ. Microbiol., **48**, no. 4, 830 – 832 (1984)
18. M. Sodeberg, O. Rossier, N. Cianciotto, J. Bacteriol., 186, 3712 – 372 (2004)
19. E. Szewczyk, (scientific editor), *Bacteriological diagnostics*, PWN, Warsaw (in polish) (2013)
20. A. Bešić, Z. Obradović, A. Dautbegović, A. Obradović, Journal of Health Science, **7**, no 1, 50 -58 (2017)

Use of EPANET solver to manage water distribution in Smart City

A. Antonowicz[1], *R. Brodziak*[2], *J. Bylka*[2], *J. Mazurkiewicz*[3], *S. Wojtecki*[1], *P. Zakrzewski*[1]

[1]Wydział Informatyki, Politechnika Poznańska, ul. Piotrowo 3, 60-965 Poznań, Polska
[2]Wydział Budownictwa i Inżynierii Środowiska, Politechnika Poznańska, ul. Piotrowo 3a, 60-965 Poznań, Polska
[3]Wydział Rolnictwa i Bioinżynierii, Uniwersytet Przyrodniczy w Poznaniu, ul. WojskaPolskiego 28, 60-637 Poznań, Polska

Abstract. Paper presents a method of using EPANET solver to support manage water distribution system in Smart City. The main task is to develop the application that allows remote access to the simulation model of the water distribution network developed in the EPANET environment. Application allows to perform both single and cyclic simulations with the specified step of changing the values of the selected process variables. In the paper the architecture of application was shown. The application supports the selection of the best device control algorithm using optimization methods. Optimization procedures are possible with following methods: brute force, SLSQP (Sequential Least SQuares Programming), Modified Powell Method. Article was supplemented by example of using developed computer tool.

1 Introduction

Water supply systems are constantly changing and developing. From ancient times until now, significant advances have been made in the field of mathematical approaches, hydraulic modelling, optimization and last monitoring. The devices have become more sophisticated thanks to advances in computers and telecommunications [1]. At present not only scientists but also practitioners need to be familiar with the latest ICT technologies supporting water distribution systems.

Particularly in recent years there has been a lot of paper taking into account modern "smart water management system" includes smart meters, smart valves, intelligent pumps and analysis models of operating in the smart cities ecosystems [2-5]. Making all these elements work together is a real challenge. There are various aspects and possibilities using smart technologies. In this context, it is necessary to continually present the possibilities of using knowledge and experience with the advanced technologies that can be operated on commonly available PC and mobile devices.

Utilization intelligent water distribution management systems is not only about convenience and quickness in decision-making, but above all the need to reduce water and energy consumption.

In 2014, there were nearly 3.9 billion people in cities and 54% in the global population, and by 2050 two thirds of the global population will live in cities, which will increase 55% of additional world water demand [6]. Water management taking into account the aspect of energy consumption is currently one of the priorities of the most cities. It's because of the reliability and water-quality performance of a city's water system which requires a lot of energy. Therefore, to solve such problems, it is proposed to use the planning and management of water supply systems in a coordinated and interdisciplinary manner [7, 8,9, 10]. One of the most important elements of this approach in Smart Cities development is the use of "ICT for sustainable growth" with the ambition to lead innovation at the worldwide scale [11]. The example of the synergy ICT and Smart City for water distribution management it was featured in this publication by application using Epanet solver.

Environmental Protection Agency [12, 13]. Epanet is commonly use for analyse municipal water distribution network. Modelling process allows determine pressures in all nodes and flow in all pipesin the network, under set of boundary conditions and parameters. During the process the set of linear mass continuity equations, non-linear energy continuity equations and are calculated. The undoubted advantage of using Epanet is the possibility to test various scenario of system operation eg. control scenarios, design variants, etc. This makes that Epanet can be use for decision support in Smart Water Network.

1.1 Research objectives

The aim of the paper is to develop a method of using EPANET solver to perform simulations allowing to choice the best scenario operation of the water supply network according to a specified criterion. For this purpose web based application was developed. Application allows remote access to the simulation model of the water distribution network developed in the EPANET environment. The application perform both single and cyclic simulations with the specified step of changing the values of the selected process variables (pump performance, valve opening rate, etc.).

2 Methodology

2.1 Tools

The created application enables simulations of hydraulic systems and was built using Python scripting with Flask framework, EPANET .DLL file (Dynamic-Link Library) and PostgreSQL PostGIS database system(Figure 1).

Flask extends Python's capabilities for Web-based solutions, but does not includes many components like database abstraction layer or validation. This issue can be solved by using one of the many extensions, which provides modules that greatly increase its capabilities. This allows for saving time needed to produce application and also reduce amount of code lines. Choosing Flask allows to place computing algorithms and scripts used to process network models on the server side (backend) and presentation layer to users web browser (frontend). Implemented PostgreSQL database is one of the object-relational database management system (ORDBMS). PostGis extension allows to support for geographic objects (Postgis Web) Simulations of water network can be performed using computing engine EPANET, developed by United States Environmental Protection Agency (EPA) Water Supply and Water Resources Division. Program computes i.e. pipes and nodes pressure, energy usage of control components, it can performs hydraulic and water quality simulations and also allows to modify parameters of water supply network elements. In this work we use EPANETTOOLS [14] which is a module prepared to use EPANET computing capabilities in Python. Hydraulic models of water supply networks in EPANET are provided in files with *.inp extension, it contains characteristic of pipes, nodes, water tanks and control components (pumps and valves). The application uses the settings of the hydraulic network stored in the database, also allowing for modification.

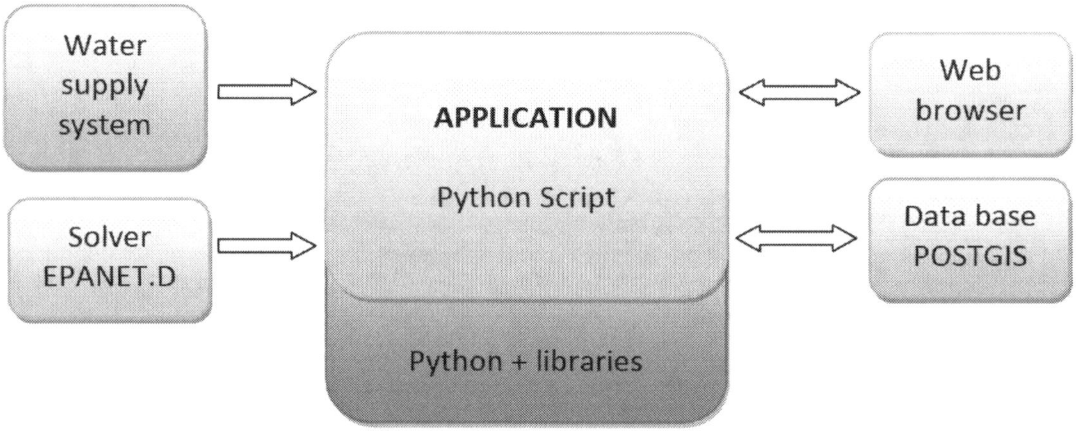

Fig. 1. Logical scheme of developed application

User communicates with application through web browser, where is possible to change basic simulation parameters (i.e. pump efficiency, energy costs) and allows to modify optimization objective (i.e. minimization of costs, minimization of energy usage). It can choose one of the implemented optimization methods:

- SLSQP (Sequential Least SQuares Programming),
- Modified non-gradient method extreme function of many variables Neldera-Meada[15, 16]
- Modified Powell method [17, 18].

The brute-force method despite of the guarantee of finding the optimum solution, because of high computational complexity and long of computation time. This method is only valid for simple and uncomplicated hydraulic models. The first method of optimization available from the application level is the Nellander-Mead Creeping Simplex method (which has been modified to take account of limitations). The next two methods of optimization with limitations are the SLSQP method [19] and Powell.

2.2 Criteria and limits for control algorithms

Various criteria and constraints must be taken into account in determining the control algorithms. Main limitations in control are due to the technical feasibility of the equipment and the water supply network. The requirements may include criteria related to environmental and economic aspects. Legal restrictions are also important. The implementation of specific requirements makes it possible to conclude that the conditions required for sustainable development are met. The work defined goal functions that indicate the main criterion for which optimization was performed. The aim is to minimize the total costs associated with the water supply network, which includes the cost of electricity purchase and operational costs (consists of environmental fee for raw water, processes of water treatment). Mathematically, this task can be written as follows:

$$min\left(Tc = \sum_{i=1}^{n} Ec_i \cdot PP_i + \sum_{i=1}^{n} Oc_i \cdot Wf_i\right) \quad (1)$$

where:
n – number of water sources,
Tc – total cost [PLN],
Ec_i – energy cost, power purchase in i-th source [PLN/kWh],
PP_i – pump power in i-th source [kWh],
Oc_i – operational cost in i-th source [PLN/m3],
Wf_i – water flow in i-th source [m3],

The purpose of the water supply system is to provide for customer the right amount of water with certain quality at the right pressure. This condition can be fulfilled through different pumps settings.

$$H_k > H_r \quad (2)$$

where:
H_k – pressure in k- node [mH2O],
H_r – required minimal water pressure in each node[mH2O]

The described criterion and limitation were taken into account in the created algorithm, whose execution indicated device settings.

3 Results

The result of the work is the computer environment for support decision-making in water supply systems control. The scope of work is limited to selecting the optimal settings of relative pumps speed. The operator need to load the file (file with Epanet model) and select the control range: minimum and maximal relative speed of pumps. Relative speed changes during the optimization process, due to selected optimization method. These values are determined in the technical characteristic of pumps. User also need to define operational cost and energy cost. Operating costs depend on water treatment technology, environmental fees, etc. Energy costs are set out in the contract with the energy provider (cost of energy may vary for different pumping stations because of the tariff). In the application also constraints for optimization needs to be determined. The operator enters the minimum pressure requirements for all nodes and relative speed pump setting step. Pump setting step is related with technical capabilities of the devices.
Before running the simulation procedure, user chooses which optimization method will be appliedThe application interface is web responsive, which ensured its mobility, legible on the different devices like smartphones, tablets, laptops.

3.1 Case study

The web application was tested on a computer module of water distribution system which consist of: 15 junction, 2 pumps, 17 pipes and two sources of water (Figure 3).

https://doi.org/10.1051/e3sconf/20183001016

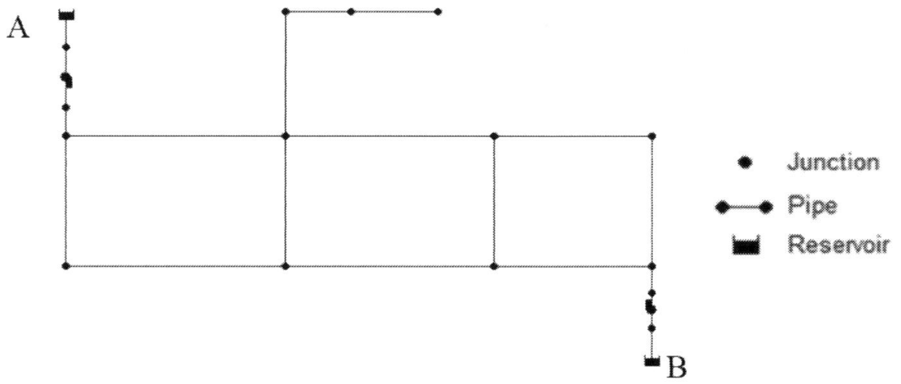

Fig. 2. The scheme of analysed water supply system model.

After the necessary data and constraints were entered the optimization procedure were performed. The result for one of optimization method is presented in figure 4. Result presents minimal operation, energy cost and sums it up. The value of selected pumps settings were presented as well.

Optimization results

Table Graph

Operation costs	Energy cost	Total cost	Pump1 setting	Pump2 setting
2.06 PLN	4.35 PLN	6.41 PLN	0.91	0.95

Fig. 3. Optimization results window – table mode.

The web application also allows analysing results on surface graphs (Figure 5). The chart contains information about: value of pump 1 setting (x-axis), value of pump 2 setting (y-axis) and value of cost (z-axis). On the graphs cut off those solutions that did not meet the optimization criteria (minimal require pressure in all nodes). The graph was made for the brute-force method.

4 Conclusion

The use of mobile application running using solver EPANET contributes to minimizing energy consumption and thus greenhouse gas emissions, without affecting the quality of drinking water. A similar system using the EPANET solver has recently been tested by other researchers who have demonstrated that its use reduces per capita water consumption and energy consumption by approximately 6-10% and by 0.5-6.2%, respectively [21].

In the paper the system for support the decision about the pump settings was presented. The system uses a well-known tool for new tasks - supporting the management of the water distribution network. Using this type of architecture allows to create complex computer tools to manage water distribution system in Smart City. Used modules of IT tools has open architectures. It also should be emphasized that all of used tools was developed under Open Source license.

Developed computer environment allows to support decision making in case when water distribution network is supplied from two sources. The main aim of task is to search the best solution for pumps control – where the cost of operations and energy is minimal. During analyzing the results of the calculations, it was noted that the result of the algorithm depended on the optimization method used.

The results of the calculation for the brute-force method and the three selected optimization method are presented in Table 1. It is easy to see that in the analyzed example, the Nelder-Meada method determined in the shortest time the same pump settings as the brute-force method.

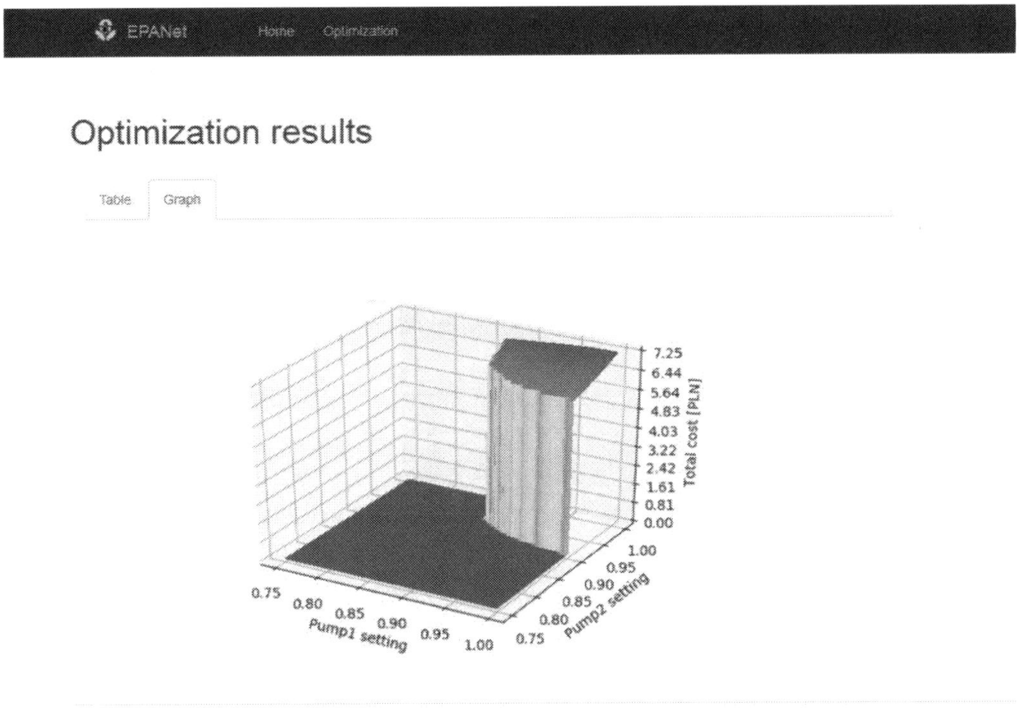

Fig. 4. Optimization results window – graph mode.

Table 1. Optimization results for various methods.

Method	Calculation time [s]	Pump setting	Total cost [PLN]
Brute-force	30.70	Pump1: 0.91 Pump2: 0.95	6.41
Nelder-Mead	3.24	Pump1: 0.91 Pump2: 0.95	6.41
SLSQP	0.12	Pump1: 0.93 Pump2: 0.97	6.58
Powell	3.10	Pump1: 0.90 Pump2: 0.99	6.43

1. Günthera M., Camhya D., Steffelbauera D., Neumayerb M., Fuchs-Hanuscha D. 2015. Showcasing a smart water network based on an

experimental water distribution system. 13th Computer Control for Water Industry Conference, CCWI 2015. Procedia Engineering 119 (2015) 450 – 457.

2. Díaz-Díaz R., Muñoz L., Pérez-González D. 2017. Business model analysis of public services operating in the smart city ecosystem: The case of Smart Santander. Future Generation Computer Systems.
3. Strzelecka A., UlanickiB.,Koop S., Koetsier L., van Leeuwen K., Elelman R. 2017. Integrating water, waste, energy, transport and ICT aspects into the smart city concept 18th Conference on Water. Distribution System Analysis, WDSA 2016. Procedia Engineering 186 (2017) 609 – 616.
4. Shahanas K.M, Sivakumar P.B. 2016. Framework for a smart water management system in the context of smart city initiatives in India. 2nd International Conference on Intelligent Computing, Communication & Convergence (ICCC-2016) Srikanta Patnaik. Interscience Institute of Management and Technology Bhubaneswar, Odisha, India. Procedia Computer Science 92 (2016) 142 – 147.
5. WIDEST Water Innovation through Dissemination Exploitationof Smart Technologies D2.3: Smart WaterGridtopicalroadmap. dostęponline: http://www.widest.eu/downloads/downloads-roadmaps (dostęp 25.06.17).
6. Arfanuzzaman Md., Atiq Rahman A. 2017. Sustainable water demand management in the face of rapid urbanization and ground water depletion for social–ecological resilience building. Global Ecology and Conservation 10 (2017) 9–22.
7. Engströma R.E., Howells M., Destouni G., Bhatt V., Baziliana M., Rogner H-H. 2017. Connecting the resource nexus to basic urban service provision – with a focus on water-energy interactions in New York City. Sustainable Cities and Society 31 (2017) 83–94.
8. Kenway S. J., Binks A., Lane J., Lant P. A., Lam K. L., Simms A. 2015. A systemic framework and analysis of urban water energy. Environmental Modelling & Software, 73(November), 272–285.
9. Scott C.A., Pierce S.A., Pasqualetti M. J., Jones A. L., Montz B. E., Hoover J. H. 2011. Policy and institutional dimensions of the water-energy nexus. Energy Policy, 39(10), 6622–6630.
10. Bazilian M., Rogner H-H., Howells M., Hermann S., Arent D. 2011. Considering the energy, water and food nexus: Towards an integrated modelling approach. Energy Policy, 39 (12) (2011), 7896–7906.
11. Gourbesville Ph. 2016. Key Challenges for Smart Water. Procedia Engineering. Volume 154, 2016, 11-18.
12. Rossman L., 2000. Epanet 2 user manual. National Risk Management Research Laboratory, Cincinnati.
13. Epanet: www: https://www.epa.gov/water-research/epanet(accessed

25.06.2017)
14. EPANETTOOLS Web 0.9.2 : Python Package Index, www: Document: https://pypi.python.org/pypi/EPANETTOOLS (accessed 25.06.2017)
15. Olsson D.M, Nelson L.S. 2012. The Nelder-Mead Simplex Procedure for Function Minimization
16. Nelder J.A., Mead R. 1965. A Simplex Method for Function Minimization. The Computer Journal 7: 308-13.
17. Powell M. J. D. 1964. An efficient method for finding the minimum of a function of several variables without calculating derivatives, Comput J.
18. Powell M J D. 2007 A view of algorithms for optimization without derivatives. 2007.Cambridge University Technical Report DAMTP 2007/NA03.
19. Kraft D. 1988. A software package for sequential quadratic programming. Tech. Rep. DFVLR-FB 88-28, DLR German Aerospace Center – Institute for Flight Mechanics, Koln, Germany.
20. Postgis Web: http://postgis.net/ (accessed 25.06.2017)
21. Santana M.V., Zhang Q., Nachabe M.H., Xie X., Mihelcic J.R. 2017. Could smart growth lower the operational energy of water supply? A scenario analysis in Tampa, Florida, USA. Landscape and Urban Planning. Volume 164, August 2017, 99–108.

Monitoring of heavy metals in selected Water Supply Systems in Poland, in relation to current regulations

Agnieszka Szuster-Janiaczyk*[1], *Piotr* Zeuschner[1], *Paweł* Noga[2], *Marta* Skrzypczak[2]

[1] Faculty of Civil and Environmental Engineering, Poznan University of Technology, Berdychowo 4, 60-965 Poznań, Poland
[2]Graduate of the Poznan University of Technology

Abstract. The study presents an analysis of water quality monitoring in terms of the content of heavy metals, which is conducted in three independent water supply systems in Poland. The analysis showed that the monitoring of heavy metals isn't reliable - both the quantity of tested water samples and the location of the monitoring points are the problem. The analysis of changes in water quality from raw water to tap water was possible only for one of the analysed systems and indicate a gradual deterioration of water quality, although still within acceptable limits of legal regulations.

Keywords: heavy metal, lead, copper, tap water, water quality monitoring

1 Introduction

The water crisis in Flint, Michigan [1], and the effects of it, has shown how important for tap water quality is stability of water, corrosion control and monitoring of heavy metals in Water Supply Systems (WSS). According to World Health Organization guidelines [2,3], monitoring of heavy metals in drinking water has become one of the elements that have to be taken into account in estimating the risk to human health, and lead, in the International Program on Chemical Safety, is one of 10 chemicals important for public health [4]. This is related to the health risks that these contaminants carry. These metals can cause cancer, promote nervous system disorders and digestive and many others [5-9].

The presence of increased concentrations of heavy metals in WDS is due to the existence of their sources in this system such as transmission pipelines, internal systems, fittings and others. Release of this metals is associated with hydraulic conditions in system (age of water, velocity) and physical, chemical and biological parameters of water influenced on stability and corrosivity of water.

To determine the presence and level of these metals in sections of Water Supply System, the proper location of the monitoring points and the determination of the quantity and frequency of sampling are needed. Research and analysis conducted in the world still didn't give a precise answer on question - how the overall monitoring of water quality

* Corresponding author: agnieszka.szuster-janiaczyk@put.poznan.pl

parameters should be. [10-13]. In the case of heavy metals monitoring, it's very important to take into account the accumulation of them in human organisms. That is why monitoring of heavy metals should be carried out in accordance with other principles which take into account the process of metal accumulation in consumer organisms [14-22]. The approach to this problem in Poland is fundamentally different from the proper approach, which may give a true picture of water quality in the network. In Polish regulation, as in the Directive [23], heavy metals are in audit monitoring, performed at a lower frequency than check monitoring, dependent on daily water production. That's way, in small systems, water quality assessment is often conducted in based on 1 water sample per year [24]. Water supply companies and sanitary inspection focus only on aspects shown if water quality meets the guidelines outlined in the Regulation of the Polish Minister of Health, the frequency and optimal localization of monitoring points aren't important for them. [15,19,21,22].

In the United States, the EPA recommends an individual approach to monitoring of heavy metals for which separate recommendations have been made. They contain precise guidelines on the amount of sampling, place of collection and corrective action [25, 15, 26, 27]. In Japan, water quality monitoring is built individually for each WSS, but must be approved by the relevant supervisory authority. That gives the opportunity to monitor the dangers contaminants which might appear in WSS [28]. In Russia, monitoring plans are elaborate in two stages, what gives the opportunity to indicating water quality parameters which should be covered by precise monitoring [29]. One of the important suggestions in the guidelines is the location of the monitoring points, where a significant part of all tested samples have to be collected in the water network.

An important element of heavy metal monitoring, which determines the reliability of the results obtained, is the quantity of water samples taken. There are several methods to design it [18, 19, 30]. In Polish regulations, the quantity of taken water samples is dependent on average of daily water production and whether a concrete water quality parameter has been qualified for check or audit monitoring. In the USA and Russia, the number of samples taken depends on the number of consumers being supplied. It is also possible to designate the necessary quantity of analysed water samples based on statistical tests. Power analysis [31], allows to determine the minimum statistically required number of samples, on the basis of which we can obtain a true picture of the analysed factor.

A reliable view of water quality can be obtained through the correct location of water quality monitoring points. There are many methods to determine a localization of sampling point. These methods are characterized by varying degrees of complexity, for example from simple random data sampling to genetic algorithms [11, 32]. They differ among themselves, among other things, the starting data needed to build them. These data can be eg. hydraulic model, water requirement, detection time, etc. Studies show that different methods give different results for localization of monitoring points in one system [11], but in every method the recommendation is that the decisive part of the monitoring point should be placed at the point of consumption of water to show real tap water quality. In Poland there is no good methodology for determining localization of sampling points for water quality assessment. The Regulation recommends the steady arrangement of monitoring points in the system, among other things, in consumers installation. Unfortunately, it's not specified, however, what part of the samples should be collected in network and what part in consumer installations .This leads to a situation in which a large proportion of the water samples analysed for heavy metal concentration comes from points located just behind the water treatment station.

1.1 Research objectives

The article presents the assessment of heavy metals monitoring conducted in Poland, on the example of three similar Water Supply Systems (WSS1, WSS2, WSS3), in relation to current legal regulations and world trends. The first two systems (WSS1 and WSS2) are stand-alone technical objects, while WSS3 is part of a larger water supply system. The characteristics of these systems are presented in Table 1.

Table 1. The characteristics of analysed Water Supply Systems.

-	WSS1	WSS2	WSS3
Number of recipients	24.150	30.201	30.493
Type of water intake	4 intakes - 13 deep wells	4 intakes - 9 deep wells	2 intakes - ground and infiltration water
Number of water treatment plants	3	3	1
Network structure	Looped-branched	Looped-branched	Looped-branched
Network length	350 km	284 km	183 km
Materials	ok. 75 % PVC; 18 % PE; 7 % cast iron; 0,15% ductile iron	72% PVC; 23% cast iron; 5% asbestos	-
Water demand	3533 m^3/d	3867 m^3/d	4000 m^3/d*)
Velocity	About 90% below 0.3m/s	About 70% below 0,5m/s	-
*) The water system plant produces 150.000 m^3/d of water, but only part of it (4000 m^3/d), supplies WSS3.			

1.2 Methodology

The results of monitoring of heavy metals for each water supply systems, from different periods were analysed in this paper: for WSS1 – years 1970 ÷ 2016; for WSS2 – years 2013 ÷ 2016; for WSS3 years 2012 ÷ 2014. The results were divided for three subdivisions of WSS: raw water samples, post-treatment samples, and samples collected in the network. In case of WSS1 there was the ability to isolated samples from internal installations – tap water quality. The total number of analysed samples of water in general monitoring and audit monitoring is presented in Table 2. Table 2 shows also the number of parameters tested as part of water quality analyses.

Table 2. Number of samples for the whole study period (general/in audit monitoring)

-	WSS1	WSS2	WSS3
raw water	17	66	551
treated water	33/2	212/32	1748/63
in network	28/8	1298/32	389/0
Number of analysed parameters	94	41	85
Number of samples with heavy metals analyzed - R/T/N*)	3/15/8	0/0/0	504/63/0
*)R/T/N- for raw water/for treated water/in network			

Samples in systems were retrieved with high variability over time. There are some years where only single results are available. Due to the lack of full data from all the water supply systems, there was no ability to produce a complete statistical analysis of the results obtained. For available results, average, minimum and maximum concentrations of the contaminant were determined. In this article the compliance of the performed monitoring with the Polish regulations on contaminants concentrations, the number of collected samples and the location of monitoring points has been analyzed. The results obtained were compared with world regulations. STATISTICA and Excel were used to develop the results.

2 Results

Table 3 summarizes the quantity of water samples taken as part of the water quality monitoring carried out. Data refers to the total number of samples and audit monitoring per year.

Table 3. Number of samples per one year (general/in audit monitoring)

-	WSS1	WSS2	WSS3
raw water	1	16.5	184
treated water	4/0.25	53/8	583/21
in network	6/1	324/8	130/0
Number of analysed parameters	94	41	85
Number of samples with heavy metals analyzed - R/T/N*)	0/0.25/1	0/0/0	168/21/0
*)R/T/N- for raw water/for treated water/in network			

The full analysis of heavy metal concentrations in the water distribution system was only possible for WSS1 (Table 4) because for WSS2 no heavy metals monitoring was carried out in the analyzed period, and in WSS3 all water quality monitoring points were located out of analysed system.

Table 4. Water quality in WSS1

Parameter	MP*)	Raw water	Treated water	From network	From taps
		mean/min÷max			
Lead µg/L	10	2/ 1÷4	1.99/ 1÷5	2.06/ 1.00÷4.00	<1/ <1÷<1
Copper mg/L	2	0.05/ 0.001÷0.002	<0.0015/ <0.001÷<0.002	0.003/ 0.0013÷0.005	<0.01/ <0.01÷<0.01
Zinc mg/L		0.048/ 0.046÷0.05	0.05	bd	Bd
Chrome µg/L	50	2/ 1÷4	2.9/ 1÷4	1.6/ 1.0÷4.0	2.7/ 1.9÷3.1
Cadmium µg/L	5	1.4/ 0.1÷1	0.22/ 0.1÷0.5	0.14/ 0.1÷0.3	0.1/ 0.1÷0.1
Mercury µg/L	1	0.08/ 0.05÷0.1	0.06/ 0.01÷0.1	0.076/ 0.03÷0.1	0.2/ 0.2÷0.2
*) Maximum permitted concentration (Regulation of the Polish Minister of Health, 2015)					

3 Discussion and conclusion

3.1 Location of the monitoring points

In all analyzed systems there is a large inconsistency in the location of monitoring points. In particular, this is evident to WSS3, where there wasn't any monitoring point for heavy metals concentration assessment in water network. The samples analyzed in this system were collected from the intake and the water treatment plant. This is due to the fact that this system is a small part of a larger water network in which all network monitoring points are located. For WSS1 the situation looks a little better - about 50% of monitoring point were located in network, but it still doesn't correspond to the recommendation, which suggest localized monitoring points also in internal installation to take tap water for analyses [15, 24, 18]. The samples for monitoring heavy metals, in this system were taken in 5 places including: 2 schools, hotel, hospital, flat. School, hospital and hotel can be considered as a good choice for localisation of monitoring points.

For WSS1 and WSS3, the amount of samples taken just after the water treatment plant was much higher than in the rest of the system, but assessment of water quality in whole system can't be based just on results from treated water, because it doesn't show the reliable metal concentrations in system.

For WSS2, the situation with monitoring of heavy metals is worst since no heavy metal monitoring points in this system is located, what is incompatible with the regulations.

For WSS2 and WSS3, in the localization of monitoring points, objects recommended in regulations, such as medical facilities and schools have been omitted.

For all analyzed WSS, their broad characteristics are available: building materials, network structure, length, velocity, water divisions, in addition to this WSS1 and WSS2 also have a full hydraulic model of network. Literature analyzes [13, 12] show that the data held by the system administrators enable the use of many different monitoring methods, so wondering is, therefore, why they can't completely use them in planning of monitoring process in analysed systems.

3.2 Number of collected samples

In accordance with the Polish legal regulations [24], the required annual quantity of water samples tested, for all systems analysed in this articles, should be 16/2 (check monitoring/audit monitoring). With respect to these values only in the WSS1 system, the quantity of water samples doesn't correspond to the provisions of the Regulation. But, if we consider that heavy metals should be included in the audit monitoring, then system WSS2 doesn't meet the requirements of the regulation, too. In the case of WSS3, which is part of the large Water Supply System (15.000 m^3/d), the total number of analysed water samples is theoretically consistent with the regulations [24], but all monitoring points are located outside the WSS3 area, so it can be stated that WSS3 is completely deprived of assessment of water quality (including heavy metals), during its distribution.

Comparing the number of samples collecting in three analysed system with global guidelines, the situation with water quality monitoring seems to be definitely worse. For example USEPA, Russian guidelines, require collecting much more samples for heavy metal monitoring. Both of these methods rely to the number of taken samples, which are calculated based on number of recipients. The amount of water samples required to obtain a reliable water quality image, can also be obtained by statistical tests of power. With regard to these tests, the number of analysed water samples, according to Polish regulations, is several to several hundred times smaller [20].

Table 5. Required quantity of samples according to different guidelines for analysed WSS

Legal acts/standards/statistical dependence	Poland	USEPA	Russia	Power analysis
Required quantity of samples	2	60	262*	14÷1042**
*) If the metals are indicated as an additional monitoring range				
**)One Proportion: Sample Size Calculation, H0: Pi = Pi0 α-0,05; Pi0-0,5; Power Goal-0,9; Pi-0,9 for 14; Pi-0,55 for 1042				

Table 5 lists the required quantities of samples for monitoring heavy metals in distribution system, for three analysed systems, based on the various available guidelines.

3.3 Water quality

In WSS1 and WSS3, the concentrations of heavy metals in treated water, just after water treatment plant, was complied with the provisions of the Polish regulation [24, 33].

Changes in the concentration of heavy metals from the intake to internal installations can be analysed only for the WSS1 system, because only in this system, the monitoring of heavy metals is carried out in the way, which allows to perform this kind of the analysis. For the whole WSS1, the metal concentrations in the water were below the limits required in the Regulation [24], but the changes in water quality during transport from the intake to the consumers were noticed (Table 4).

In the water from the network, collected in water mains, there was no increase in the concentration of lead, in relation to the concentrations of this factor indicated in treated water, while double increase in copper concentration in the network was observed, compared to the average concentration indicated in treated water - from 0.0015 mgCu/L in water treatment plant, to 0.003 mgCu/L in network. In addition, a further increase in the average concentration of copper, compared to the average concentration in the water network, was noted in internal installations - from 0.003 mgCu/L to <0.01 mgCu/L.

There was also an increase in the average concentration of chromium in water from installation installations, with respect to treated water - from 1.6 µgCr/L to 2.7 µgCr/L.

In the analysed system, there was noticed also a successive increase in mercury concentrations from 0.06 µgHg/L (for treated water) to 0.076 µgHg/L (for water from network), to 0.2 µgHg/L (for internal installations).

3.4 Conclusion

In the investigated systems, water quality monitoring was conducted with high variability over time and with low regularity, in relation to the number of tested samples. Moreover, the location of the monitoring points focused mainly on treated water, which allows us to assume that in the investigated systems there isn't credible monitoring of heavy metals in the water, although monitoring is carried out in accordance with applicable regulations.

Based on the data from WSS1, it can be stated that the analysed water, at each monitoring point, complied with guidelines included in Polish regulations, but it should be presumed, that this is not a complete picture of water quality in this system, because the points in internal installations concerned utilities (schools, hospital, hotel), which are under constant supervision of the sanitary inspection, hence the administrators of these facilities are likely to take special care of their technical condition [33]. Others studies conducted in Poland by Postawa [34] show that, with more detailed monitoring, where the location of the monitoring points was determined by the RDT method, for 1440 samples, there was network zone in which about 8.4% of water samples had exceeded 10 µg/L, and in the other zone there were about 60% of samples with excess lead concentration.

In future studies, the description of the analysed systems, should be supplemented by a detailed description of the internal installation - age of installations, used materials, the water sampling method, and the way the installations are used by the customers, because all these factors are very important in determining the tap water quality [35, 33].

Symptomatic is, that even small water supply systems, such as the ones analysed in this article, have often very good tools, in the form of hydraulic models, to design reliable water quality monitoring plans in the system, but they don't use them to prepare effective monitoring of heavy metals. Probably it will change when the law regulations in this area will changes and administrators of water supply systems will have to build reliable water quality monitoring plans for monitor metals concentrations in water.

It should be emphasized, that according to the Polish regulations, a very serious problem with reliable water quality monitoring, have the small water supply systems where the measurement of heavy metal concentrations is once a year or less often and just in the treated water, not in the network. However, this is a problem that affects small waterworks also in other European countries and in the world.

References

1. K. Pieper, M. Tang, M.A. Edwards, Environ. Sci. & Technol., **51**(4), 2007-2014 (2017)
2. World Health Organization (WHO), Lead in drinking-water. Background document for development of WHO Guidelines for Drinking-Water Quality. Geneva, World Health Organization (WHO/SDE/WSH/03.04/09; http://www.who.int/water_sanitation_health/dwq/chemicals/lead.pdf)
3. World Health Organization (WHO), Manual for the public health management of chemical incidents, Geneva, Switzerland (2009), Available online: http://www.who.int/environmental_health_emergencies/publications/Manual_Chemical_Incidents/en/,
4. World Health Organization (WHO). Ten Chemicals of Major Public Health Concern, (2010) Available online: http://www.who.int/ipcs/assessment/public_health/chemicals_phc/en/ (assessed on May 2017)
5. M.A. Edwards , S. Triantafyllidou, D. Best, Environ. Sci.& Technol., **43** (5), 1618-1623 (2009)
6. N. Zhang, S. Zang, Q. Sun, Ecotoxicology **23**, 518–526 (2014)
7. V. Demir, T. Dere, S. Ergin, Y. Cakır, F. Celik, Water Resources **42**, 508–516 (2015)
8. N.H.A.Razak, S.M. Praveena, A.Z. Aris, Z. Hashim, J. Epidemiol. Glob. Health **5**, 297-310 (2015)
9. V. Elumalai, K. Brindha, E. Lakshmanan, Water **9**, 234, www.mdpi.com/journal/water, (2017)
10. G. Dorini, P. Jonkergouw, Z. Kapelan, F. di Pierro, S.T. Khu, D. Savic, An efficient algorithm for sensor placement in water distribution systems, 8th Annual Water Distribution Systems Analysis Symposium, Cincinnati, Ohio, USA, August 27-30, (2006)
11. A. Ostfeld, J.G. Uber, E. Salomons, J.W. Berry, W.E. Hart, C.A. Phillips, J.P. Watson, G. Dorini, P. Jonkergouw, Z. Kapelan, F. di Pierro, S.T. Khu, D. Savic, D. Eliades, M. Polycarpou, S.R. Ghimire, B.D. Barkdoll, R. Gueli, J.J. Huang, E.A. McBean, W. James, A. Krause, J. Leskovec, S. Isovitsch, J. Xu, C. Guestrin, J. VanBriesen, M. Small, P. Fischbeck, A. Preis, M. Propato, O. Piller, G.B. Trachtman, Z.Y. Wu, T. Walski, J. Water Res. Plan. Man., **134** (6), 556-568 (2008)
12. S. Rathi, R. Gupta, Procedia Engineering nr **89**, 181-188 (2014)
13. E. Hołota, B. Kowalska, JCEEA, **32** (62), 81-98 (2015)
14. E.J. Hoekstra, C.R. Hayes, R. Aertgeerts, A. Becker, M. Jung, A. Postawa, L. Russel, S. Witczak, Guidance on sampling and monitoring for lead in drinking water, European Commission JRC Scientific and Technical Reports EUR 23812 EN, (2009)
15. Lead and Copper Rule Monitoring and Reporting Guidance for Public Water Systems,

USEPA, (2010)

16. Lead and copper corrosion control in new construction, The report of the research project no. 4164, Water Research Foundation (AWWA RF),US EPA, (2011)
17. USEPA (U.S. Environmental Protection Agency), (2014). USEPA Integrated Risk Information System (IRIS) online database. Available at: http://www.epa.gov/iris (Accessed: Nov. 2015).
18. A. Postawa, S. Witczak, *Metale i substancje towarzyszące w wodach przeznaczonych do spożycia w Polsce*, Akademia Górniczo-Hutnicza im. Stanisława Staszica w Krakowie, Kraków (2011)
19. B. Clark, S. Masters, M.A. Edwards, Environ. Sci.& Technol.,**48**(12), 6836-6843 (2014)
20. C.R. Hayes, N. Croft, E. Phillips, S. Craik, M. Schock, J.of Water Supply: Research and Technology -AQUA, **65**(5), 373-383 (2016)
21. S. Chowdhury, M.A.J. Mazumder, O. Al-Attas, T. Husain, Science of the Total Environment, **569–570,** 476–488 (2016)
22. S. Masters, J. Parks, A. Atassi, M.A. Edwards, Environ Monit Assess, **188**(3):177 (2016)
23. Directive 98/83/EC of the November 1998 on the quality of water intended for human consumption-Official Journal of European Communities L 330/32 of 5.1298
24. Regulation of the Polish Minister of Health of 27 November 2015, The Regulation on the quality of water intended for human consumption Dz. U. z 2015 r. poz. 139 i 189
25. Safe Drinking Water Act, *Water Infrastructure Improvement for the Nation Act*, P.L. 114-322, (2016)
26. Grassroots Rural and Small Community Water Systems Assistance Act, P.L. 114-98, enacted 11 of December, (2015)
27. M. Tiemann, *Safe Drinking Water Act (SDWA): A Summary of the Act and Its Major Requirements*, Congressional Research Service 7-5700, (1 March 2017)
28. *Water Resources Management in Japan*, Policy, Institutional and Legal Issues, Environment and Social Development, East Asia and Pacific Region The World Bank, (April 2006)
29. Standard of 2.1.4.1074-01, Drinking water. Hygienic requirements for water quality from centralized water supply systems. Quality control, Russian Federation, (2002)
30. S. Masters, M.A. Edwards, Environ. Eng. Sci. **32**(5), 361-369 (2015)
31. Power analysis of statistical tests - http://www.statsoft.pl/textbook/stathome_stat
32. D. Kowalski, B. Kowalska, M. Kwietniewski, Ochrona Środowiska, **35**(3), 45-48 (2013)
33. A. Szuster-Janiaczyk, J. Bylka, T. Schiller, M. Skrzypczak, Jakość wody w instalacjach wewnętrznych w odniesieniu do procesów kształtujących tę jakość, na podstawie wybranego Systemu Zaopatrzenia w Wodę, in *Instalacje wodociągowe i kanalizacyjne: projektowanie, wykonanie, eksploatacja*, Vol. 3, red. Z. Heidrich, A. Melasińska, J. Chudzicki, Wyd. Seidel-Przywecki sp. z o.o., Warszawa, 2017
34. A. Postawa, J.of Water Supply: Research and Technology-AQUA, **64**(1):85 (2015)
35. K. Wątor, A. Mika, A. Postawa, Wpływ materiału instalacji oraz czasu stagnacji wody na stężenia wybranych metali w wodzach przeznaczonych do spożycia in *Zaopatrzenie w wodę, jakość i ochrona wód*, red. Z. Dymaczewski, J. Jeż-Walkowiak, Poznań, (2016)

The influence of synthetic hyetograph parameters on simulation results of runoff from urban catchment

Karolina Mazurkiewicz[1*], and *Marcin* Skotnicki[1]

[1]Institute of Environmental Engineering, Poznań University of Technology, Pl. M. Skłodowskiej-Curie 5, Poznań, Poland

Abstract. The paper presents the results of analysis of the influence of the maximum intensity (peak) location in the synthetic hyetograph and rainfall duration on the maximum outflow from urban catchment. For the calculation Chicago hyetographs with a duration from 15 minutes to 180 minutes and peak location between 20% and 50% of the total rainfall duration were design. Runoff simulation was performed using the SWMM5 program for three models of urban catchment with area from 0.9 km^2 to 6.7 km^2. It was found that the increase in the rainfall peak location causes the increase in the maximum outflow up to 17%. For a given catchment the greatest maximum outflow is generated by the rainfall, which time to peak corresponds to the flow time through the catchment. Presented results may be useful for choosing the rainfall parameters for storm sewer systems modeling.

1 Introduction

Rainfall input data for modeling the outflow from urban catchment can be represent by the historical recorded rainfalls, synthetic hyetographs or hyetographs developed with stochastic models [1, 2]. Synthetic hyetographs are mainly used in the design of new sewerage systems, including determining the level of flood protection. They are useful especially where there is no access to historical rainfall data. Synthetic rainfalls can be developed on the basis of direct analysis of recorded rainfall data or IDF (or DDF) curves. This form of data is very often the only source of information about the characteristics of rainfall in the area. IDF curve allows to specify the average rainfall intensity as a function of the rainfall duration and the frequency of occurrence. DDF curve shows a similar relationship in relation to the cumulative rainfall depth. With the use of IDF or DDF curves block rainfalls can be determined, i.e. the rainfalls with constant intensity for considered rainfall duration. This kind of data is insufficient for the rainfall - runoff transformations in the simulation model [3]. It's because they do not take into account the one of the main rainfall characteristics - the variability of intensity over time.

[*]Corresponding author: *karolina.mazurkiewicz@put.poznan.pl*

Basic rainfall characteristics obtained from the IDF or DDF curves are transformed into synthetic rainfall with variable intensity or depth in time. One of the most popular synthetic rainfall in Germany and Poland is Euler (type II) rainfall [1], characterized by the location of the maximum intensity (intensity peak) in the 30% of the total rainfall duration. The popularity of this hyetographs is a result in large part of simple design. Location of the rainfall intensity peak for real rainfalls is not constant [4, 5, 6]. Synthetic rainfall should represent the characteristics of real rains, so the construction of the synthetic hyetograph should enable the change of rainfall intensity peak location. This possibility is given by Chicago hyetographs [7]. The development of synthetic rainfall by this method, however, is relatively complex, especially when compared with Euler rainfalls.

The use of synthetic rainfall is sometimes criticized [8]. The primary complaint is the lack of linearity of the rainfall - runoff transformation which causes that rainfall and runoff frequency of occurrence are not equal [9]. The rainfall frequency of occurrence considered in the IDF or DDF curves concerns only one rainfall characteristic, which is the average intensity or cumulative depth. The outflow is much more complex, characterized for example by volume and intensity peak. The rainfall frequency of occurrence can be assigned to one of the outflow characteristics. In the simulation analysis of sewerage systems both the free surface flows and pressure flows are taken into account. If there is no flooding, runoff is generated from impervious area. Runoff from pervious area does not occur or is negligibly small [10, 11]. In this situation, if the parameters of the catchment are constant for concerned task, rainfall with given depth will generate the same runoff volume, regardless of the form of synthetic hyetograph. This outflow characteristics can be compared with the frequency of rainfall occurrence. In formulating the assumption that for a given depth of rainfall and it's duration synthetic hyetograph generates the greatest outflow peak, it can be assumed that the frequency of rainfall occurrence can be identified with the frequency of occurrence of maximum outflow. This justifies the use of synthetic rainfalls, especially in the case of a limited access to other rainfall data.

The value of the frequency of rainfall occurrence for a given task are specified in regulations, for example European Standards [12]. The rainfall duration is usually determined according to individual decisions. In the literature recommended rainfall durations are mostly related to the flow time through the catchment and are equal to multiply of this value [13]. The flow time by the catchment, identified with the time of concentration, is derived from the traditional methods of sewerage systems design based on The Rational Formula [14]. For constant rainfall intensity maximum outflow will occur at a time when the whole catchment will be involved in generating the outflow. To meet this condition the rainfall must last so long that storm water from the farthest point of the catchment will reach the considered cross section. In the case of rainfalls with parameters variable at the time, the greatest outflow appears after the pulse of rainfall with a maximum intensity or depth. Therefore, it should be expected that the peak location is related to the area of the catchment, represented by the flow time. Mentioned issue is raised in very small number of publications [1,13], so it was consider worth investigating.

The aim of the analysis presented in the paper is the assessment of the impact of the synthetic hyetograph parameters (rainfall duration and peak location) on the maximum outflow. Obtained results can be used to unify the determination of the synthetic hyetograph parameters for the verification of sewerage systems requirements according to the EN 752 standards.

2 Materials and methods

In presented analysis the synthetic hyetographs, whose design is based on the assumptions of the Chicago method [7], are used. The location of the maximum intensity (the peak) is specified by the ratio r - the dimensionless time to peak related to the rainfall duration. A detailed description of the construction of this hyetograph can be found in the literature [15, 16]

The rainfall intensities were calculated with the use of Bogdanowicz-Stachý Formula, which is a form of DDF curve for Polish conditions [17]. To achieve the purpose of this analysis the hyetographs for frequency of occurrence of 2 years were developed. This value is recommended for storm sewer design according to the standards EN-752. Outflows caused by these rainfalls should be the free surface flows and the outflow hydrographs should not be distorted as a result of the pressure flow. The synthetic rainfalls with a durations T_D from 15 min to 180 min were analyzed (the durations increased of 5 min). The ratio r was established on the basis of the literature review and analysis of rainfall data registered in the city of Poznań [15]. The values of ratio r varied from 0.2 to 0.5 (the increment was 0.1). In total, 136 synthetic rainfalls were used for storm sewer system simulations. The examples of hyetographs developed for T_D = 90 min are shown in the figure 1.

The decrease in the maximum rainfall intensity with the increase in the value of the ratio r, can be assign to properties of used hyetographs. The functions describing the variation of the rainfall intensity before the peak and after the peak are dependent on the value of the ratio r [7]. The higher the r value, the smaller increases of rainfall intensity in the following time steps. For values of r = 0.5, functions describing the rainfall intensities before and after the peak are equal and the hyetograph is symmetric.

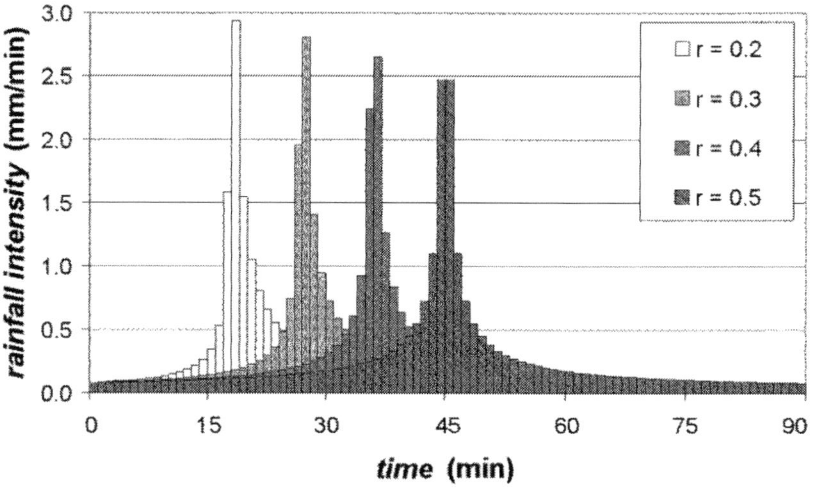

Fig. 1. The examples of synthetic hyetographs for rainfall duration T_D = 90 min and different values of the ratio r

Determined synthetic hyetographs were used to perform outflow simulations from three real catchments. Their basic characteristics are shown in table 1. Catchments described by A and B are located in the city of Bydgoszcz, the catchment described by C is located in the city of Poznań. Catchment models have been developed in SWMM5 [18].

The storm water from all catchments is collected into drainage system of circular cross-sections (fig. 2). In concerned drainage systems there are no special objects e.g. retentions reservoirs or pumping station, which could affect the hydrographs. The rainfall - runoff transformation is a result only of the surface runoff and flow in the channels.

 https://doi.org/10.1051/e3sconf/20183001018

Table 1. The basic characteristics of the catchments used in the analysis

Catchment description	**Total area** *F*(ha)	**Percent of impervious area** ψ **(%)**	**Longest flow time** T_F **(min)**
A	89	26	21
B	172	26	30
C	670	29	50

To determine the relationship between the catchment characteristics and the parameters of synthetic rainfalls generating the maximum outflow for each catchment, the longest time flow T_F through the channels for each catchment was calculated (table 1). The values of T_F were calculated using the Chezý formula with the Manning coefficient for cross-sections and slopes of sewers the same as in the catchment models, assuming full flow conditions.

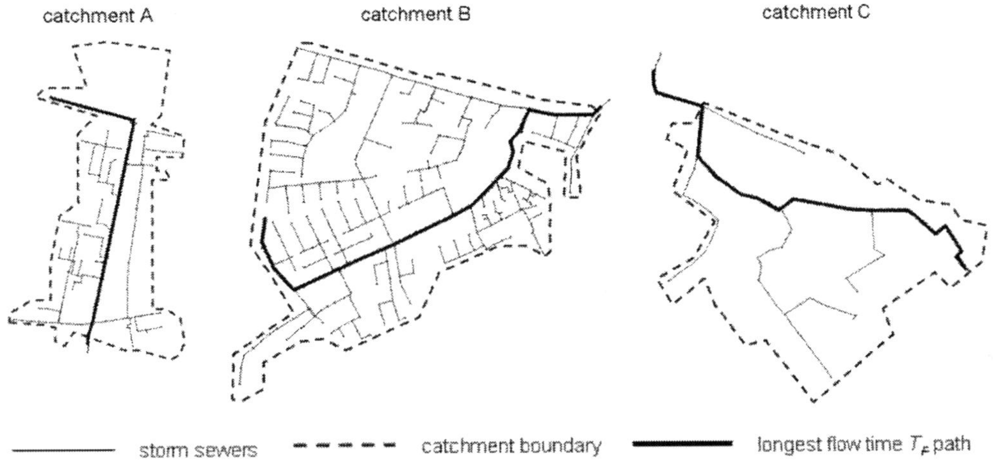

Fig. 2. The schemes of analyzed catchments (figures without proportions between catchments)

Concerned catchments have a monitoring system of rainfalls and outflows. On the catchments A and B rain gauges and flow meters are operated by MWiK in Bydgoszcz. Measurements on the catchment C are performed by Poznań University of Technology. Data provided by MWiK in Bydgoszcz shows that for the real rainfalls catchment models represent the maximum outflow with an accuracy of ±15% and the differences of the times to peak do not exceed 5 min. A similar accuracy were obtained for the catchment C [19]. Thus the accuracy of simulation results should be considered as acceptable.

The influence of rainfall duration T_D and rainfall peak location (represented by the ratio *r*) on maximum outflow Q_P were examined. The maximum outflow is the greatest outflow computed for specific rainfall parameters: T_D and *r*. Outflow hydrographs were determined for the cross – sections located at the end of the longest flow time path (fig. 2). Due to the different areas of the individual catchments, the calculated values of the maximum outflow vary significantly. To allow the comparison of values Q_P computed for each catchment, a dimensionless relative maximum outflow δQ was used:

$$\delta Q = \frac{Q_P}{Q_B}(-) \quad (1)$$

where:
Q_P –maximum outflow computed for a specified rainfall duration T_D and ratio r (dm^3/s)
Q_B - base maximum outflow computed for the duration T_D = 180 minutes and r = 0.3 (dm^3/s)

The value of ratio r for rainfall generating base maximum outflow was taken according to the Euler hyetograph type II, where the peak is located at 30% of the total rainfall duration. In presented analysis the longest rainfall duration was chosen, as for long rainfall durations maximum outflows take the greatest values. The base maximum outflow is therefore the greatest maximum outflow generated by the rain with ratio r equal to 0.3, and this value is constant for the catchment area.

3 Results and discussion

For a specific rainfall duration T_D, the increase in ratio r causes an increase in maximum outflow (fig. 3a, c, e). These changes are similar for all catchments - the increments of the relative maximum outflow δQ vary from 0.02 to 0.17. The average value of the increment δQ changes from 0.09 for catchment C to 0.14 for the catchment B and is not related to the catchment area. Changes of Q_P may be partly related to the changes of the maximum rainfall intensity, which are a result of changes in the ratio r.

It should be noticed that the increase in the maximum outflow occurs in spite of decrease in the maximum rainfall intensity (fig. 1). This is the result of the catchment capacity to buffer short rainfalls with high intensities. Assumed time step of discretization equal to 1 minute allows to accurate describe the rainfall variability within rainfall duration. However, high temporary rainfall intensities do not generate sufficiently great outflows. Buffering maximum rainfall intensities is not related to the total area of the catchment area. It takes place at the stage of transformation the rainfall into the surface runoff in the subcatchments.

For large catchment areas (catchments B and C), maximum outflows for rains of short durations (less than 30 min) are clearly smaller than the value Q_P calculated for the longer rains. This is the result of surface retention of the catchment. In case of short rains, the rainfall depth before the peak is used to fill the surface retention. The less rainfall depth is transformed into the outflow, what in turn reduces the outflow peak.

For constant ratio r, the increase in rainfall duration T_D causes the increase in maximum outflow (fig. 3b, d, f). Change of relative maximum outflow δQ vary from 0.07 to 0.21. The average increase of value δQ ranges from 0.09 to 0.19, respectively for the catchment A and C. For considered rainfall durations and values of ratio r, the influence of the rainfall duration T_D on the maximum outflow is slightly greater than the influence of the ratio r. Average change of the maximum outflow increases with an increase in the catchment area.

When a certain value of rainfall duration, called threshold time T_{DT}, is reached, the maximum outflow remains almost constant (fig. 3b, d, f). Values of T_{DT} evaluated for each catchment were compared with the longest flow time T_F through a network (tab. 2). In addition, for a given rainfall time to peak T_P for hyetograph was specified. Time to peak T_P was obtained by multiplication rainfall duration T_D and ratio r. On the basis of a comparison of an average values of time to peak for hyetograph T_P and flow time T_F for a considered catchments, it can be concluded that these values are similar.

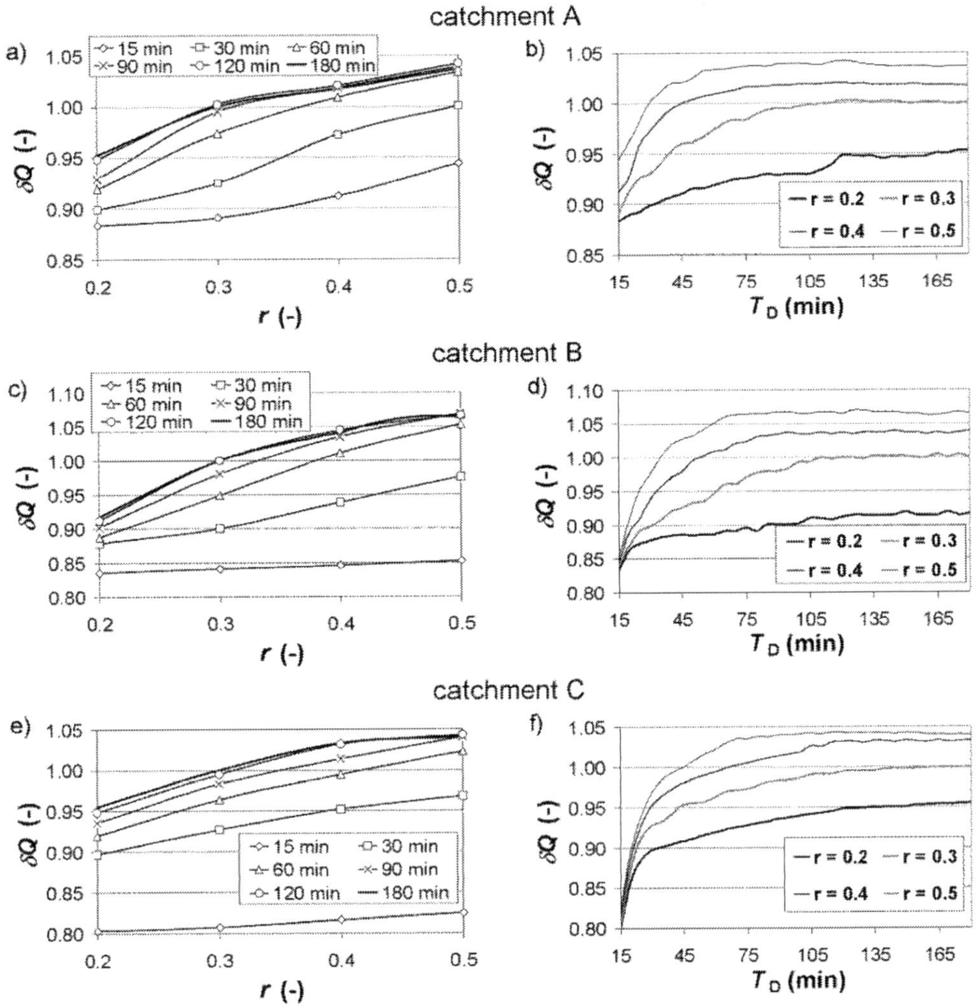

Fig. 3. The relative maximum outflow δQ as a function of the ratio r (a)(c)(e) and as a function of the rainfall duration T_D (b)(d)(f)

In order to complete the analysis the outflow hydrographs were compared. The hydrographs calculated for catchment A and rainfalls with equal time to peak T_P were taken into account (fig. 4).

All considered rainfalls generate maximum outflows of comparable value and in the same time (for the same time to peak of maximum outflow). The maximum difference Q_P does not exceed 8% and times to peak of maximum outflows do not differ by more than 2 min.

Presented results indicate that the maximum outflow from the catchment is a function not only of the total rainfall duration and the corresponding average rainfall intensity or depth (according to the IDF or DDF curve), but also of the location of the peak in the hyetograph. The increase in rainfall depth as a result of increase in rainfall duration changes the shape of outflow hydrograph after the peak (fig.4).

According to recommendations presented in the literature, the maximum outflow from the catchment is generated by the rainfall with a duration of at least four times larger than the flow time [13]. Cited analysis was performed with the use of Euler hyetograph type II, whose peak is at 30% of the total rainfall duration (time to peak T_P is therefore 0.3 T_D).

After extending the rainfall duration T_D to the value of the 4 T_F, time to peak T_P of 1.2 T_F is obtained. This value corresponds to the results presented in the paper. Increase in the time to peak T_P can be obtained not only by extending the total rainfall duration but also by changing the rainfall peak location. In case Chicago hyetographs increase in the value of ratio *r* leads to the slightly increase in the maximum outflow.

Table 2. Threshold times T_{DT} of rainfalls for analyzed catchments

Ratio *r* (-)	**Catchment A** (T_F = 21 min)		**Catchment B** (T_F = 30 min)		**Catchment C** (T_F = 50 min)	
	T_{DT} **(min)**	T_P **(min)**	T_{DT} **(min)**	T_P **(min)**	T_{DT} **(min)**	T_P **(min)**
0.2	120	24.0	125	25.0	175	35.0
0.3	100	30.0	110	33.0	165	49.5
0.4	70	28.0	95	38.0	135	54.0
0.5	55	27.5	70	35.0	90	45.0
	Average	27.4	Average	32.8	Average	45.9

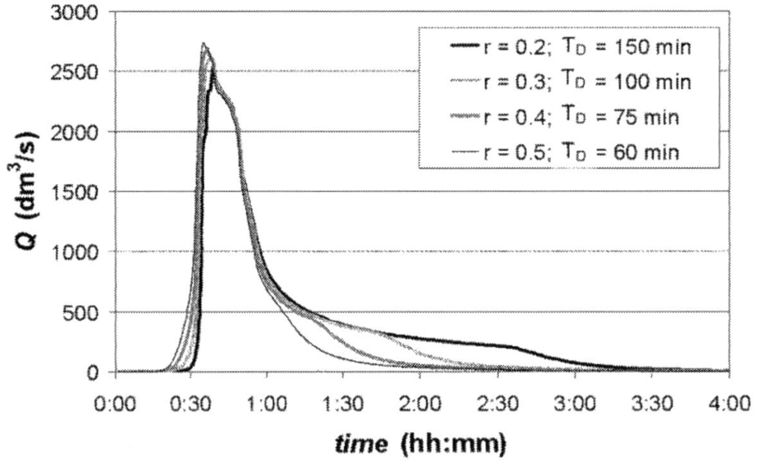

Fig. 4. Examples of outflow hydrographs for catchment A and rainfalls of T_P = 30 min

4 Conclusions

Results of the presented analyses allow to formulate the following conclusions:

1. The increase in the ratio *r* causes the increase in the maximum outflow.
2. For a specific value of ratio *r* maximum outflow increases with an increase in rainfall duration until the threshold time is reached. After exceeding the threshold time maximum outflow does not change its value.
3. The threshold time decreases with an increase in ratio *r*.
4. The comparable maximum outflows can be generated by synthetic rainfalls with a specific time to rainfall peak, regardless of the value of the total rainfall duration and ratio *r*.

5. The greatest maximum outflow is caused by the hyetograph with the location of the rainfall peak similar to the longest flow time through the drainage network.

The authors thank the company Miejskie Wodociągi i Kanalizacja in Bydgoszcz for sharing the catchment models and measurement data.
Work made from the Poznań University of Technology resources (01/13/DSPB/0856).

References

1. T. G. Schmitt: *Kommentar zum Arbeitsblatt A 118 "Hydraulische Bemessung und Nachweis von Entwässerungssystemen"*. DWA, Hennef (in German) (2000)
2. P. Licznar, J. Łomotowski, D. E. Rupp: Random cascade driven rainfall disaggregation for urban hydrology: An evaluation of six models and a new generator. Atmospheric Research **99**, 563–578 (2011)
3. L. Alfieri, F. Laio, P. Claps: A simulation experiment for optimal design hyetograph selection, Hydrological Processes **22**, 813-820 (2008)
4. P. Urcikán, J. Horváth: Synthetic design storm and its relation to Intensity-Duration-Frequency Curves. Water Science and Technology Vol. **16** No 8-9, 69–83, (1984)
5. V. Arnell: *Rainfall data for the design of sewer pipe systems; with the supplement: Description and Validation of the CTH-Urban Runoff Model*, Report Series A:8, Chalmers University of Technology, Dept. of Hydraulics, Goteborg, Sweden (1982)
6. N. El-Jabi, S. Sarrat: Effect of Maximum Rainfall Position on Rainfall-Runoff Relationship. Journal of Hydraulic Engineering, **117 (5)**, 681-685 (1991)
7. C. J. Keifer, H. H. Chu: Synthetic rainfall pattern for drainage design. ASCE Journal of the Hydraulics Division, **83** (HY4), 1-25 (1957)
8. B. J. Adams, C. D. D. Howard: Design storm pathology, Canadian Water Resources Journal/Revue canadienne des ressources hydriques, **11:3**, 49-55 (1986)
9. A. Vigilone, G. Blöschl: On the role of storm duration in the mapping of the rainfall to flood return period. Hydrology and Earth System Sciences, **13**, 205-216, (2009)
10. J. B. Stall, M. L. Terstriep: *Storm sewer design – an evaluation of the RRL method*. Office of Research and Monitoring US EPA, Washington (1972)
11. M. J. Boyd, M. C. Bufill, R. M. Knee: Predicting pervious and impervious storm runoff from urban drainage basins. Hydrological Sciences Journal, **39:4**, 321-332 (1994)
12. EN 752: *Drain and sewer systems outside buildings*. PKN, Warszawa (2008)
13. B. Kaźmierczak, A. Kotowski: *Weryfikacja przepustowości kanalizacji deszczowej w modelowaniu hydrodynamicznym*. Prace Naukowe Instytutu Inżynierii Ochrony Środowiska Politechniki Wrocławskiej 92, Seria: Monografie 57, Wrocław (in Polish) (2012)
14. ASCE: *Design and Construction of Urban Stormwater Management Systems*. ASCE Manuals and Reports of Engineering Practice no. 77, WEF Manual of Practice FD-20 (1992)
15. K. Mazurkiewicz: *Wyznaczanie charakterystyki opadu obliczeniowego dla potrzeb modelowania odpływu ze zlewni miejskiej*. Rozprawa doktorska, Politechnika Poznańska (in Polish) (2016)
16. A. L. L. da Silveira: Cumulative equations for continuous time Chicago hyetograph method. Revista Brasileira de Recursos Hidricos/Brasilian Journal of Water Resources, Porto Alegre, v. **21**, n. **3**, 646-651 (2016)
17. E. Bogdanowicz, J. Stachý: *Maksymalne opady deszczu w Polsce. Charakterystyki projektowe*. Materiały badawcze, seria: Oceanologia i Hydrologia, IMGW, Warszawa (in Polish) (1998)

18. L. A. Rossman: *Storm Water Management Model User's Manual Version 5.1*, 09.2015, www.epa.gov/water-research/storm-water-mamagement-model-swmm (2015)
19. M. Skotnicki, M. Sowiński: Dokładność odwzorowania stopnia uszczelnienia zlewni cząstkowych w modelowaniu odpływu ze zlewni miejskiej. Gaz Woda i Technika Sanitarna, **7-8**, 276-279 (in Polish) (2011)

Contamination content introduced with rain water to the rivers after they have been cleaned in separators of petroleum compounds

Dawid Łapiński [1*], *Józefa* Wiater [1]

[1] Department of technology in engineering and environmental protection, Faculty of Civil and Environmental Engineering, Bialystok University of Technology, Wiejska 45A, Poland

Abstract. The article presents the results of rainwater quality studies. The rainwater flowing from the surface of urbanized areas of the city of Białystok discharged into the river after they have been cleansed with separators petroleum compounds. Also discussed are issues related to with rain water and the problem to need develop them. Contaminated flushes from sealed surfaces such as street squares, sidewalks, parking lots, etc., during periods of rain or thawing, pose a serious threat to the natural environment. The work presents analyzes of such pollutants as general slurry, heavy metals, chlorides, BOD_5, COD, etc., which once penetrate rainwater into Biała River. Five measurement and control points located in a typical urban drainage basin were selected for the study. The tests were taken in the spring of 2017 in two measuring series. The first series was taken during the long-term precipitation, while the second was after the almost one-month rain-free period, at the time of the first precipitation.

Keywords: Rain water; Precipitation; heavy metals; separators; White River.

ISSUES RAINWATER

The development of civilization and the increase of the urban areas has forced the need for rainwater runoff from the surface. The priority is to collect them as quickly as possible from squares, streets, sidewalks and roofs, and next drain to the rain drain or land-water receiver[1,20]. Contaminated flushes from sealed surfaces such as street squares, sidewalks, parking lots, etc., during periods of rain or thawing, pose a serious threat to the natural environment[2,13]. In urban areas, it is a disorder of the hydrological cycle resulting from the growth surface of the impermeable and significant reduction in retention capacity and infiltration of rainwater. Watercourses in urbanized catchments, especially in big cities, become the main receivers of sewage and highly contaminated surface runoff, which causes their chemical and hygiene-sanitary conditions to deteriorate[11]. The search for a rational way of treating rainwater is one of the most urgent challenges in water management.

* Corresponding author: *d.lapinski@pb.edu.pl*

Rainwater is already contaminated during rainfall as a result of contact with contaminated atmospheric air. Raindrops capture in atmosphere contained it inter alia, volatile dust, SO_2, NO_2, CO_2, fumes and chemicals from industrial areas[8]. The next stage of further pollution of the precipitation is surface runoff. The pollutants washed from the catchment area such as streets, squares, roofs, parks and pavements include sand, gravel, dust, rubbish, salt from the prevention of black ice, tire abrasion products, leaves and other parts of plants, fuel spills, etc. [15,16].The content of impurities, both organic and non-organic, in rain waters is very diverse. The amount and type of pollutants entering rainwater depends mainly on the type of catchment and the way it is managed, as well as the time between successive rainfalls[20]. On the composition of the water is mainly influenced: type of surface of streets, squares and sidewalks, density of road network and traffic intensity, organization and method of street cleaning, method of anti-glaze control, participation of green areas in the city, intensity and duration of rain, as well as many other factors[3,4].

The main pollutants and their describing parameters occurring in rainwater can include:

- heavy metals (zinc, chromium, cadmium, lead, nickel, copper)
- aromatic hydrocarbons (benzo [a] pyrene, pyrene, naphthalene)
- total suspension, BOD, COD, nitrogen, phosphorus, anions, cations, conductivity, pH

The content of pollutants changes during rainfall, in general their concentration is the highest in the initial stage of rainfall, because the first runoff washes off the terrain surfaces[2]. This is especially important after long periods without precipitation. This phenomenon is caused by the pollution not only from the surface, but also from impurities deposited in the dewatering devices[3,18]. One of the main sources of rainwater and snowmelt pollution is motorization. Runoffs from roads are loaded with significant concentrations of general suspended solids, COD, ether extract, petroleum substances and heavy metals. Heavy metals are found in rainwater in the form of dissolved salts, ions or insoluble compounds[1,18]. Higher metal concentrations are usually observed in runoff from industrial areas, rather than from residential areas.

Looking at literature, both Polish and foreign, related to the problem of rainwater, one can come across various definitions and formulations of rainwater or rainwater. In the Polish legal system, the terms "rainwater" and "meltwater" are not clearly defined[1,3,7,11]. However, in various legal acts, the definition of rainfall sewage and meltwater sewage appears. Pursuant to the Environmental Protection Law Act of 27 April 2001, by means of sewage is meant, among other things, rainwater or snowmelt, entered into open or closed sewage systems, originating from contaminated surfaces with permanent surface, in particular from cities, ports, airports, industrial, commercial, service and storage areas, transport bases, roads and parking lots[22]. This definition was repeated in the Act of 18 July 2001 on Water Law. The same entry is also made in the Act of June 7, 2001 on collective water supply and collective sewage disposal[23]. The basic legal acts talking about the possibilities of development and quality standards for rainwater are the aforementioned Water Law from July 18, 2001. with d. and Regulation of the Minister of the Environment on July 24, 2006 on the conditions to be met when introducing sewage into waters or into the ground, and on particularly harmful substances. This regulation defines only the values of suspension and hydrocarbons in rainwater discharged to the receiver and states that they should not contain contaminants in the amount of 100 mg • dm^{-3} total suspension and 15 mg • dm^{-3} petroleum substances[24].However, other dangerous substances are omitted, such as heavy metals, which due to their bioaccumulation potential may have a negative impact on the environment, in particular on rainwater receivers.

Rainwater from urbanized areas can be drained in a conventional way by collective

https://doi.org/10.1051/e3sconf/20183001019

sewage systems or, alternatively, managed at the place of their creation. The main method of pre-treating rainwater, before they are discharged to the water receiver, are devices based on physical methods mainly using sedimentation and filtration processes[11]. The most important parameter allowing to determine the negative impact of rainwater on the water receiver is the in which the accumulation of most pollutants takes place. Even up to 90% of the total suspension is a mineral slurry in the form of sand, gravel and soil particles, which absorbs other impurities on its porous surface. Strict relations were found between the amount of suspension and the chemical and biochemical oxygen demand as well as the concentration of heavy metals[5,6]. The most commonly used devices for preclining rain watercourses include various types of separators, settling tanks, absorbent wells, sediment ponds, soil filters, etc.[7].

Subject and methodology of research

Within the framework as part of the research carried out at the Department of Technology in Engineering and Environmental Protection, an analysis of the composition of rainwater pollutants from a selected catchment in the agglomeration of the city of Białystok was carried out. A catchment located in the south-eastern part of the city, characterized by diversified buildings (high- and medium-sized housing estates, low-detached, single-family neighborhoods, service and recreation areas) and the presence of the main communication glazing was selected for the research. Five measurement and control points have been selected as shown in the figure below.

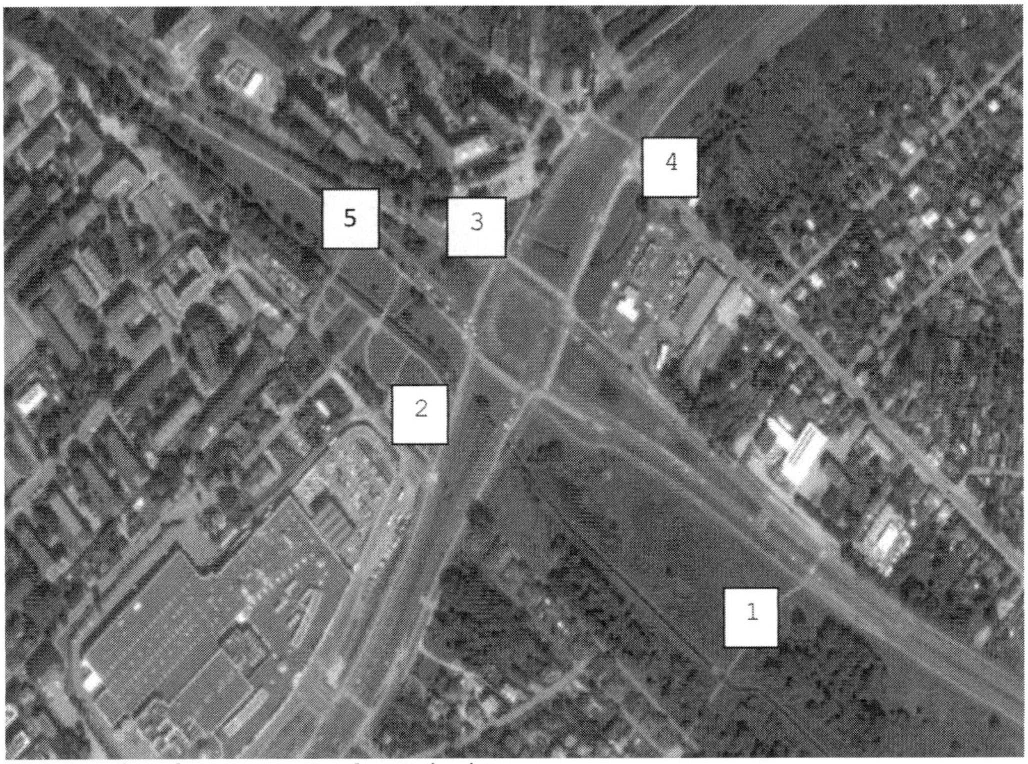

Fig 1. Location of measurement and control points.

Research samples were effluents from the separators of petroleum compounds located in the city of Białystok as well as water taken from the Biała River. The Biała River is a fourth-order river, a left tributary of Supraśl. It takes its origin not far from the Protasy village and its length is 32.7 km. Constant tributaries of Biała are: the foot from the Upper Dojlid, Dolistówka and Bażantarka. The river is strongly regulated, some of its sections are rectilinear with technically formed banks and outlets of storm water drainage[9].

Samples were taken in April and May 2017. It was the period of spring rainfall characterized by various types of intensity and their occurrence time. It was important to take samples in the first 20 minutes of atmospheric precipitation because, according to literature data, the highest concentration of contaminants rinsed off the land surface occurs [17,19,20].Samples for testing were taken in the coastal zone, at the outlet of separators (example drawing 2) and from the water depth of the river. The collected samples without fixation were immediately transported to the laboratory in order to determine the selected indicators. In the water collected, the following were determined: total suspension by the weigh methodt, BOD_5 determined by respiromotor method using the OXI TOP kit, COD by the dichromate method using a HACH LANGE DR / 5000 spectrophotometer, anions and cations using a Thermo Scientific ICS 5000+ ionograph, heavy metals by the ASA method on atomic absorption spectrometer with Thermo Scientific iCE3400 electrothermal atomization and pH and conductivity by potentiometric method. Determinations were made according to current regulations or on the basis of the commonly used and recommended by the scientific literature analytical methodologies.

Fig 2. Measurement and control point no. 3, Outlet from petroleum separator.

Research results

Tables 1,2,3,4,5 and 6 show the results of the water samples tested. Results are the arithmetic mean of the three repetitions. The conducted research showed a very varied content of individual parameters and confirmed the information on high variability of the pollutant indicators in the case of rainwater studied in the literature[2,10,12,18]. Measurement points 1 and 5 were water taken from the Biała River before supplying rainwater from separators and supplying rainwater. The points 2,3,4 were sewage from the outlet selected for the analysis of separators.

Table 1. Basic physicochemical parameters of tested. First series of research.

Parameter	Unit	1	2	3	4	5
pH	-	8,14	8,23	8,11	8,09	8,12
O_2	mg/l	10,06	9,25	8,35	9,25	10,11
Temperature	°C	17,70	17,30	17,90	17,60	17,00
Conductivity	μS/cm	517,00	987,00	1208,00	1031,00	618,00
Turbidity	mg/l	26,00	20,00	13,50	7,20	21,00
	FAU	73,00	19,00	28,00	22,00	62,00
	NTU	10,50	7,65	11,50	9,19	6,51
Color	mg Pt/l	64,00	57,67	24,00	22,83	56,00
Basicity Zm	mval/dm^3	4,00	5,50	5,60	3,60	4,40
Overall acidity	mval/dm^3	2,20	2,60	3,30	1,76	2,86
Hardness	mval/dm^3	5,00	6,48	6,84	4,40	5,44
Total suspension	mg/l	163,00	115,00	87,00	72,00	120,00
BOD_5	mg/l	4,00	5,00	2,00	0,00	1,00
COD	mg/l	53,50	55,50	44,00	28,00	47,00

Table 2. Basic physicochemical parameters of tested. Second series of research.

Parameter	Unit	1	2	3	4	5
pH	-	8,03	7,38	7,72	7,68	7,88
O_2	mg/l	8,56	8,17	8,13	8,06	7,92
Temperature	°C	20,30	20,90	20,80	20,50	20,70
Conductivity	μS/cm	434,00	527,00	1127,00	674,00	617,00
Turbidity	mg/l	20,40	398,90	214,30	220,10	56,40
	FAU	34,60	526,00	254,00	253,00	75,00
	NTU	25,40	357,00	162,00	159,00	51,10
Color	mg Pt/l	55,00	330,00	216,00	190,00	105,00
Basicity Zm	mval/dm^3	3,40	2,80	4,50	2,60	3,50
Overall acidity	mval/dm^3	1,32	2,64	1,76	1,54	1,48
Hardness	mval/dm^3	4,60	7,80	8,12	4,00	5,28
Total suspension	mg/l	349,00	640,00	324,00	321,00	182,00
BOD_5	mg/l	5,00	50,00	25,00	31,00	12,00
COD	mg/l	106,00	453,00	317,00	242,00	140,00

Table 1 and 2 shows significant differences in the individual quality parameters of the waters tested. They result from the aforementioned fact that the first series was collected during the rain but during the few days of rain that washed away most of the contaminants in the previous days, while the second series was carried out after a long rainless period and during the sampling we managed to collect the so-called 'first' wave of "pollution carried along with precipitation.

The pH during the whole research period remained at a similar level and ranged from 7.38 to 8.23. Similarly, the temperature of the tested water and wastewater from separators was constant in individual series. After the dry season there was a drop in dissolved oxygen at the control and control point 1 and 5 located on the river. Large differences were observed in the color and turbidity, especially for points 2.3 and 4, ie at the outlet of separators. The values between the first and second measurement series differed by an order of magnitude. As an example, point no. 3 in which the color and turbidity value expressed in FAU has changed from about 20 to 200. Similarly as described in the literature, there was also observed a relationship between the increase in the amount of total suspension and the biological and chemical oxygen demand. For these parameters, there was also a change in the order of magnitude between the first and the second series of tests. Analyzing all the parameters presented in Table 1 and 2, a slight deterioration of water status in the Biała River was observed. However, the measurement and control points were located on a fairly short section of the examined river and to fully illustrate the impact of the urban agglomeration on its condition, it would be necessary to examine the water before and after Białystok. The content of dissolved cations and anions in water in both measurement series remained at a similar level. It can be concluded that the increase in the total suspension content does not affect dissolved cations and anions. Small changes can be observed only in the case of phosphates, nitrates and ammonium ions. This is probably due to the flushing of fertilizers from adjacent green areas and garden plots. Results of the studied anions and cations are presented in Table No. 3 and 4.

Table 3.Content of cations and anions in tested samples. First series of research.

Parameter	Unit	1	2	3	4	5
F^-	mg/l	0,13	0,10	0,11	0,12	0,13
Cl^-	mg/l	29,52	135,05	175,81	175,02	45,11
Br^-	mg/l	0,00	0,03	0,05	0,03	0,00
NO_3^-	mg/l	3,31	7,92	23,22	29,04	9,13
NO_2^-	mg/l	0,03	0,06	0,41	0,13	0,13
PO_4^{3-}	mg/l	0,01	0,02	1,93	1,91	0,07
SO_4^{2-}	mg/l	30,38	50,60	47,78	51,11	38,43
Na^+	mg/l	29,70	118,20	176,40	204,60	69,90
NH_4^+	mg/l	0,40	1,20	3,20	2,10	1,30
Mg^{2+}	mg/l	22,40	35,40	29,60	23,60	24,90
K^+	mg/l	8,80	12,90	66,90	26,80	11,50
Ca^{2+}	mg/l	159,50	212,90	227,00	143,50	176,30

Table 4. Content of cations and anions in tested samples. Second series of research.

Parameter	Unit	1	2	3	4	5
F^-	mg/l	0,13	0,12	0,10	0,07	0,10
Cl^-	mg/l	22,92	72,54	185,27	104,01	56,68
Br^-	mg/l	0,00	0,26	0,00	0,00	0,01
NO_3^-	mg/l	2,06	11,65	20,33	13,91	11,71
NO_2^-	mg/l	0,03	0,36	0,41	0,29	0,35
PO_4^{3-}	mg/l	0,08	0,64	1,70	0,53	0,20
SO_4^{2-}	mg/l	25,37	22,06	38,90	34,89	37,94
Na^+	mg/l	23,30	71,50	177,80	118,30	56,30
NH_4^+	mg/l	0,40	5,80	1,80	3,80	1,40
Mg^{2+}	mg/l	19,30	12,00	25,10	14,50	22,90
K^+	mg/l	9,10	17,20	42,60	19,50	14,60
Ca^{2+}	mg/l	132,20	116,20	198,90	111,90	158,70

Large differences were observed while the contents of individual elements and heavy metals. The results of tests on heavy metals and macronutrients are summarized in the Table No. 5 and 6.

Table 5. Content of macroelements and heavy metals in tested samples First series of research

Parameter	Unit	1	2	3	4	5
Na	mg/l	19,50	66,05	36,70	74,55	15,90
Mg	mg/l	6,55	13,05	6,68	7,73	6,60
K	mg/l	4,79	7,34	28,89	12,99	5,81
Mn	µg/l	59,50	127,50	430,50	64,00	67,50
Fe	µg/l	641,00	416,50	474,50	390,50	410,50
Cr	µg/l	0,00	5,00	0,00	0,00	5,00
Ni	µg/l	5,50	21,00	8,50	7,50	7,00
Cu	µg/l	5,50	26,50	20,50	12,00	10,00
Zn	µg/l	50,00	156,50	150,00	203,00	67,50
Pb	µg/l	0,00	5,50	2,00	0,00	1,26
Cd	µg/l	0,00	0,00	0,00	0,00	0,00

Table 6. Content of macroelements and heavy metals in tested samples Second series of research

Parameter	Unit	1	2	3	4	5
Na	mg/l	35,43	295,90	243,10	250,80	85,90
Mg	mg/l	17,00	19,40	13,90	12,60	22,10
K	mg/l	14,35	49,47	33,02	26,99	20,51
Mn	μg/l	98,10	835,33	818,82	403,90	208,21
Fe	μg/l	1900,15	14693,68	7076,07	3348,35	2063,06
Cr	μg/l	3,00	29,53	19,02	8,51	6,01
Ni	μg/l	1,75	17,52	9,51	7,01	4,50
Cu	μg/l	8,01	105,11	98,60	54,05	17,52
Zn	μg/l	127,38	1158,16	660,66	500,00	223,22
Pb	μg/l	11,26	49,05	116,53	17,52	14,01
Cd	μg/l	65,07	349,00	718,72	254,25	190,19

The content of heavy metals in the tested samples remained at level the microgram in liter and was similar to the values presented by other authors[14,18,19,20]. By using the previously quoted literature information, it can be concluded that most heavy metals are accumulated in the suspension, which explains the differences between their content in the first and second research series. This is most evident in the case of cadmium and lead which in the first series were practically absent or at the background level. The second research series confirms that rainwater discharged from squares and streets contains large amounts of heavy metals. When comparing heavy metal contents at 1 and 5 measurement and control points, small changes can be observed on the studied river section, despite their significant amounts supplied with rainwater, but it should be noted that according to other authors, heavy metals very quickly accumulate in sediments river[10,17,18].

Conclusions

1. On the basis of the conducted research, it can be concluded that the inflow of rainwater changes the values of physical and chemical indicators of water in the Biała River.
2. The duration of rain has a decisive influence on the amount of pollutants in rainwater as the pollution carried in the first 15-20 minutes of precipitation is much higher.
3. On the basis of the conducted tests, it was found that water collected from the same points, at different dates and in weather conditions, is characterized by a different degree of pollution.

The research work was carried out within the framework of the work of its own W / WBiIS / 10/2015 funded that funds for learning.

References

1. G. Sakson, M. Zawilski, E. Badowska, A. Brzezińsk, Rainwater waste pollution as the basis for choosing how to use it, *Czasopismo Inżynierii Lądowej, Środowiska I Architektury*, JCEEA, t. XXXI, z. 61 (3/I/14), lipiec-wrzesień **11**,(2014)
2. E. Eriksson, A. Baun, P.S. Mikkelsen and A. Ledin: Selection of stormwater priority pollutants, 10th International Conference on Urban Drainage, Copenhagen/Denmark, 21-26 August (2005)
3. E. Ociepa, Assessment of pollution of rain sewage to sewage systems, *Inżynieria i Ochrona Środowiska,* t. 14, nr 4, **7**, (2011)
4. J. Królikowska , Pre-treatment of rainwater sewage using a hydroseparator, *Wodociągi i Kanalizacja*, 4, **2**. (2009)
5. S.L. Dąbkowski, K. Górska, J. Górski, B.Szeląg, Wstępne wyniki badań ścieków deszczowych w jednym z kanałów w Kielcach. *Gaz, Woda i Technika Sanitarna*, nr 6, **4**. (2010)
6. K.L. Rule, S.D.W. Comber, D. Ross, A. Thornton , C.K. Makropoulos, R. Rautiu, Diffuse sources of heavy metals entering an urban wastewater catchment, *Chemosphere*, 63,**8**, (2006)
7. A. Wałęga, K. Chmielowski, W.z Miernik, Seminatural systems for draining and purifying rainwater from urbanized areas - legal and technical aspects, *POLSKA AKADEMIA NAUK*, Oddział w Krakowie, **10**, Nr 6/2009,
8. G . Mangani, A . Berloni, F .Bellucci, F .Tatano, M. Maione, Water, Air and Soil Pollut, 160,**15**, (2005)
9. A. Wysocka - Czubaszek, W. Wojno, Seasonal variability of water chemistry in a small river in an urbanized basin, *Przegląd Naukowy – Inżynieria i Kształtowanie Środowiska* nr 63,**12**, (2014)
10. S. Ballo, , M. Liu, i J. Chang,. Pollutants in stormwater runoff in Shanghai (China): Implications for management of urban runoff pollution. Progress in Natural Science, 19,**7**, (2009)
11. J. Królikowska, A. Królikowski, Rainwater. Discharge, development, pretreatment and use, Publisher Seidel-Przywecki, Piaseczno (2012)
12. E. Zębek, M. Szwejkowska, Evaluation of the impact of pre-treated rainwater on the number of cyanobacteria in the downtown Jeziorak lake, small at various precipitation altitudes, *Ochrona środowiska*, vol. 36, nr 1,(2014)
13. E. Ociepa, A. Kisiel i J. Lach, Protection of surface water against rainwater effluents from sewage systems, *Proceedings of ecopole*, vol. 3, no. 1, (2009)
14. M. Niemiec, B. Wiśniowska - Kielian, Contamination with heavy metals flowing down from roads taken from evaporation tanks, inżynieria ekologiczna nr 20, (2008)
15. M. Kayhaniana, C. Suverkroppb, A. Rubyb, K. Tsayc, Characterization and prediction of highway runoff constituent event mean concentration. *Journal of Environmental Management*, 82, **16**, (2007)
16. E. Burszta - Adamiak, J. Łomotowski, Discharge of rainwater in areas with dispersed buildings. *Infrastruktura i Ekologia Terenów Wiejskich* , nr 31(1), **12**, (2006)
17. A. Graham, R.L. Parker, Vertical distribution of heavy metals in Springwood Lake sediments, Richmond, Indiana. *Geological Soc. of Am*., 35, 6, (2003)
18. Ł. Bąk, J. Górski, K. Górska, B. Szeląg, Content of suspensions and heavy metals in selected waves of rain sewage in the urban catchment, Ochrona Środowiska, 2,3, (2012)
19. K.L. Rule, S.D.W. Comber, D. Ross, A. Thornton , Makropoulos C.K., Rautiu R.:

Diffuse sources of heavy metals entering an urban wastewater catchment, Chemosphere, 63, **8**, (2006)

20. T. Molenda, Dynamics of changes of selected pollutants in surface runoffs of the urban catchment, Infrastructure and ecology of rural areas. Nr 4/3/2006, *POLSKA AKADEMIA NAUK*, Oddział w Krakowie, **7**,(2006)
21. The Act of 27 April 2001 on Environmental Protection Law (consolidated text: Journal of Laws of 2001 No. 62, item 627, as amended).
22. Act of 18 July 2001. Water law (consolidated text: Journal of Laws of 2001 No. 115, item 1229,
23. Regulation of the Minister of Environment of 24 July 2006 on conditions to be met when introducing sewage into waters or into the ground, and on substances particularly harmful to the aquatic environment (Journal of Laws of 2006 No. 137, item. 984 with later changes).

Stability evaluation of modernized bank protections in a culvert construction

Mariusz Cholewa [1*], *Karol* Plesiński [1], *Katarzyna* Kamińska[1] and *Izabela* Wójcik [1]

[1]University of Agriculture in Krakow, Faculty of Environmental Engineering and Lands Surveying, Department of Hydraulic Engineering and Geotechnics, al. Mickiewicza 24/28, 30-059 Kraków, Poland

Abstract. The paper presents stability evaluation of the banks of the Wilga River on a chosen stretch in Koźmice Wielkie, Małopolska Province. The examined stretch included the river bed upstream from the culvert on a district road. The culvert construction, built over four decades ago, was disassembled in 2014. The former construction, two pipes that were 1.4 m in diameter, was entirely removed. The investor decided to build a new construction in the form of insitu poured reinforced concrete with a 4 x 2 m cross section. Change of geometry and different location in relation to the river current caused increase in the flow velocity and, as a consequence, erosion of both protected and natural banks. Groundwater conditions were determined based on the geotechnical tests that were carried out on soil samples taken from the banks and the river bed. Stability calculations of natural slopes of the Wilga River and the ones protected with riprap indicate mistakes in the design project concerning construction of the river banks. The purpose of the study was to determine the stability of the Wilga River banks on a selected section adjacent to the rebuilt culvert. Stability of a chosen cross section was analysed in the paper. Presented conclusions are based on the results of geotechnical tests and numerical calculations.

Introduction

Road culverts built in the last century often did not have any design projects. Lightweight traffic was assumed depending on the local economic, ground and technical conditions. It mainly concerned district and local roads where traffic was low at that time. Over the years traffic density increased, often the type of the road also changed. Modernization and rebuilding of the road infrastructure were necessary.

Nowadays it is required to fulfil the conditions set out in appropriate legal acts [1], detailed regulations of the Minister of Transport [2] and standards. Some of the constructions built several decades ago do not fulfil the current standards. The managers are trying as far as possible to make the necessary changes or rebuild aging constructions.

* Corresponding author: m.cholewa@ur.krakow.pl

1. Purpose

The paper presents stability evaluation of the banks of the Wilga River on a chosen stretch in Koźmice Wielkie, Małopolska Province. The examined stretch included the river bed upstream from the culvert on a district road. The culvert construction, built over four decades ago, was disassembled in 2014. The former construction, two pipes that were 1.4 m in diameter, was entirely removed (Fig. 2b). The investor decided to build a new construction in the form of insitu poured reinforced concrete with a 4 x 2 m cross section (Fig. 3). Change of geometry and different location in relation to the river current caused increase in the flow velocity and, as a consequence, erosion of both protected and natural banks (Fig 4a,b). Groundwater conditions were determined based on the geotechnical tests that were carried out on soil samples taken from the banks and the river bed. Stability calculations of natural slopes of the Wilga River and the ones protected with riprap indicate mistakes in the design project concerning construction of the river banks. The purpose of the study was to determine the stability of the Wilga River banks on a selected section adjacent to the rebuilt culvert. Change of geometry and different location in relation to the river current caused increase in the flow velocity which lead to erosion of the banks. Stability of a chosen cross section was analysed in the paper. Presented conclusions are based on the results of geotechnical tests and numerical calculations.

2. Characterization of the watercourse

The source of the Wilga River is located in Pawlikowice (Wielicki Foothills). The river starts at the altitude of 370 m above sea level and after about 21 km it flows into the Vistula River at the altitude of 200 m above sea level. The beginning of the river meanders through stone formations and after that it flows only through finegrained cohesive soils (Fig. 1a). The outlet stretch was regulated and embankments were built so the urban area is protected against the backwater from the Vistula River. The Wilga River has an irregular flow. During floods it rises, the flow velocity increases and the river intensely erodes its bed creating often vertical slops that are 3 m high (Fig. 1b). When the water levels are low they do not exceed a few centimetres.

Fig. 1. Natural bed of the Wilga River a) upstream and b) downstream from the culvert.

3. Characterization of the road culvert

3.1 Condition of the construction before rebuilding

The culvert was built in 1970s for lightweight horse and vehicle traffic. At that time the road was rather quiet and unpaved. In the following years the road was paved using metallurgical slag, later an asphalt surface was placed (Fig. 2a). Two reinforced concrete pipes with an internal diameter of 1.4 m were used as elements carrying water. The length of each pipeline was 10.3 m (Fig. 2b). The culvert was built without any design project since it was not necessary at the abovementioned economic conditions.

Fig. 2. a) Localization of the culvert b) condition of the construction before rebuilding

3.2 Current condition of the construction

Former culvert pipe was entirely disassembled. On the ground level where geotechnical parameters were appropriate the base of a new culvert was built. It is a reinforced concrete culvert poured on site; it is a closed frame based directly on solid clayey soils. The front walls are situated parallel to the road axis. The total length of the construction is 11.74 m. The horizontal diameter is 4.00 m, the vertical 2.00 m (Fig. 3) and the total carriageway is 10.34 m wide.

Fig. 3. Culvert on the upstream side and the view of damages

3.3 Bank protection failure

In the first year (2015) after the construction was finished the bank protection on the right side formed from riprap slid along with the soil from the slope fig. below.

Fig. 4. The view of damages that occurred after the first year of exploitation a) from the downstream side and b) upstream side

In the following year (2016) the improved protection was also gradually destroyed. Progressive failure of the bank is presented in fig. below.

Fig. 5 The view of damages that occurred after the second year of exploitation from the upstream side

4. Stability evaluation

Slope stability evaluation can be performed using different methods [3, 4, 5, 6, 7]. Bishop's method, where the failure surface is circular-cylindrical, was used in the presented calculations. An attempt was made to determine whether the change of water conditions would influence the values of the factor of safety of a slope with a relatively dangerous inclination of 1.5:1. Two calculation models were assumed, starting with the one when the water level is low and then the one during the flood season in July. Surface load on the top of the road from traffic was also taken into account.

https://doi.org/10.1051/e3sconf/20183001020

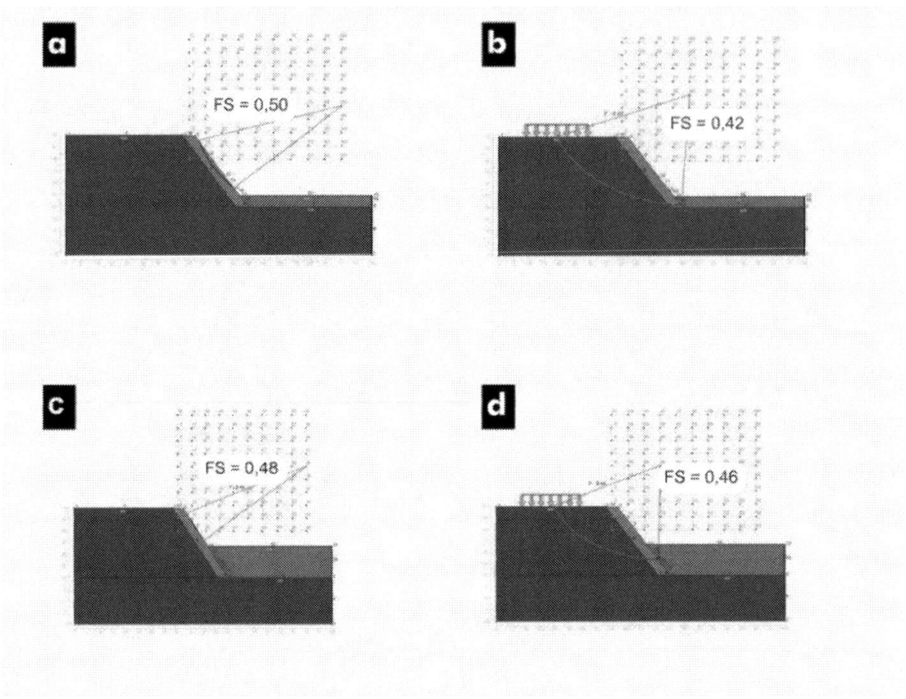

Fig. 6 Calculation scheme for the first model – at low water level a) without any load, b) vertical load of 1000 kN. Calculation scheme for the second model – at high water level c) without any load, d) vertical load of 1000 kN

Scheme presented in Fig. 6a shows the results of calculations at low water level – parameters of the soil (Tab. 1) and geometry of the slope resulted in the value of the global stability factor of safety at 0.50.

Table 1. Geotechnical parameters of the ground and slope soils

No	Name	I_L	ρ [Mg/m³]	c [kPa]	Ø [°]	Water
1	soil of the slope - plastic clay	0.50	2.00	26.00	16.00	yes
2	consolidated clayey ground	0.00	2.10	40.00	10.00	yes
3	riprap					

Apart from total saturation of the slope, the scheme in Fig. 6b assumes that there is a heavy transport vehicle on the top of the road, which was presented on the scheme as vertical load of 1000 kN. The factor of safety is 0.42. Failure surface goes deeply into the road.
When the water level is high (Fig. 6 c,d) the values of the factor of safety do not differ significantly from the first model. Adequately it was 0.48 for the road without any load and 0.46 when there is a surface load of 1000 kN.

Table 2. Results of the stability calculations

Vertical load on the top of the slope	1st model low water level	2nd model high water level
	Value of the factor of safety	
No load	0.50	0.48
1000 kN	0.42	0.46

All obtained results of the factor of safety are below the limit value of 1.0 (Tab. 2). When the water level is low it undermines the base of the slope and at the same time the water filtrates through the ground from the opposite side of the road embankment. During the low and medium flow, in the hole scour, the average values of velocity at measurement points were in range from V = 0,210 m/s to V = 0,333 m/s. The maximum values at vertical measurements points were equal from V = 0,223 m/s to V = 0,370 m/s. Then, the values measured above the river bed were in range from V = 0,060 m/s to V = 0,32 m/s. The river bed is a natural drainage element for the adjacent area, so water supply to the river banks has to be taken into consideration [8, 9, 10]. When the water level is high the hydraulic pressure is compensated by the water pressure. Although another unfavourable phenomenon connected with water uplift pressure occurs (decrease in the value of density in the calculation scheme) [11, 12, 13]. It should also be noticed that high current velocity washes out and lifts the soil material.

Summary

The carried out tests and calculations gave consistent results concerning stability of the protected banks. With unfavourable geometry of the slope and groundwater conditions the calculated values of the factor of safety are below 1. Observations and tests were carried out during both low and high water levels. An important piece of information was the behaviour of the protected banks during drawdown – the hydraulic pressure contributed to the stability failure [14, 15].

Such results are influenced by the values of mechanical properties of the saturated soil in the protected banks [16, 17]. The tests carried out in a shear box apparatus showed that the angle of internal friction is the same when the moisture content was close to the highplastic consistency. Uplift pressure of the filtrating water also has a negative influence in the tested crosssections. Although primarily it was the force connected with filtrating water that caused the bank protection to slide toward the flow cross-section.

The presented test object gives us the possibility to observe filtration phenomena – the critical ones – that result in reduction of the useful crosssection of the river. When the water level is high it infiltrates to the slope and it saturates the soil, reducing strength parameters in the saturated zone [18]. The main reason for the unfavourable values of the factor of safety is the infiltrating ground water as well as the velocity and uplift pressure of the flowing water.

References

1. Ustawa z dnia 21 marca 1985 r. o drogach publicznych (Dz. U. Nr 14, poz. **60**) z pózniejszymi zmianami.
2. Rozporzadzenie Ministra Transportu, Budownictwa i Gospodarki Morskiej z dnia 25.04.2012 w sprawie ustalenia geotechnicznych warunków posadowienia obiektów budowlanych (Dz. U. z 2012 r., poz. **463**).
3. R. Handy, M. Spangler. Geotechnical Engineering. Soil and Foundation Principles and Practice. McGrawHill Professional New York, USA. (2006)
4. C. Venkataramaiah. Geotechnical Engineering. New Age International Publishers, Daryaganj, Delhi. (2006)
5. M. Borys, K. Mosiej. Wytyczne wykonywania ocen stanu technicznego i bezpieczeństwa wałów przeciwpowodziowych. Wydawnictwo IMUZ, Falenty. (2005)
6. R. Kulma, R. Zdechlik. Modelowanie procesów filtracji, Wydawnictwa AGH, Kraków. (2009)
7. L. Rembeza. Nieustalony przepływ filtracyjny przez wał przeciwpowodziowy z uszczelnieniami. Gosp.Wod. **9**, 380–382. (2007)
8. T. Strzelecki, S. Kostecki, S. Żak. Modelowanie przepływów przez ośrodki porowate, Dolnośląskie Wydawnictwo Edukacyjne, Wrocław. (2008)
9. J. Chai, N. Miura, S. Shen. Performance of embankments with and without reinforcement on soft subsoil. Canadian Geotechnical Journal, **39**(4), 838-848. (2002)
10. I. E. Alvarez, R. Rubio, H. Ricalde. Beach Restoration with Geotextile Tubes as Submerged Breakwaters in Yucatan, Mexico. Geotextiles and Geomembranes, **25**(4-5), 233–241. (2007)
11. F. Gassner. Field Observation of GCL Shrinkage at a Site in Melbourne Australia. Geotextiles and Geomembranes, **27**(5), 406–408. (2009)
12. C. C. Huang, Y. S. Tzeng, C. J. Liao. Laboratory Tests for Simulating Abrasion Damage of a Woven Geotextile. Geotextiles and Geomembranes, **25**(4-5), 293–301. (2009)
13. A. Koffler, M. Choura, A. Bendriss, E. Zengerink. Geosynthetics in Protection Against Erosion for River and Coastal Banks and Marine and Hydraulic Construction. Journal of Coastal Conservation, **12**(1), 11–17. (2008)
14. Ch. Huang, Ch. Liao. Abrasion damage of geogrids induced by turbid flow. Geotextiles and Geomembranes, **25**(2), 128-138. (2007)
15. S. Lukman, J. A. Otun, D. B. Adie, A. Ismail, I. A. Oke. A Brief Assessment of a Dam and Its Failure Prevention. Journal of Failure Analysis and Prevention, **11**(2), 97–109. (2011)
16. X. Junqiang, T. Li, X. Li, X. Zhang, Q. Zong. Daily Bank Erosion Rates in the Lower Yellow River Before and After Dam Construction. Journal of the American Water Resources Association (JAWRA), **50**(5), 1325-1337. (2014)
17. W. T. Oh, S. K. Vanapalli. Influence of Rain Infiltration on the Stability of Compacted Soil Slopes. Computers and Geotechnics, **37**(5), 649–657. (2010)
18. M. Cholewa, K. Plesiński, K. Kamińska, P. Baran. Warunki hydrodynamiczne przepływu rzeki Wilgi w okolicy przepustu drogowego w miejscowości Koźmice Wielkie. Acta Scientiarum Polonorum seria Formatio Circumiectus, **15**(3), 45 – 56. (2016)

E3S Web of Conferences **30**, 02001 (2018) https://doi.org/10.1051/e3sconf/20183002001
Water, Wastewater and Energy in Smart Cities

Composting of sewage sludge with solid fraction of digested pulp from agricultural biogas plant

Wojciech Czekała[1],* *Jacek* Dach[1], *Jacek* Przybył[1], *Jakub* Mazurwiekiwcz[1], *Damian* Janczak[1], *Andrzej* Lewicki[1], *Anna* Smurzyńska[1], *Kamil* Kozłowski[1]

[1]Institute of Biosystems Engineering, Poznan University of Life Sciences, ul. Wojska Polskiego 28, 60-637 Poznan, Poland

Abstract. Sewage sludge management is an important element of environmental protection. Composting and anaerobic digestion are the biological conversion methods for sewage sludge management. Mass and volume reduction is a result of a properly composted process. Solid fraction of digested pulp can be use as co-substrate, because it is good structural material. The aim of the study was to determine the possibility of composting sewage sludge with a solid fraction of digestate. The compost mix consisted of 25 kilograms of sewage sludge and 20 kilograms solid fraction of digestate in fresh mass. The experiment was carried out in laboratory conditions. Bioreactors of 165 dm^3 volume were used. The experiment included two stages. Stage I took place in bioreactors and lasted until the cooling phase of the compost was complete. Stage II included compost maturation for a period of eight months (to 287 day of composting). The reduction of mass obtained at the end of Stage I amounted 30.2%. At the end of Stage II, it was 86.7% relative to the initial weight of the compost. The maximum value of temperature was 75.1°C. Studies have shown that sludge with a solid fraction of digestate can be a suitable substrate for composting with sewage sludge.

Abbreviations
AD – anaerobic digestion
DM – dry matter
FM – fresh matter
MM – mineral matter
OM – organic matter in Total Solids
SF – solid fraction of digested pulp
SL – sewage sludge

1 Introduction

One of the most important environmental issues is the management of waste, including by-products of wastewater treatment processes such as sewage sludge (SL) [1,2]. Management of waste can take place using technologies compatible with sustainable development [3]. SL can be a problematic product for environment and requires management [4]. The primary reason is the amount of their formation and the content of odorogenic and harmful substances [5]. Therefore, it is necessary to look for technologies that will help them to be used for the benefit of the environment [6]. The most commonly management method of SL in last years was storage. Legal regulations that entered into force on January 1, 2016 prohibit the storage of SL. There is an order for sewage treatment plants to take measures to manage them in a way that is safe for the environment [7].

Composting and anaerobic digestion (AD) can be mentioned as part of the biological conversion process [8,9]. Composting is a controlled process of aerobic decomposition of organic matter with microorganisms, carried out in prisms or reactors [10]. Among the advantages of this method are the significant reduction in the weight and volume of composted mixtures and the final product, compost, which is an organic fertilizer [11]. SL is a substrate often used for the compost production [12,13]. This is largely due to the high degree of hydration and high nitrogen content. By adjusting the composition e.g. by adding straw it is possible to create the right proportion of substrate.

Methane fermentation takes place at sewage treatment plants, landfills and in the agriculture biogas plants [14]. In each of these cases, biodegradation of the substrates rich in organic matter occurs. The main product of the anaerobic digestion is biogas [15,16]. It is a mixture of gases, dominated by methane and carbon dioxide. Ammonia, hydrogen sulphide, hydrogen and other gases are present in smaller amounts. The second product is digested pulp (digestate), which is a residue from the decomposition process [17]. Digestate is mainly composed of mass of non-fermented compounds and bacterial biomass. Its quantity is closely related to the mass of the feedstock to the biogas plant. However, chemical composition, is largely conditioned by the type of substrates used in the biogas production process. The digested pulp is in the liquid form. Dry matter is usually 2 to 10%. An important element in the digestate management is the separation process [18]. The result of the process are at least two fractions - solid and liquid [19]. Due to its abundance in organic matter, nitrogen and high humidity, it is possible to select the appropriate proportion of the composted mixture. The produced compost is an environmentally safely product and is desirable due to its fertilizer properties [20].

Considering the high volume of nitrogen in SL in on the one hand, and the SF in carbon on the other, it is possible to use them in the composting process. is As a result of oxygen decomposition processes compost hygienisation occurs. This is an important aspect of this process, since sludge is often a source of microbiological contaminants and helminths' eggs.

The purpose of the study was to determine the possibility of composting SL in a mix with a SF. The paper shows the origin and properties of the substrates used for composting, presents composting methodology and discusses the changes related to the particular parameters during composting. Moreover, the authors have done the discussion to compare the results of their own research with those obtained by other scientists.

2 Methodology

2.1 Substrates and composting mixtures

SL used in the experiment came from one of the wastewater treatment plants near Poznan. The substrate was dehydrated using presses. Dehydrated municipal substrate was composted with a SF of the digested pulp from an agricultural biogas plant, located in Poland. Primarily corn silage, fruit pulp and distillery were used for the production of biogas in this installation. The compost mix consisted of 25 kilograms SL and 20 kilograms of SF in fresh mass. The dry weight of the SL was 14.56% and the SF was 32.52%. The organic matter content in dry matter was 75.58% for sludge and 92.94% for digested pulp, respectively. During composting the temperature changes, carbon dioxide emissions, dry matter and organic matter, pH, conductivity and bulk density were monitored (tab. 1).

Table 1. Participation and selected properties of the components used in the composting process.

Substrates	Properties				Participation of substrates
	DM [%]	OM [%]	pH [-]	cond. [$mS \cdot cm^{-1}$]	FM [%]
SF	32.52	92.94	9.07	1.27	20
SL	14.56	75.58	7.09	0.24	25

The nitrogen content in the compost mixture was 24.82 $g \cdot kg^{-1}$ DM and phosphorus 16.66 $g \cdot kg^{-1}$ DM. The experiment included two stages. Stage I took place in bioreactors and lasted until the cooling phase of the compost was complete. Stage II included compost maturation for a period of eight months (to 287 day of composting).

2.2 System for laboratory composting

The experiment was carried out in laboratory conditions in the Laboratory of Ecotechnologies (Poznan University of Life Sciences). Bioreactors of 165 dm^3 volume were used for the experiment equipped with equipment for monitoring the decomposition of organic materials in aerobic and anaerobic conditions (fig. 1). Due to unique construction it is possible to continuously control the process going on inside the reactor.

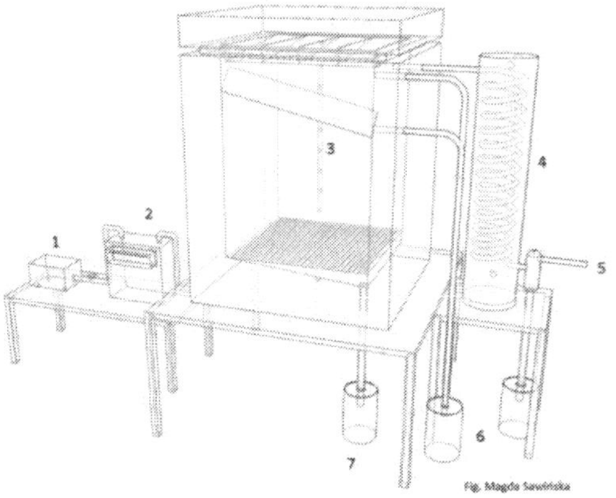

Figure 1. Schematic diagram of the system for laboratory composting 1 – air supply pomp, 2 – flow meter and flow regulator, 3 – chamber of reactor, 4 – air cooler, 5 – cold water, 6,7 – condensate and leachate collectors.

2.3 Composting process

During the experiment the air flow was regulated in the range of 0.8 to 2.7 $dm^3 \cdot min^{-1}$. The initial value of the air injected into the chamber was 2.5 $dm^3 \cdot min^{-1}$. Twice a day correctness was checked by rotameter. When approaching 5% oxygen in the air, the airflow was increased in order to prevent the occurrence of anaerobic conditions.

2.4 Physical and chemical analysis

In order to investigate physico-chemical parameters changes of the compost, samples were taken at the beginning, during the opening of the chambers (turning) and at the end of the experiment. Turning consisted of an opening of the bioreactor and mixing of the total contents of the chamber to aerate the compost and improve the structure of the composting material. The material for analysis was taken from the five different places in chamber and averaged sample was created. At the beginning and at the end of the experiment as well as during the turning, the mass and volume changes measurements were made using a laboratory scale. This allowed to determine the weight loss and volume during the experiment. Based on the changes in the tested parameters the bulk density of fresh compost was calculated.

The construction of the reactors allowed to collect gravity leaks from the compost. Special cooling system allows to condensate the vapor escaping from the chamber and from the radiator. DM of samples was determined by drying method at 105°C for 24 h. Dried samples (each in three replicates) were incinerated to define OM and MM. Organic matter was determined in the furnace (temp. 550°C, 3h). Measurement was done in triplicates. Measurement of pH and conductivity was done using Elmetron CX-401 multifunction device.

2.5 Gaseous measurements

During the composting process, air from the compost chambers was tested using a GeoTech GA5000 gas analyzer. The device allowed analysis of five gases in the concentration range: CH_4 from 0 to 100%, CO_2 from 0 to 100%, O_2 from 0 to 25%, NH_3 from 0 to 1000 ppm, H_2S from 0 to 10000 ppm. Once a week, calibration of gas analyzer was done.

2.6 Temperature measurement

Temperature measurement was done using stick that tests a temperature of 20 centimetres from the bottom of the chamber. These measurements were recorded on the hard disk of the computer every 8 hours.

3 Results

3.1 Characteristics of physical and chemical parameters changes

The mass reduction obtained at the end of Stage I amounted 30.2%. At the end of Stage II, it was 86.7% relative to the initial weight of the compost. This was due to the evaporation of water and the reduction of organic matter. After maturation, the dry matter content was 85.63%.

The highest content of OM occurred on the first day of the experiment - 93.45%. After turning, its content decreased to 81.38%, and at the end of Stage I was 73.32% (tab. 2). During maturation of compost (over five months), the OM level was slightly changed eventually reaching 74.72% on two hundred nineteenth day.

Bulk density increased from 444 $kg{\cdot}m^{-3}$ at the beginning of the experiment to 557 $kg{\cdot}m^{-3}$ (34 day) and 620 $kg{\cdot}m^{-3}$ on the sixty-nine day. Along with the duration of the process, the pH has changed from 7.65 to 8.5 (34 days) to 8.7 (51 days). At the end of Stage I (69 days), the level was 7.53. The maturation phase contributed to a decrease in the reaction, reaching a value of 6.55. Conductivity increased from 0.79 $mS{\cdot}cm^{-1}$ at the beginning of the experiment to 3.88 $mS{\cdot}cm^{-1}$, an increase of 391%.

Table 2. Selected compost properties during the process

	Phase of composting	Day	FM [kg]	DM [%]	OM [%]
Compost	Start	0	45	32.67	93.45
	Turning	34	36.1	19.71	81.38
	Finish of Stage I	69	31.4	17.25	73.32
	Finish of Stage II	287	6.0	85.63	74.72

3.2 Temperature changes

The temperature of the composted mixture during the experiment ranged between 18 and 75.1°C. The maximum value defining the thermophilic phase was demonstrated on the third day of the experiment. On the thirty-fourth day of the experiment, when the compost temperature reached a similar level of ambient temperature (22.6°C), turning was made. After this step, which was associated with oxygenation of the mixture, the temperature increased to a maximum of 54.7°C. In the following days there was a gradual decline in temperature to the sixty-ninth day, which was the end of the composting process in chamber (fig. 2).

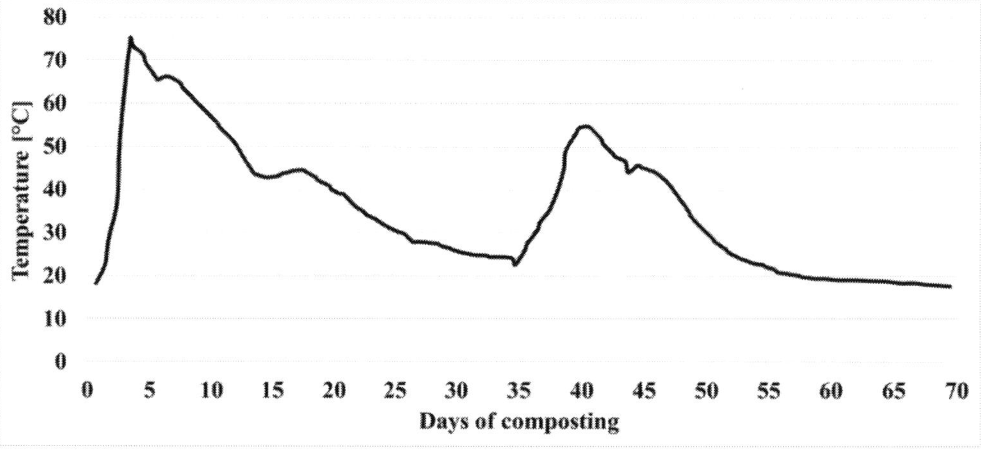

Figure 2. The temperature of the composted mixture during the experiment

3.3 Gaseous emission changes

The minimum concentration of oxygen at 9.1% level occurred on the third day of the study. Along with the compost being cooled down, the microbial activity has weakened, resulting in an oxygen increase to 17-20% level observed on the eleventh day, and lasted until the thirty-fourth. Decomposition of organic compost caused the decrease of oxygen content in the compost up to forty-first day. On forty-ninth day an oxygen stabilization started in chamber to the level of 20%, until the end of Stage I of the experiment.

Carbon dioxide is generally negatively correlated with the concentration of oxygen during composting process. The increase in its concentration causes a direct proportional loss of oxygen, and their sum should be close to that occurring naturally in atmospheric air. Maximum values of carbon dioxide concentration of 11.7% were reported with the highest compost temperatures.

Emission of ammonia from the compost heap reached its maximum of 989 ppm on the fourth day of the experiment, decreasing to less than 100 ppm only on the eleventh day of composting. At the same time, there was no effect of turning on ammonia emissions. Methane and hydrogen sulfide levels were also monitored. Emission of these gaseous proves the presence of anaerobic conditions. Hydrogen sulfide was observed in both compartments only in the first days of composting, but its level was marginal. H_2S maximum emission occurred on the third day, 28 ppm.

Composting of SL with SF is relatively new. Straw is the most popular substrate for composting with SL. In the study [8] the research on composting SF as monosubstrate was carried out. The addition of sewage sludge in own research allowed to achieve higher temperatures (fig. 2). This was due to the supply of SL with significant amounts of nitrogen and improved C/N ratio. Digestate was use in composting in research Franke-Whittle et al. [21]. The results of this authors also confirm the possibility of using digestate for composting.

4 Conclusion

Studies have shown that SL with a SF can be a suitable substrate for composting. The proof of that are for example change of parameters and temperatures during the process. This was proven among others by the reduction of organic matter from 93.45% at the beginning of the experiment to 78.32% after 69 days of the process. Knowing that sewage sludge is rich in nitrogen, it is important to keep in mind the appropriate proportions between components so as to avoid excessive ammonia emissions.

Information about sources of financing

This work was supported by Human Capital Operational Programme, Sub-measure 8.2.2, co-financed by the European Union under the European Social Fund coordinated by Dr. Eng. Wojciech Czekała within the period of 2012-2013 (POKL 8.2.2/30-404-12/13) and 2013-2014 (PO KL 8.2.2/30-355-13/14). The research project "Scholarship support for doctoral students specializing in majors strategic for Wielkopolska's development".

References

1. J. Yuan, D. Chadwick, D. Zhang, G. Li, S. Chen, W. Luo, L. Du, S. He, S. Peng. Effects of aeration rate on maturity and gaseous emissions during sewage sludge composting. Waste Management **56**, 403-410 (2016)
2. J.B. Bień, A. Grosser, E. Neczaj, M. Worwąg, P. Celary. Co-Digestion of Sewage Sudge with Different Organic Wastes – Review. Polish Journal of Environmental Studies Volume **2**, 24-30 (2010)
3. D. Wojcieszak, J. Przybył, A. Lewicki, A. Ludwiczak, A. Przybylak, P. Boniecki, K. Koszela, M. Zaborowicz, K. Przybył, K. Witaszek. Use of neural image analysis methods in the process of determine the dry matter content in the compost. Proc. of SPIE Vol. **9631** 963118-1 (2015)
4. J. Havukainen, M.T. Nguyen, L. Hermann, M. Horttanainen, M. Mikkilä, I. Deviatkin, L. Linnanen. Potential of phosphorus recovery from sewage sludge and manure ash by thermochemical treatment. Waste Management **49**, 221-229 (2016)
5. M. Michałkiewicz, J. Jeż – Walkowiak, Z. Dymaczewski, M.M. Sozański. Dezynfekcja ścieków. Inżynieria Ekologiczna **24**, 38-51 (2011)
6. M. Kacprzak, E. Neczaj, K. Fijałkowski, A. Grobelak, A. Grosser, M. Worwag, A. Rorat, H. Brattebo, A. Almås, B.R. Singh. Sewage sludge disposal strategies for sustainable development. Environmental Research **156**, 39-46 (2017)
7. D. Kosicka-Dziechciarek, A. Wolna-Maruwka, J. Mazurkiewicz. The Danger of Pathogenic Organisms in Sewage Sludge and Methods of Their Reduction. Archives of Waste Management and Environmental Protection **17(4)**, 127-138 (2015)
8. W. Czekała, J. Dach, R. Dong, D. Janczak, K. Malińska, K. Jóźwiakowski, A. Smurzyńska, M. Cieślik. Composting potential of the solid fraction of digested pulp produced by a biogas plant. Biosystems Engineering **160**, 25-29 (2017)
9. M. Cieślik, J. Dach, A. Lewicki, A. Smurzyńska, D. Janczak, J. Pawlicka-Kaczorowska, P. Boniecki, P. Cyplik, W. Czekała, K. Jóźwiakowski. Methane fermentation of the maize straw silage under meso- and thermophilic conditions. Energy **115(2)**, 1495-1502 (2016)

10. D. Janczak, K. Malińska, W. Czekała, R. Cáceres, A. Lewicki, J. Dach. Biochar to reduce ammonia emissions in gaseous and liquid phase during composting of poultry manure with wheat straw. Waste Management **66**, 36–45 (2017)
11. M. Zaborowicz, D. Wojcieszak, K. Górka, S. Kujawa, R.J. Kozłowski, K. Przybył, N. Mioduszewska, P. Idziaszek, P. Boniecki. Determination of dry matter content in composted material based on digital images of compost taken under mixed visible and UV-A light. Proc. SPIE 10033, Eighth International Conference on Digital Image Processing (ICDIP 2016), **100332G** (August 29, 2016) (2016)
12. M.C. Gutiérrez, A. Serrano, J.A. Siles, A.F. Chica, M.A. Martín. Centralized management of sewage sludge and agro-industrial waste through co-composting. Journal of Environmental Management **196**, 387-393 (2017)
13. L. Meng, W. Li, S. Zhang, C. Wu, K. Wang. Effects of sucrose amendment on ammonia assimilation during sewage sludge composting. Bioresource Technology **210**, 160-166. (2016)
14. R.L. Grando, A.M. de Souza Antune, F.V da Fonseca, A. Sánchez, R. Barrena, C. Font. Technology overview of biogas production in anaerobic digestion plants: A European evaluation of research and development. Renewable and Sustainable Energy Reviews 80, 44-53 (2017)
15. C. Zhao, X. Cui, Y. Liu, R. Zhang, Y. He, W. Wang, C. Chen, G. Liu. Maximization of the methane production from durian shell during anaerobic digestion. Bioresource Technology **238**, 433-438 (2017).
16. K. Kozłowski, A. Lewicki, P. Sołowiej, M. Neugebauer, A. Smurzyńska, 2016. Usage of waste whey as mono-substrate in continuous fermentation process. Energy And Clean Technologies Conference Proceedings, SGEM Vol. III: 345-350 (2016)
17. M.J. Stolarski, M. Krzyżaniak, K. Warmiński, J. Tworkowski, S. Szczukowski, E. Olba–Zięty E., Gołaszewski J. Energy efficiency of perennial herbaceous crops production depending on the type of digestate and mineral fertilizers. Energy **134**, 50-60 (2017)
18. W. Czekała, S. Bartnikowska, A. Lewicka, A. Bugała, Z. Zbytek, A. Lewicki. Economic and energy efficiency of the solid biofuels produced from digested pulp. MATEC Web of Conferences **60**, 04005 ICCBS (2016)
19. W. Czekała, J. Dach J. Przybył, P. Boniecki, A. Lewicki, P.C.R. Carmona, D. Janczak, H. Waliszewska. The energetic efficiency of solid fraction of digestate pulp from biogas plant in production of solid and gaseous biofuels - a case study of 1 MWel biogas plant in Poland. Proceedings of the 2nd International Conference on Energy & Environment: Bringing Together Engineering and Economics: 547-553 (2015)
20. I. Sigurnjak, C. Vaneeckhaute, E. Michels, B. Ryckaert, G. Ghekiere, F.M.G. Tack, E. Meers. Fertilizer performance of liquid fraction of digestate as synthetic nitrogen substitute in silage maize cultivation for three consecutive years. Science of The Total Environment 599–600, 1885-1894 (2017)
21. I.H. Franke-Whittle, A. Confalonieri, H. Insam, M. Schlegelmilch, I. Körner. Changes in the microbial communities during co-composting of digestates. Waste Management **34**, 632–641 (2014)

Define of internal recirculation coefficient for biological wastewater treatment in anoxic and aerobic bioreactors

Volodymyr Rossinskyi[1,*]

[1] Department of Ecobiotechnology and Bioenergy, National Technical University of Ukraine «Kyiv Polytechnic Institute» , 37, Prospect Peremohy, 03056, Kyiv-56, Ukraine

Abstract. The biological wastewater treatment technologies in anoxic and aerobic bioreactors with recycle of sludge mixture are used for the effective removal of organic compounds from wastewater. The change rate of sludge mixture recirculation between bioreactors leads to a change and redistribution of concentrations of organic compounds in sludge mixture in bioreactors and change hydrodynamic regimes in bioreactors. Determination of the coefficient of internal recirculation of sludge mixture between bioreactors is important for the choice of technological parameters of biological treatment (wastewater treatment duration in anoxic and aerobic bioreactors, flow capacity of recirculation pumps). Determination of the coefficient of internal recirculation of sludge mixture requires integrated consideration of hydrodynamic parameter (flow rate), kinetic parameter (rate of oxidation of organic compounds) and physical-chemical parameter of wastewater (concentration of organic compounds). The conducted numerical experiment from the proposed mathematical equations allowed to obtain analytical dependences of the coefficient of internal recirculation sludge mixture between bioreactors on the concentration of organic compounds in wastewater, the duration of wastewater treatment in bioreactors.

1 Introduction

The biological oxygen demands (BOD_5) in municipal wastewater usually are 100-300 mgO_2/l. The municipal wastewater usually contains suspended solids 147.3-468.2 mg/l, phosphor compounds 0.2-15.8 mg/l, nitrogen compounds 13.2-63.7 mg/l, synthetic detergents 8-25 mg/l [1, 2].

Synthetic detergents as amphiphilic xenobiotics, have a negative impact on the process of oxidation of organic compounds in units for biological wastewater treatment through localization substrate, dispersing the sludge, breach of oxygen regime and inhibition of

* Corresponding author: wrossin@live.com

activated sludge microorganism activity [3]. The synthetic surfactants with membranotropic properties affect microorganisms [4, 5].

The anoxic and aerobic biological treatment technology is mainly used for the effective removal of organic compounds and nutrients from municipal wastewater [6].

The assessment oxidation velocity (kinetic parameters) of organic compounds in aerobic and anoxic bioreactors with activated sludge depending on the initial concentration of organic contaminants is carried out by Monod equation [6-8].

To ensure denitrify bacteria the necessary substrate, nitrates and nitrites, treatment of wastewater is carry out in anoxic bioreactors with organizing recycle of sludge mixture from aerobic bioreactors.

The rate of internal recirculation of sludge mixture between anoxic and aerobic bioreactors is 50-500% [9]. Biological process of wastewater treatment operate with internal recycle of 200% [10]. The change rate of internal recycle of sludge mixture between the aerobic and anoxic bioreactor leads to a redistribution of the concentration of dissolved organic compounds in wastewater and change hydrodynamic regimes in bioreactors.

The effect of internal recirculation of sludge mixture between aerobic and anoxic bioreactors on efficiency of wastewater treatment from organic compounds is an important scientific issue for choice technological parameters of biological wastewater treatment.

2 Materials and methods

The experimental research of define internal recirculation coefficient was conducted for wastewater treatment from organic compounds in anoxic and aerobic bioreactors (Fig. 1).

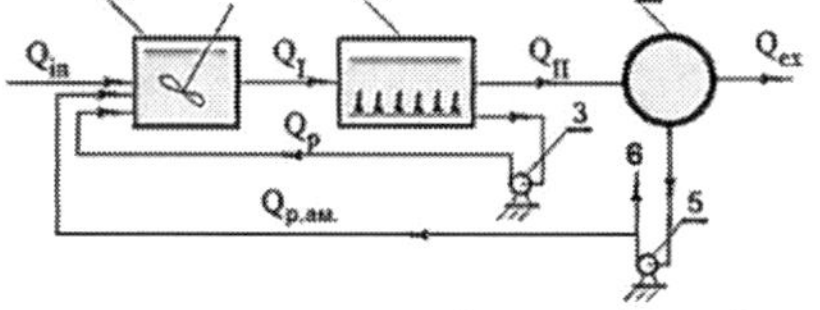

Fig. 1. Scheme of biological wastewater treatment in anoxic and aerobic bioreactors: 1 – anoxic bioreactor; 2 – aerobic bioreactor; 3 – recirculation pump; 4 – secondary clarifier; 5 – recirculation pump of activated sludge; 6 – disposal of excess activated sludge; Qin – inlet wastewater flow to anoxic bioreactor; Q_I – outlet wastewater flow from anoxic bioreacor; Q_{II} – wastewater flow from aerobic bioreactor to secondary clarifier; Qp – internal recirculation flow (nitrate recycle); $Q_{p.am.}$ – recirculation flow of activated sludge; Q_{ex} – flow of treated wastewater.

Internal recirculation flow capacity (Q_p, m^3/hour) can be defined:

$$Q_p = n_H \cdot Q_{in} \quad (1)$$

where n_H is a coefficient of internal recycle of sludge mixture, unit fractions.

Considering that the oxidation of organic compounds in anoxic bioreactors perform facultative bacteria using oxygen of nitrites and nitrates, the concentration of organic compounds in sludge mixture after anoxic bioreactor can be defined:

$$C_I^{ex} = C_I^{in} - [t_I(0,\ t_I^I)] \cdot dC_I/dt \quad (2)$$

where C_I is concentration of organic compounds in sludge mixture in anoxic bioreactor, $mgBOD_{total}/l$; t_I is a duration of wastewater treatment in the anoxic bioreactor, hour.

It is obvious that concentration of organic contaminants in sludge mixture at the outlet of the anoxic bioreactor (C_I^{ex}, $mgBOD_{total}/l$) numerically equal to the concentration of organic contaminants in sludge mixture inlet the aerobic bioreactor (C_{II}^{in}, mgBOD$_{total}$/l) (Fig. 1).

The oxidation of organic compounds carried by aerobic microorganisms in the aerobic bioreactors, so the concentration of organic contaminants in sludge mixture after aerobic bioreactor can be expressed:

$$C_{II}^{ex} = C_{II}^{in} - [t_{II}(0,\ t_i^{II})] \cdot dC_{II}/dt \tag{3}$$

where C_{II} is concentration of organic compounds in sludge mixture in aerobic bioreactor, mgBOD$_{total}$/l; t_{II} is a duration of wastewater treatment in the aerobic bioreactor, hour.

The renovation existing aerotanks by releasing some useful volume of aerotanks under anoxic bioreactor duration (t_I, hour) of wastewater treatment in anoxic bioreactor without nitrate recycle can be defined:

$$t_I = W_I / Q_{in} \tag{4}$$

where W_I is useful volume of anoxic bioreactor, m^3.

The duration of wastewater treatment (t_I^P, hour) in the anoxic bioreactor can be expressed given the internal recirculation flow:

$$t_I^P = W_I / (Q_{in} + Q_p) \tag{5}$$

where Q_p is internal recirculation flow capacity, m^3/hour.

The velocity of passing an elementary volume of sludge mixture in the anoxic bioreactor can be defined considering changing of internal recirculation flow capacity:

$$\theta_I = t_I^P / t_I = 1/(1 + n_н). \tag{6}$$

The balance equations system for determinate of organic compounds concentration change by phased wastewater treatment in anoxic and aerobic bioreactors with internal recirculation of sludge mixture considering activated sludge recirculation from secondary clarifier can be presented:

$$\begin{cases} n_н = Q_p / Q_{in};\quad C_I^{in} = \dfrac{C_{mix} + n_н \cdot C_p}{(1 + n_н)};\quad C_{mix} = \dfrac{C_{in} + C_{ex} \cdot R}{1 + R}; R = \dfrac{a_i}{\dfrac{1000}{J_i} - a_i}; \\ C_I^{ex} = C_I^{in} - \dfrac{1}{(1 + n_н)} \cdot \dfrac{dC_I}{dt} \cdot t_I(0, t_i^I);\quad C_{II}^{in} = C_I^{ex}; C_{II}^{ex} = C_{II}^{in} - \dfrac{1}{(1 + n_н)} \cdot \dfrac{dC_{II}}{dt} \cdot t_{II}(0, t_i^{II}) \end{cases}, \tag{7}$$

where C_{in} is a inlet concentration of organic compounds to anoxic bioreactor, mgBOD/l; C_{ex} – concentration of organic compounds in treated wastewater, mgBOD/l; C_p – concentration of organic compounds in internal recirculation flow, mgBOD/l; C_{mix} – concentration of organic compounds in mixture of activated sludge and wastewater, mgBOD/l; $n_н$ – internal recirculation coefficient, fraction of 1; C_I – concentration of organic compounds in anoxic bioreactor, mgBOD/l; t_I – duration of wastewater treatment in anoxic bioreactor, hour; C_{II} – concentration of organic compounds in sludge mixture in aerobic bioreactor, mgBOD/l; t_{II} – duration of wastewater treatment in aerobic bioreactor, hour; J_i – index of activated sludge, cm^3/g; R – coefficient of activated sludge recirculation, unit fractions.

The complex computational experiment for definition of internal recirculation coefficient was conducted for wastewater containing biodegradable organic compounds with concentration C_{in} = 300 mgBOD/l by dose of activated sludge a_i=2 g/l, dissolved oxygen concentration in anoxic bioreactor 0.1 mg/l, dissolved oxygen concentration in aerobic bioreactor 2 mg/l, concentration of nitrate 5 mgN-NO_3^{-1}/l, index of activated sludge 80 cm^3/g.

3 Results and discussions

Decrease of concentration of organic compounds from 300 mgBOD/l to 15 mgBOD/l with treatment duration of wastewater in anoxic bioreactor 0.5 hours and aerobic bioreactor 3 hours is provided with a coefficient of internal recirculation of sludge mixture of 180% (Fig. 2). Reduced treatment time in the aerobic bioreactor up to 1.5 hours needs an increase the coefficient of internal recirculation to 420%. An increase in the coefficient of internal recirculation can reduce the useful volume of the aerobic bioreactor, but requires the selection of a more productive recirculation pump.

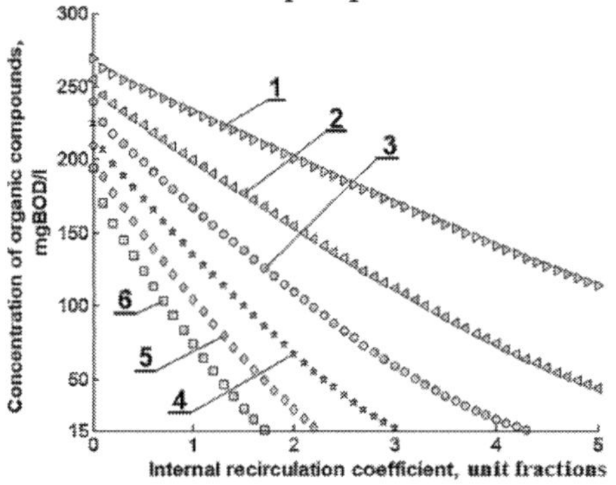

Fig. 2. Dependences of the concentration of biodegradable organic compounds (BOD_{total}) from internal recirculation coefficient and duration of wastewater treatment in anoxic (0.5 hour) and aerobic bioreactors (hour): 1 – 0,5; 2 – 1; 3 – 1,5; 4 – 2; 5 – 2,5; 6 – 3.

The rate of organic compounds removal also depends on the wastewater treatment duration in anoxic condition. For decrease of concentration of organic compounds from 300 mgBOD/l to 15 mgBOD/l with wastewater treatment duration in anoxic bioreactor 0.5 hours and aerobic bioreactor 2 hours is provided with a coefficient of internal recirculation of sludge mixture of 380% (Fig. 3). The reduced treatment time in the anoxic bioreactor up to 1 hours needs an increase the coefficient of internal recirculation to 340% for treatment time in the aerobic bioreactor is 2 hours.

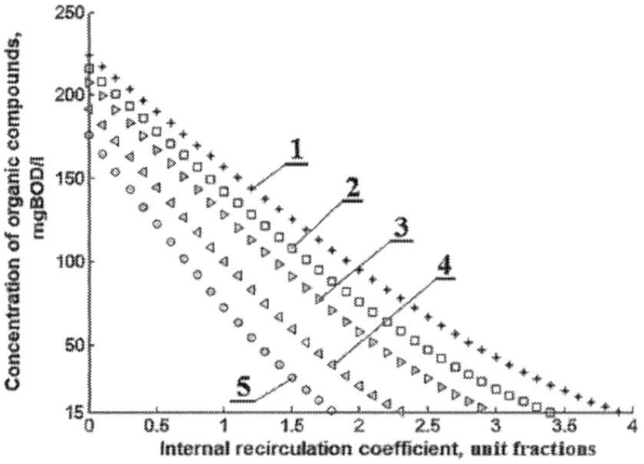

Fig. 3. Dependences of the concentration of biodegradable organic compounds (BOD_{total}) from internal recirculation coefficient and duration of wastewater treatment in aerobic (2 hour) and anoxic bioreactors (hour): 1 – 0,5; 2 – 1; 3 – 1,5; 4 – 2.

The twice decreasing the duration of treatment of wastewater in aerobic bioreactors requires an increase in the coefficient of internal recirculation by 57%. However, twice decreasing the duration of treatment of wastewater in anoxic bioreactors requires an increase in the internal recirculation coefficient to 10%. Regulation of useful volumes of bioreactors enables to minimize electricity consumption for recycling sludge mixture.

The synthetic detergents inhibit the processes of oxidation of organic compounds in anoxic and aerobic bioreactors. The assess a rate of removal of organic compounds carried by the numerical experiment for the nitrate concentration in sludge mixture 5 $mgN\text{-}NO_3^{-1}/l$ and change the concentration of synthetic detergents in the wastewater from 0 to 20 mg/l. The semi-saturation constant for synthetic detergents was adopted 30 mg/l [11].

For decrease of concentration of organic compounds from 300 mgBOD/l to 15 mgBOD/l with treatment duration of wastewater duration of wastewater treatment in anoxic bioreactor 0.5 hour and aerobic bioreactor 2 hour and without synthetic detergents is provided with a coefficient of internal recirculation of sludge mixture of 380%. Decrease of concentration of organic compounds from 300 mgBOD/l to 15 mgBOD/l with treatment duration of wastewater duration of wastewater treatment in anoxic bioreactor 0.5 hour and aerobic bioreactor 2 hour is provided with a coefficient of internal recirculation of sludge mixture of 500% for wastewater contains synthetic detergents with concentration of 5 mg/l (Fig. 4). The increasing concentrations of synthetic detergents from 0 to 10 mg/l needs to increase the coefficient of the internal recirculation coefficient to 45%.

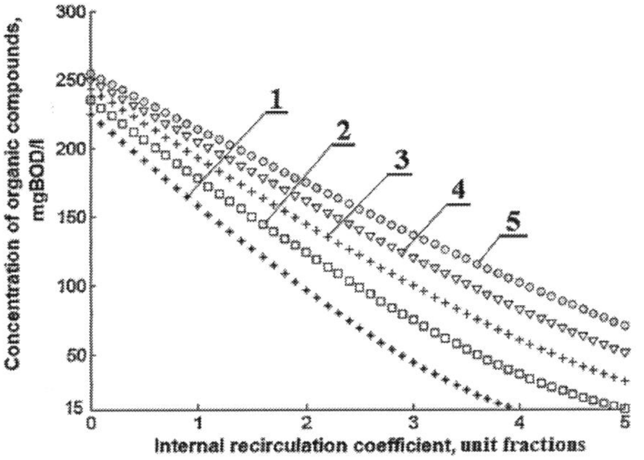

Fig. 4. Dependences of the concentration of biodegradable organic compounds (BOD_{total}) from internal recirculation coefficient, duration of wastewater treatment in anoxic (0.5 hour) and aerobic bioreactors (2 hour) and concentration of synthetic detergents in sludge mixture (mg/l): 1 – 0; 2 – 5; 3 – 10; 4 – 15; 5 – 20.

4 Conclusion

The internal recirculation coefficient for biological wastewater treatment in anoxic and aerobic bioreactors is important for choosing the exploitation of technological parameters of bioreactors with recirculation of sludge mixture.

Although for biological wastewater treatment it is recommended to take internal recirculation coefficient of 50-500% [9] and 200% [10], define of internal recirculation coefficient can be determined by iterative computation by the balance equations system (7). The increase the efficiency of biological wastewater treatment from organic compounds achieved through the increase of the internal recirculation coefficient considering inhibition biological treatment processes by synthetic detergents in bioreactors with different oxygen conditions.

The full assessment of wastewater treatment of organic compounds in bioreactors with different oxygen conditions along with the internal recirculation rate requires consideration of the recirculation rate of activated sludge from the secondary clarifier.

References

1. M. H. Gerardi, *Wastewater Bacteria* (2006)
2. O. Vasylenko, O. Polishchuk, L. Vasylenko, Ecological safety and Environmental Sciences, **15** (2014)
3. V. Rossinskyi, Bulletin of National University of Water Management and Nature Resources Use, **3** (2016)
4. S. A. Ostroumov, *Biological Effects of Surfactants* (CRC Press. Taylor & Francis. Boca Raton, 2006)
5. A. Dereszewska, S. Cytawa, R. Tomczak-Wandzel, K. Medrzycka Pol. J. Environ. Stud. **24**, (2015)
6. M. Henze, C. P. L. Grady, W. Gujer, G. V. R. Marais, T. Matsuo, *Activated Sludge Model No. 1.* (London, 1987)
7. V. M. Rossinskyi, L. A. Sablii, Issues of Water Supply, Water Disposal and Hydraulics, **27** (2016)
8. O. J. Oliynyk, T. S. Hayrapetyan, Reports of the National Academy of Sciences of Ukraine, **5** (2015)
9. M. Henze et al., *Wastewater Treatment: Biological and Chemical Processes* (Springer: Berlin, Heidelberg, 2013)
10. T. Yoona, H. Lee, C. Kima. Journal of Membrane Science, **242** (2004)
11. V. M. Rossinskyi, L. A. Sablii, *The 8th Eastern European Young Water Professionals Conference : Leaving the Ivory Tower Bridging the Gap between Academia, Industry, Services and Public Sector* (Oficyna Wydawnicza Politechniki Wrocławskiej, 2016)

E3S Web of Conferences **30**, 02003 (2018) https://doi.org/10.1051/e3sconf/20183002003
Water, Wastewater and Energy in Smart Cities

A comparative analysis of selected wastewater pretreatment processes in food industry

Katarzyna Jaszczyszyn[1,*], *Wojciech* Góra[1], *Zbysław* Dymaczewski[1], and *Robert* Borowiak[1]

[1]Faculty of Civil and Environmental Engineering, Poznan University of Technology, 5 Piotrowo, 60-965 Poznan, Poland

Abstract. The article presents a comparative analysis of the classical coagulation with the iron sulphate and adsorption on bentonite for the pretreatment of wastewater in the food industry. As a result of the studies, chemical oxygen demand (COD) and total nitrogen (TN) reduction were found to be comparable in both technologies, and a 29% higher total phosphorus removal efficiency by the coagulation was observed. After the coagulation and adsorption processes, a significant difference between mineral and organic fraction in the sludge was found (49% and 51% for bentonite and 28% and 72% for iron sulphate, respectively).

1 Introduction

Industrial pollution are a large and various group of contaminants. According to the NACE Rev. 2 – Statistical classification of economic activities in the European Community and in Poland - Polish Classification of Activities [1-2], in the Classification scheme, food industry is included in the Section C – Manufacturing. The section named: Manufacture of food products and Manufacture of beverages contains two categories of food industry. The first is: processing of products from agriculture, forestry, hunting and fishing in the food and drink for humans or animals and activities which result in a variety of products, such as: meat, fish, fruit and vegetables, fats and oils, dairy products, grain mill products, pet food, other groceries. And the second one contains: production of soft drinks, mineral water, alcohol, beer and wine. Poland is one of the largest agri-food industry producer in Europe [3]. In 2013 in Poland the share of sold production of the agri-food industry in the sold production of the entire processing industry exceed 22%, it is about PLN 215.7 billion, EUR 51.48 billion [4].

In Poland, the largest share in the wastewater production (over 85%) comes from electricity, gas, steam and hot water supply and about 9.4% from manufacturing. Wastewater from manufacturing of food products and of beverages in the total amount of industrial wastewater is about 1.2%, and in the Section C – Manufacturing it contains, about 12.1% [5].

In industries water is used for different purposes - during the production process, washing, diluting, cooling and other. Compared to other industries sectors, the food industry uses a relatively large amount of water per unit of product. The consumption of

* Corresponding author: katarzyna.jaszczyszyn@put.poznan.pl

water in the food industry is related to the required high quality of the product and hygienic considerations.

One of the major consumers (24%) of freshwater among food and beverage processing facilities on the global market is the meat processing industry [6]. Typical water usage in this industry is about 1.5 to even 60 m^3/ton of product [7-8].

Wastewater in the food industry generally contain high concentration of organic pollutant. These are, above all, macromolecular protein complexes, carbohydrates, fats and parts of fruits/vegetables, bones etc. [9-11]. Wastewater from meat industry contain high concentration of suspended solids, including pieces or fat, feathers, flesh, manure or undigested feed [10, 12]. Blood as one of the major dissolved pollutants in slaughterhouse wastewater has high chemical (COD) and biological (BOD_5) oxygen demand. Meat industry wastewater is very harmful to the environment. Effluent discharge from slaughterhouse has caused deoxygenation of rivers and contamination of groundwater [12].

Table 1. General characteristics of selected industrial wastewater [6, 10-11, 13-14].

		Slaughterhouse	Bakery
COD	mg/L	500-15900	1.5-18 kg/tonne
BOD_5	mg/L	150-4635	155-8500
TN	mg/L	50-841	N/D*
TP	mg/L	25-200	N/D
TSS	mg/L	270-6400	130-5700 (13430)
Fats/grease	mg/L	160-2500	60-1200 (4490)
pH	-	4.9-8.1	4.7-8.4

*N/D-no data

The bakery industry is another example of the food industry with a high water consumption, which is about 9% of freshwater consumption in beverage and food industries [6]. More than a half of the water in bakery plants is discharged as wastewater. Wastewater is primarily generated from cleaning operations including equipment cleaning and floor washing. It may contain flour, sugar, salt, yeast, water, and oil/fat. During the manufacturing processes, hot water mixed with detergents is used in Cleaning-In-Place systems [13].

The different types of industrial wastewater contamination requirements causes a variety of strategies for treatment technologies. About 65-90% of the organic matter in food industry wastewater is colloidal or particulate and it can be reduced to a high degree by the chemical pretreatment of raw wastewater [15]. Most of the solids can be removed using simple sedimentation techniques with the solids recovery as slurry or sludge. However, classical pretreatment methods for industrial wastewater which based on mechanical and physical-chemical processes, are often insufficiently effective, and also waste streams are difficult to manage. Coagulation process using inorganic coagulants may result in the production of huge volume of sludge and aluminium or iron salts that may be retained in the treated water [16]. According to that, new and highly effective pretreatment methods that will take into account the residual pollutions as well as an impact of applied processes on the sludge management, are being developed [17-18]. An exemplary method that meets the current criteria is an adsorption with mineral sorbents application combined with flocculation [19].

Clay minerals show outstanding adsorption properties for numerous organic products [20-22].

This paper presents a comparative research results of two different wastewater pretreatment methods in three industrial plants - meat (IP1, IP2) and bakery industry (IP3).

2 Materials and methods

Averaged wastewater samples from three industrial plants (IP): slaughter and meat processing (IP1 and IP2) and bakery industry (IP3) were analysed. The choice of the industrial plant type was followed by, among others, the ease of their collection with technical equipment and the diversity in the composition of pollution.

All of the samples for analysis were taken from the wastewater treatment line of each plant as averaged and raw wastewater. Pretreatment processes with iron sulphate and natural Na-bentonite, as a mineral sorbent, were carried out.

The natural bentonite samples used for the experiments was obtained from Hekobentonity Ltd., Poland. Bentonite was applied as a 5% water suspension.

The specific surface area of bentonite was determined by the nitrogen adsorption–desorption method at 77K on ASAP 2420 analyser. BET surface area for the bentonite was 58.517 m^2/g. Elemental composition of raw bentonite was analysed by SEM equipped with EDS detector (QUANTA 250 FEG, FEI). Bentonite chemical composition (in %): Na_2O – 3.60, MgO – 2.41, Al_2O_3 – 24.92, SiO_2 – 63.49, K_2O – 0.69, CaO – 1.65, Fe_2O_3 – 3.24.

The iron sulphate coagulant ($Fe_2(SO_4)_3$) was obtained from Kemipol Ltd., Poland. Polyelectrolytes were delivered by Korona JV., Poland. Hydrochloric acid 0,1M (Chempur, Poland) and sodium hydroxide 0,1M (Chempur, Poland) were used for the pH adjustment.

All chemicals were of analytical grade and used as received. In the experiments, distilled water was used for all experiments to preparing the solutions and suspensions.

Batch experiments was conducted in laboratory conditions in 1dm^3 reactors using a Kemira Flocculator equipped with 6 glass beakers connected to a tachometer controlled stirrers. Kemira Flocculator allows to control both processes coagulation and flocculation. Depending on the specifics of the process, a different process time was programmed for quick mixing, slow mixing, and sedimentation. Mixing times and speeds have been selected to illustrate the processes that take place on a technical scale in a classical coagulation and flocculation system.

The turbidity measurement was performed on the Turb 350IR nephelometer, WTW.

The tests were performed in triplicate for each test to obtain a more reliable result.

The scope of physical-chemical analysis included determining the basic parameters of raw and purified (pre-treated) wastewater (COD, total nitrogen and total phosphorus concentration and dry matter in the obtained sediment).

All chemical analysis of wastewater and sludge were conducted in accordance with the applicable standards and standard methods. COD analysis was prepared with ISO 15705:2002 - Determination of the chemical oxygen demand index (ST-COD).

Determination of dry matter and water content on a mass basis in sediment and sludge was also conducted. For this purpose, the dry weight of the sludge was tested after drying in drying oven thermostatically controlled with forced air ventilation, maintaining a temperature of (105±5)°C.

The sludge was subjected to a drying process in accordance with the guidelines in EN 12880: 2004 - Determination of dry residue and water content. As the final result, the average of three samples was taken. The same crucibles were used for the subsequent loss on ignition measurement, ignited at 550°C in the muffle furnace.

3 Experimental results

Initially, concentration of chemical oxygen demand (COD), total nitrogen (TN) and total phosphorus (TP) in untreated wastewater samples from three industrial plants (IP1, IP2, IP3) were investigated. The results are compiled in Table 2.

Table 2. Untreated (daily averaged) wastewater samples characterization.

		Sample ID		
		IP1	IP2	IP3
COD	mg/L	2960 ± 62.2	4730 ± 125.3	9530 ± 357.6
TN	mg/L	228 ± 6.9	206 ± 5.3	66.5 ± 4.5
TP	mg/L	25.6 ± 0.8	28.8 ± 1.3	12.8 ± 1.0

IP1, IP2 - slaughter and meat processing, IP3 - bakery industry.

On the basis of preliminary research, the flocculant that works best with the given coagulant on a given wastewater test was selected to further analysis.

The first stage of the research was coagulant dose selection. There were carried out six series in every stage of the research. Dosages in the range of 0.15 - 1.2 g/L for iron sulphate and 0.3 - 1.8 g/L for bentonite were tested.

The final dosage of the reagents was determined on the basis of the turbidity of the supernatant liquid.

As the best dose, it was found the smallest turbidity of the supernatant liquid. Fig. 1. and Fig. 2. illustrate the experimental results of the coagulant dosage selection.

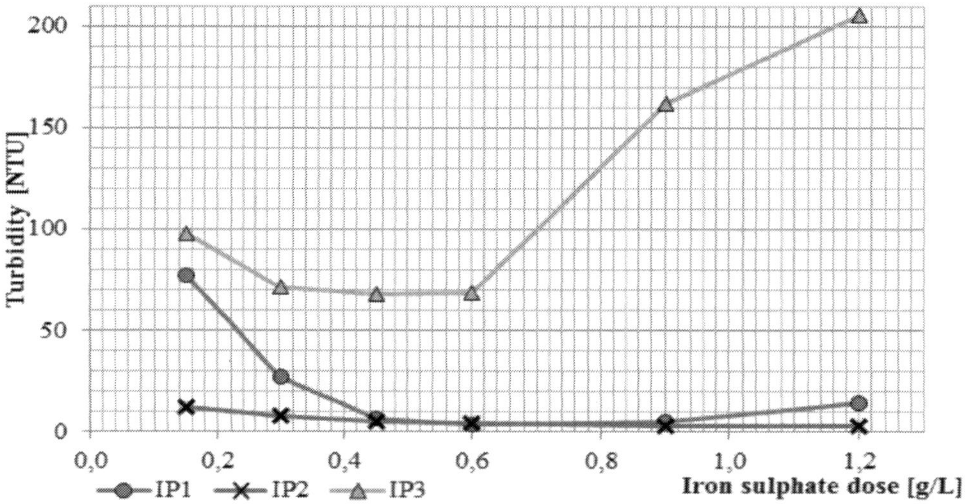

Fig. 1. Influence of doses of iron sulphate on wastewater turbidity.

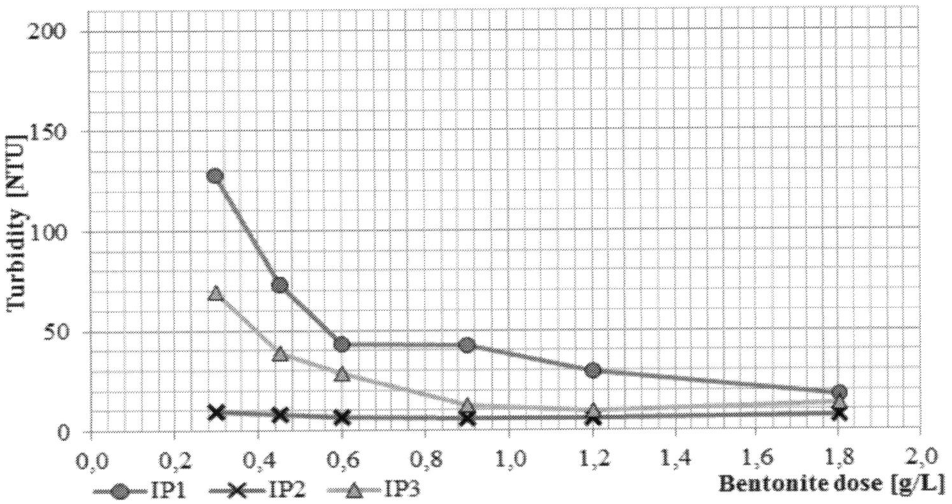

Fig. 2. Influence of doses of bentonite on wastewater turbidity.

Reagent doses selected to the further researches: iron sulphate IP1 – 0.6g/L, IP2 – 0.9g/L, IP3 – 0.45g/L and for bentonite IP1 – 1.8g/L, IP2 – 0.9g/L, IP3 – 0.9 g/L.

For selected reagent doses pre-treated wastewater samples and obtained sediments have been analysed. The effectiveness results of the COD, Nitrogen and Phosphorus reduction are compiled in Table 3. Effectiveness of the reduction of selected pollutants are also shown in the Fig. 3.

Table 3. Effectiveness of the reduction of selected pollutants.

		Bentonite			**Iron sulphate**		
		IP1	**IP2**	**IP3**	**IP1**	**IP2**	**IP3**
COD	%	84.4	84.7	71.0	93.4	95.3	79.4
TN	%	83.1	77.4	82.7	90.0	72.1	88.0
TP	%	53.9	69.4	82.8	98.4	98.6	96.9

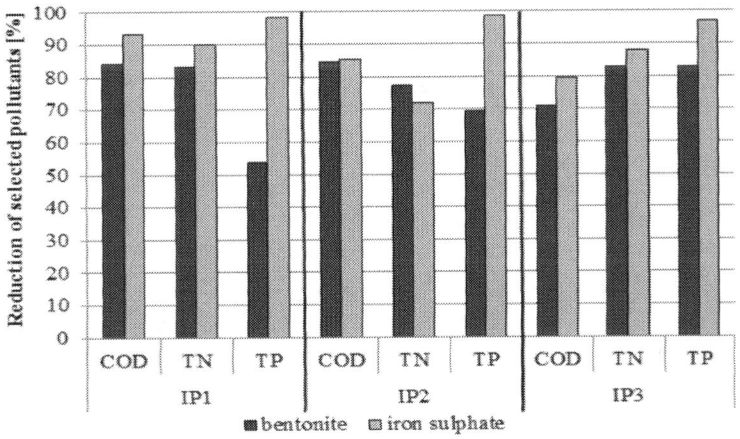

Fig. 3. Effectiveness of the reduction of COD, total nitrogen and total phosphorus [%]

Percentage content of the fixed solids and volatile solids contained in the sediments are compiled in Table 4. and shown in the Fig. 4.

Table 4. Percent of dry matter content in the sediment sludge.
(FS – fixed solids = mineral matter fraction, VS – volatile solids = organic fraction)

		Bentonite			Iron sulphate		
		IP1	**IP2**	**IP3**	**IP1**	**IP2**	**IP3**
FS	%	63.0	39.7	44.3	23.9	27.9	25.0
VS	%	37.0	60.3	55.7	76.1	72.1	75.0

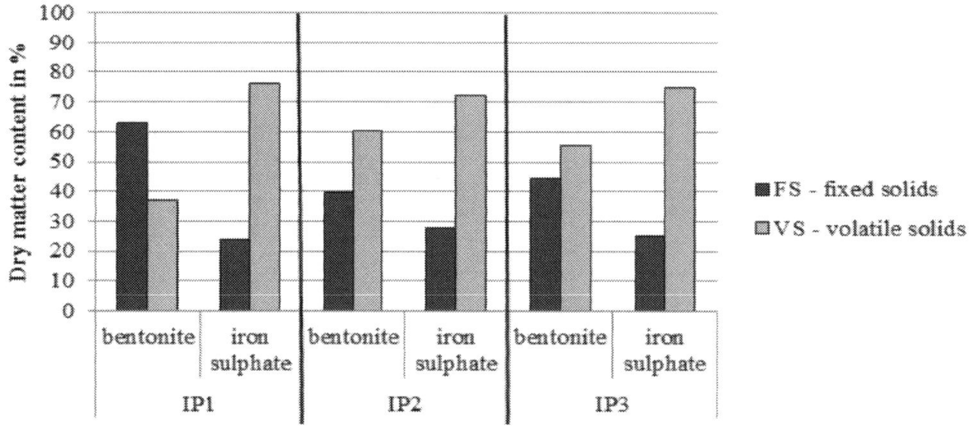

Fig. 4. Dry matter content (in%) in the sediment sludge after pretreatment.

4 Discussion

Initially, concentration of chemical oxygen demand (COD), total nitrogen (TN) and total phosphorus (TP) in daily averaged untreated wastewater samples from three industrial plants (IP1, IP2, IP3) were investigated. The results are compiled in Table 2.

Similar concentrations of total nitrogen and total phosphorus were found in wastewater samples from plants IP1 (slaughter) and IP2 (meat processing). COD was higher in the sample of wastewater from the meat processing plant (about 60%).

Wastewater from meat industry is suitable for biological treatment [7]. For IP1 and IP2 the ratio of C: N: P is 100: 7.7: 0.9 and 100: 4.4: 0.6 respectively, which nearly ensures the balance of carbon compounds and nutrients for biological processes in untreated wastewater. In the case of IP3 sample (from bakery), COD value was more than twice as high as in the meat industry samples, while significantly lower concentrations of both total nitrogen and total phosphorus, which is typically for the bakery industry (see Table 1.). The ratio of C: N: P in the wastewater sample from the bakery (IP3) is 100: 0.7: 0.1 and is not balanced for biological treatment. In the case of biological treatment using following pretreatment processes, it is appropriate to proportional reduce the concentration of pollutants from the meat industry plants (IP1, IP2) and to maximize the reduction of COD in the case of wastewater from the bakery (IP3).

In the pretreatment process of wastewater from meat industry (IP1 and IP2) with iron sulphate as a coagulant, the efficiency of total phosphorus removal was 98-99%. Similar results were obtained by some authors with application of other coagulants - polyaluminum chloride or alum as the reagent - 99.9 and 98% TP removal, respisevely [6]. By using alum, ferric chloride and ferric sulfate Amuda and Alade achieved significantly lower efficiency: 45, 32 and 39% TP removal, respectively [23].

Other hand, by application of the bentonite the efficiency of TP removal for IP1 and IP2 was 54 and ~70%. In wastewater from bakery industry (IP3), the total phosphorus removal efficiency was 97% with iron sulphate, while with bentonite ~83%.

For all investigated wastewater samples, a lower total phosphorus removal efficiency was obtained by the use of bentonite. This is most likely related to the chemical precipitation of orthophosphates using iron sulphate [24].

The use of bentonite may be beneficial to preserve the increased phosphorus content after pretreatment processes to balance the composition of effluent for the biological processes as an appropriate treatment degree.

For all samples, total nitrogen removal using both agents: iron sulphate and bentonite has yielded comparable efficacy, and on average 83% and 81%, respectively. The COD reduction was most affected by the use of iron coagulant - 79-93%, and 71-85% for bentonite.

From the values obtained during the determination of dry matter content and dry residue, the largest share of organic matter content was obtained by using iron coagulant (above 70% for each type of industry). The content of organic matter in the bentonite-treated sediment was varied and oscillate between 37% and 60.3%.

The high content of mineral fraction in the sediments after pretreatment with bentonite affects the potential methods of its management. Considering the iron content in natural bentonite (Fe_2O_3 - 3.24%) and the iron content in the coagulant - $Fe_2(SO_4)_3$, it is expected to considerably lower iron content in the sediment sludge after bentonite application. This fact may also be important for some sediment management methods.

5 Summary

For both methods - iron sulphate coagulation as well as adsorption with bentonite, comparable COD and total nitrogen removal efficiencies were obtained (the slightly better results were obtained for the COD removal using iron sulphate). In the coagulation, the efficiency of phosphorus removal was 29% higher than in the adsorption. Other hand, in the adsorption process, about a 20% higher of the mineral matter content in the sludge was achieved.

Application of bentonites for the pretreatment of food processing industry, may be an alternative method to the classical coagulation process.

Analysing the directions of sludge management, depending on the applied pretreatment technology, significant differences in sediment composition obtained in both methods should be taken into account.

References

1. Central Statistical Office of Poland, *Polish Classification of Activities PKD*; in Polish (2007)
2. Eurostat, *N.A.C.E. Rev.2.* (Office for Official Publications of the European Communities, Luxemburg, 2008)
3. Global Agricultural Information Network, *Food Processing Sector Report* (Warsaw, 2016)
4. Polish Information and Foreign Investment Agency, *Food sector in Poland* (Economic Information Department, 2015)

5. Central Statistical Office of Poland, *Statistical Yearbook of Industry 2016* (Warsaw, 2017)
6. C.F. Bustillo-Lecompte, M. Mehrvar, J Environ Manage **161** (2015)
7. World Bank Group, *Meat processing and rendering. Project Guidelines: Industry Sector Guidelines*, in: Pollution Prevention and Abatement Handbook (1998)
8. A. Steinhoff-Wrześniewska, A. Rajmund, J. Gwozdoń, Inżynieria Ekologiczna **32**; in Polish (2013)
9. F. Meinck, H. Stoff, H. Kohlschuter, *Ścieki przemysłowe* (Arkady, Warsaw, 1975), Polish translation of: F. Meinck, H. Stoof, H. Kohlschütter, Industrieabwasser (1968)
10. H. Rüffer, K.H. Rosenwinkel, *Oczyszczanie ścieków przemysłowych: Poradnik.* (Oficyna Wydawnicza Projprzem-Eko, 1998), Polish translation of: K.H. Rosenwinkel, Taschenbuch der Industrieabwasserreinigung (Oldenbourg, 1991)
11. C.J. Banks, Z. Wang, *Treatment of meat wastes*. In: Waste Treatment in the Food Processing Industry, in: L. K. Wang, Z. T. Hung, H. H. Lo, C. Yapijakis, Waste treatment in the food processing industry (Taylor & Francis Group, Florida, 2006)
12. D.I. Massé, L. Masse, Canadian Agricultural Engineering **42,** 3 (2000)
13. J.P. Chen, L. Yang, R. Bai, Y.T. Hung, Bakery Waste Treatment, in: L.K. Wang, Y.T. Hung, H.H. Lo, C. Yapijakis, Waste treatment in the food processing industry (Taylor & Francis Group, Florida, 2006)
14. B. Bartkiewicz, K. Umiejewska, *Oczyszczanie ścieków przemysłowych*; in Polish (Wydawnictwo Naukowe PWN, 2010)
15. M.H. Sorour, A.G. Aboulnour, A.A. Mostafa, H.F. Shaalan, M.H. El-Sayed, and H.A. Hani, World Applied Sciences Journal **26**, 8 (2013)
16. T. Kazi, A. Virupakshi, IJIRSET **2**, 8 (2013)
17. Ministry of the Environment, *Najlepsze dostępne techniki BAT - wytyczne dla branży spożywczej: owocowo-warzywnej (soki i nektary, przetwory, mrożonki;* in Polish (ENVIRON Poland Sp. z o.o., Warsaw, 2004)
18. A.P. Vanerkar, S. Satyanaryan, S. Satyanaryan, IJCPS **2**, Special Issue (2013)
19. E. Worch, *Adsorption technology in water treatment: fundamentals, processes, and modelling* (de Gruyter, Berlin/Boston, 2012)
20. Z. Lizhong, Z. Jianying, L. Yimin, S. Xueyou, Q. Wenbin, Journal of Environmental Sciences **8**, 3 (1996)
21. J. Kaleta, D. Papciak, A. Puszkarewicz, Rocznik Ochrona Srodowiska **15**, 3; in Polish (2013)
22. T.P.A. Shabeer, A. Saha, V.T. Gajbhiye, S. Gupta, K.M. Manjaiah, E. Varghese, Environ Technol. **35** (2014)
23. O.S. Amuda, A. Alade, Desalination **196.1** (2006)
24. E. Nassef, ESTIJ **2**, 3 (2012)

Model Studies on the Effectiveness of MBBR Reactors for the Restoration of Small Water Reservoirs

Agata Nowak[1], Robert Mazur[2,], Ewa Panek[1], Joanna Chmist[3]*

[1] Department of Geoinformation, Photogrammetry and Remote Sensing of Environment, AGH University of Science and Technology in Krakow, Poland
[2] Department of Environmental Management and Protection, AGH University of Science and Technology, Poland
[3] Department of Ecology and Environmental Protection, Poznań University of Life Sciences, Poland

Abstract. The authors present the Moving Bed Biofilm Reactor (MBBR) model with a quasi-continuous flow for small water reservoir restoration, characterized by high concentrations of organic pollutants. To determine the efficiency of wastewater treatment the laboratory analysis of physic-chemical parameters were conducted for the model on a semi-technical scale of 1:3. Wastewater treatment process was carried out in 24 h for 1 m3 for raw sewage. The startup period was 2 weeks for all biofilters (biological beds). Approximately 50% reduction in COD and BOD_5 was obtained on average for the studied bioreactors. Significant improvements were achieved in theclarity of the treated wastewater, with the reduction of suspension by 60%. The oxygen profile has improved significantly in 7 to 9 hours of the process, and a diametric reduction in the oxidative reduction potential was recorded. A preliminary model of biological treatment effectiveness was determined based on the conducted studies. In final stages,the operation mode was set in real conditions of polluted water reservoirs.

1 Introduction

The primary goal of Water Frame Directive is to achieve a good state of uniform bodies of surface water. Many large and small water reservoirs, heavily transformed by humans still require urgent corrective action in order to achieve the main aim of WFD. The process of water quality monitoring is crucial to prepare the proper environmental risk assessment (ERA). Environmental risk assessment procedures are currently being promoted by the EU as a necessary tool for further environmental risk management actions. The US EPA published Guidelines for Ecological Risk Assessment [1], nevertheless, this document contained very general guidelines for the ERA process [2]. In the EU has developed a number of specific algorithms of ERA for the surface water, and one of the most commonly applied methods is a TRIAD, developed in the UK[2]. The results of the ERA are adatabase to take further action under the environmental risk management procedure (ERM). The ERM algorithms require the

*Corresponding author: mazur@agh.edu.pl

involvement of legal, political, economic and engineering instruments. Often these procedures are cyclical, with a constant modification of the main objectives and with control of stages andend effects. The ERA results are also a database for selecting appropriate methods for restoration of degraded water reservoirs, already at the planning stage of the ERM procedure. In order to choose thebest types of lake reclamation methods can be used multi-criteria analysis such as AHP, Rembrandt or Promethe[3, 4]. The cost of water treatment of polluted water bodies is very high, especially when they are characterized by high loads of organic pollution.

The scientific progress in Environmental Engineering provides us with informationabout new effective biotechnologies of water treatment and purification from different types of pollution. The best prospects give us interdisciplinary research in this field of environmental biotechnologies [5]. Researchers in their studyhave been looking for effective strains and consortiums of microorganisms which are able to maintain an effective biodegradation and biotransformation of different types of pollution[6, 7, 8, 9, 10, 11]. The above-mentionedstudy directions aimto develop dedicated biopreparations and high-efficiency wastewater treatment technologies [6]. Most of wastewater treatment technologies require the maintenance of controlled and homogeneous bioreactors operating conditions [12]. However, for two decades, the market of WWTP observed thedynamic development of MBBR technology, in which the main objective is to minimize power consumption and volume of the reactor while increasing the degree of wastewater treatment. This types of reactors are used in wastewater treatment from communal and industrial sources in global scale [13]. The MBBR reactors are characterizedby wide tolerance for changes in physicochemical factors (temp., pH, redox, conductivity, organic loads, etc.) [14]. In the two decades, over 450 large and nearly 110,000 smaller WWTP have been developed around the world using a high-efficiency biofilm bed. There are therefore justified scientific grounds for the possibility of adapting this technology to the reclamation of open water bodies in temperate climate zone [13, 15]. MBBR reactors require an energy input to maintain the aeration process and flow of the medium through the bed. The application of RES (renewable energy sources) in power supply of field installation allows for energy self-sufficiency of this type of technology.

It is also possible to use traditional sources of electricity and prepare the installation in astationary mode with the blower station located on the reservoir bank [16].

2 Methodology

The authors conducted a study on the biological treatment efficiency of municipal wastewater in MBBR model reactors designed in laboratory scale 1:3. Four MBBR reactors were constructed with cylindrical geometryand dimensions: diameter φ = 400 mm, L_{tot} = 90 cm, h_{basket} = 70 cm, $V_{act.}$ = 0,088 m^3, $V_{tot.}$ = 0,133 m^3 (Fig. 1.). The bed mediaofreactor were 3 cm long carriers made of an electrical conduit (PE) and thetotal active area was 30,8 m^2. Each reactor was operating in plastic containers with volume V = 1 m^3, fully submerged, 5 cm below the surface of wastewater. The aeration process was carried out by fine bubble disc diffuser (φ =240 mm) with compressed air suppliedby rubber pipes and generated bya blower (model AT-80). Inorder to ensure a smooth adjustment of the supplied air into reactor system, two valves were mounted: control and relief valve. A rotameters were used to measure the flow of supplied air into the bed of MBBR and Mammoth reactors.Established air flow was $Q_p = 2{,}2\ m^3{\cdot}h^{-1}$, at which the flow and velocity of the wastewater through the reactor was obtained, respectively:$Q_w = 8\ m^3{\cdot}h^{-1}$and $v = 0{,}38\ m{\cdot}s^{-1}$.The above air flow ensured the full circulation of the carriers into the bed and the biofilm on the colliding carriers wasn't intensively detached from theexternal surface. The startup period for biological bed was 2 weeks, after which it was obtained a stable process for the wastewater treatment. The bioreactors were operated on the wastewater after primary mechanical treatment by rotary mechanical coarse in Left-side WWTP in Poznan. The average loads of CODwere in the range 500 ÷ 600 mg $O_2 \cdot dm^{-3}$ in raw wastewater. Nutrients concentration was recorded respectively in ranges: PO_4^{-3} 7 – 14 mg·dm^{-3}, NH^{4+} 27 ÷ 84 mg·dm^{-3}, NO^{3-} 1,2 ÷ 2,6 mg·dm^{-3}, NO^{2-} 0,08 ÷ 1,4 mg·dm^{-3} in raw sewage.

The hydraulic retention time was: $$RT = \frac{V_{act}}{Q_D}[d] = \frac{0.088\ m^3}{1\frac{m^3}{d}} = 0{,}088\ [d] \qquad (1)$$

where: V_{act} –volume of thereactor, Q_d – influent flow rate

The operation mode was divided into 1,5 h of aeration (nitrification phase) and 1,5 h without aeration (denitrification phase).The floating carriers provide adequate biomass suspension of biofilm in the denitrification phase, no additional mixing is required. The samples of treatedwastewater were taken after treatment process, before its discharging from model reactor chambers.

Table 1 presents laboratory technics used during analysis of physical and chemical parameters of studied wastewater (raw and after treatment).

Table 1. Laboratory procedures and methods of studied physical and chemical parameters.

No.	Method	Used procedures
1	COD	Procedure Merckmillipore in NOVA 60 Spectroquant®, 114555 - CODcuvette test, method for range: 100 - 1000 mg $O_2 \cdot dm^{-3}$
2	BOD_5	According to the procedure Oxi-top (WTW company) in 5 days period (in dark glass bottles with continuous stirring and temperature 20 °C).
3	NH^{4+}	Spectrophotometric method, by spectrophotometer Spectroquant NOVA 60 by Merck. Method PB-06, issue 2 from 1 0.05.2007.
4	NO^{3-}	Spectrophotometric method, by spectrophotometer Spectroquant NOVA 60 by Merck. Method PB-04 issue 2 from 10.05.2007.
5	NO^{2-}	Spectrophotometric method, by spectrophotometer Spectroquant NOVA 60 by Merck. Method PB-05 issue 2 from 10.05.2007.
6	PO_4^{3-}	Spectrophotometric method, by spectrophotometer Spectroquant NOVA 60 by Merck. Method PB-07 issue 2 from 10.05.2007.
7	SO_4^{2-}	Spectrophotometric method, by spectrophotometer Spectroquant NOVA 60 by Merck. Method (064) 114548 5 – 250 mg/l SO4 Bariumsulfate, turbidimetric

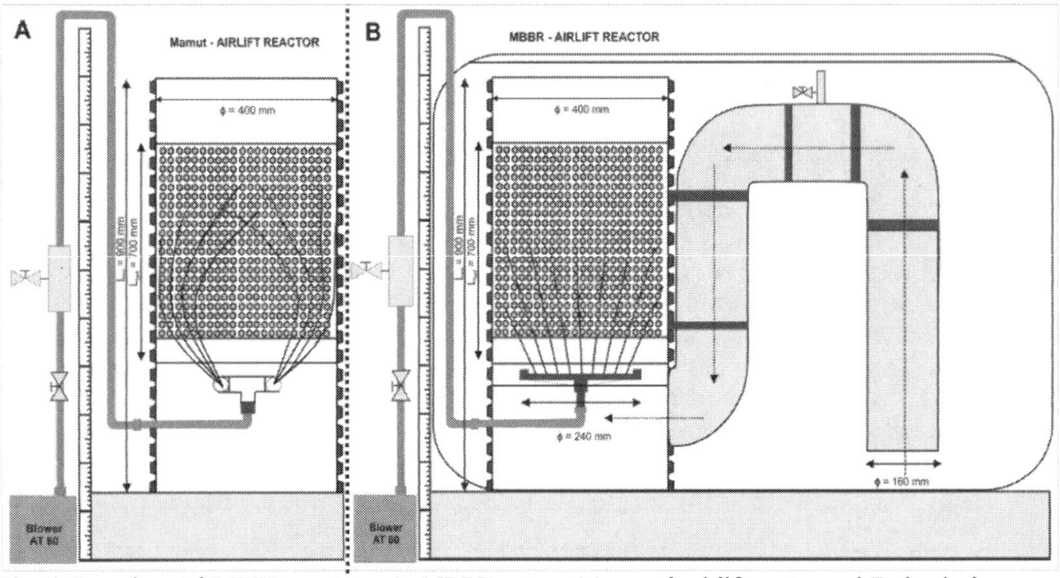

Fig. 1. Experimental MBBR reactors: A. MBBR – type Mammoth airliftreactor and B.classical MBBR airliftreactor in scale 1:3 in Left-side WWTP in Poznan.

3 Results

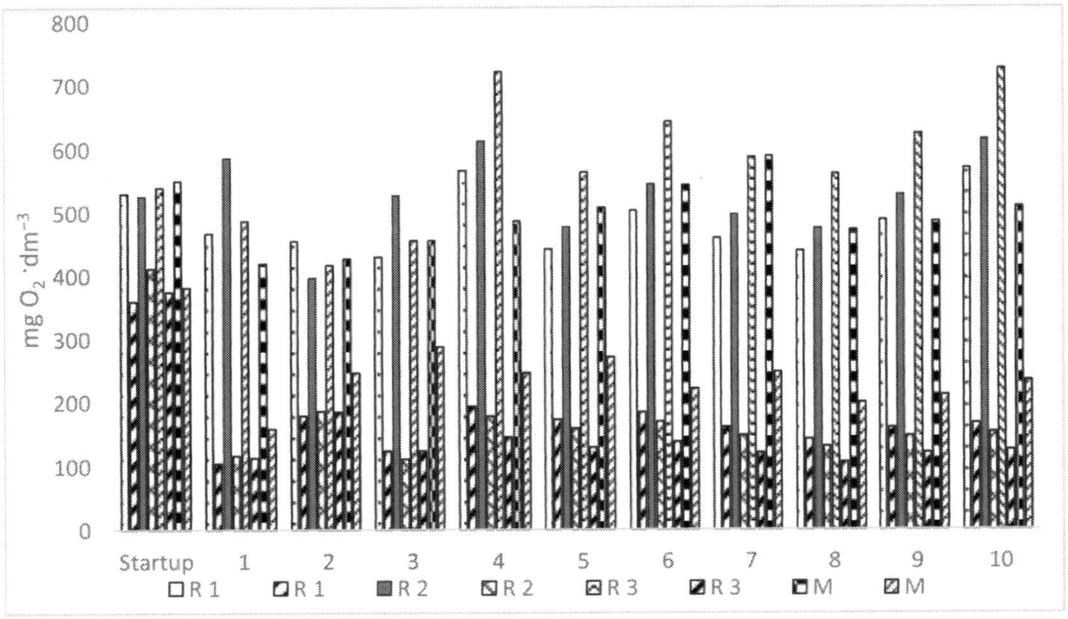

Fig. 2. Reduction of COD after 24h wastewater treatment [mg $O_2 \cdot dm^{-3}$].

Figure2 and 3 present changes in COD and BOD_5 after 24 h of wastewater treatment process in bioreactors MBBR (R1, R2, R3) and Mammoth (M). The startup period is characterized by worst treatment results in comparison to further experiments after 2 weeks later, in which the efficiency of the treatment has improved significantly. The article presents results for 10 randomly chosen treatment processes in two months period of theexperiment.

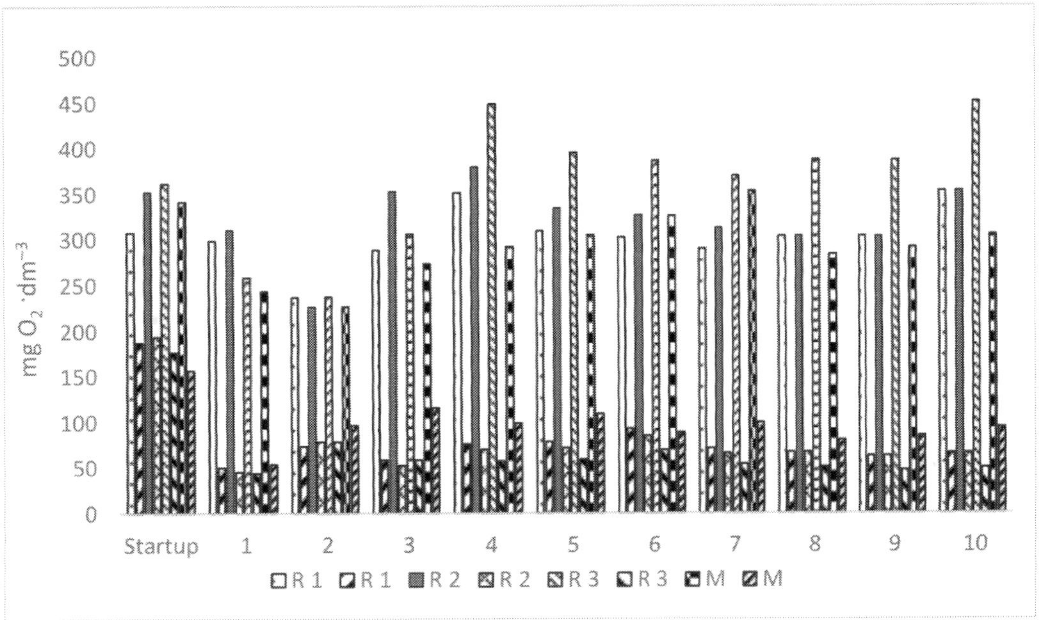

Fig. 3. Reduction of BOD_5 after 24h wastewater treatment process [mg $O_2 \cdot dm^{-3}$].

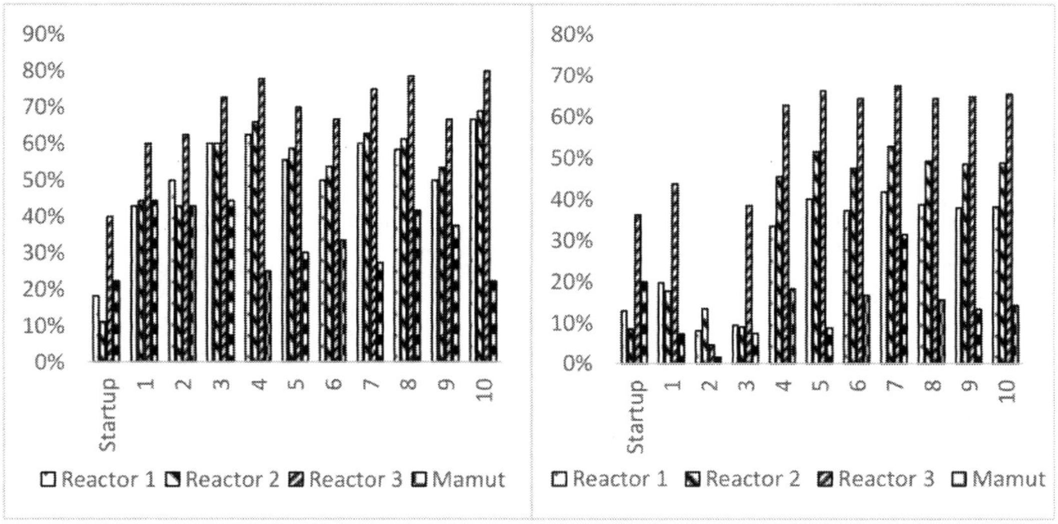

A. B.

Fig. 4. Reduction of nutrients after 24h MBBR operation: A. PO_4^{3-} and B. NH_4^+[in %].

The changes in nutrients concentration PO_4^{3-}, NH_4^+ and NO_3^- (Fig. 4 and 5) were statistically significant due to the initial level with exception of changes in NO_2^- ions levels. The concentration of this ions was highly variable depending on whether aerobic or anaerobic conditions were provided. Due to the accepted sampling procedure during the aeration phase, the NO_2^- concentrations in the experimental samples were characterized by significant non-specific changes.

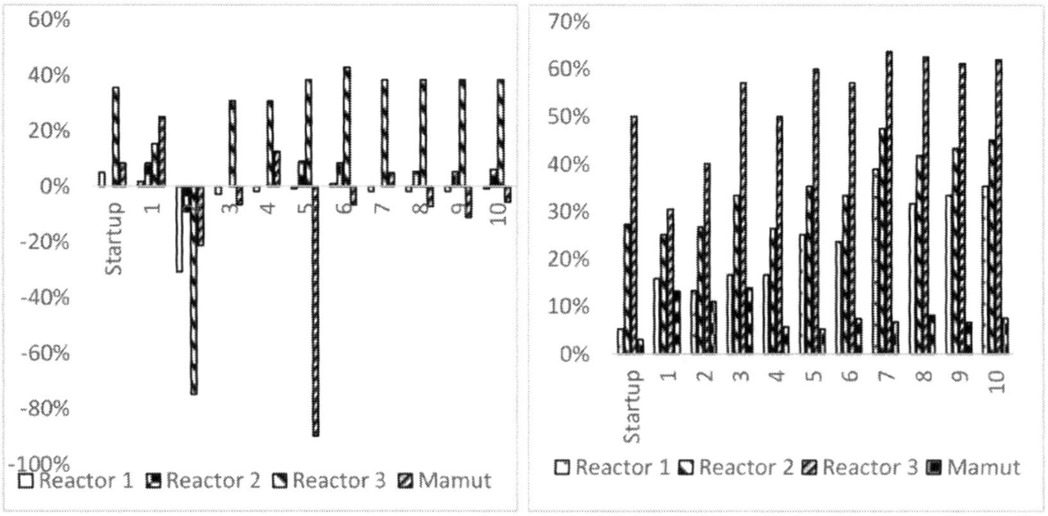

A. B..

Fig. 5. Reduction of nutrientsafter 24h MBBR operation: A. NO_2^- and B. NO_3^- [in %].

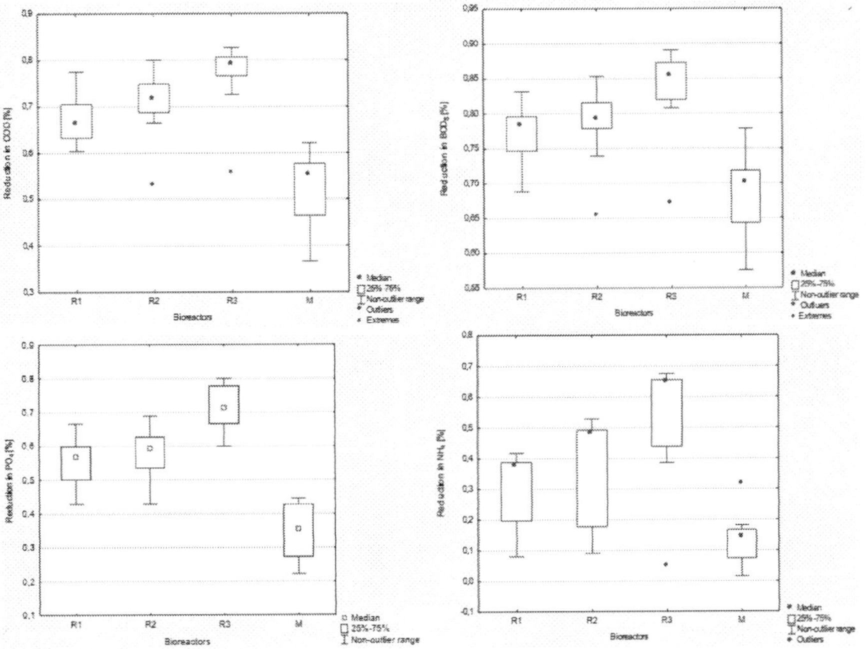

Fig 6. Statistical analysis of theeffectiveness of reduction of selected pollutants in investigated reactors.

4 Discussion

The experiment was conducted during the summer months since 1st of July till 10th of September, thedaily temperature of treated wastewater were ranged in 20 – 28° C and were conducive to biochemical processes. The concentration of oxygen in raw sewage in the bed dropped to level 0,5 mg $\cdot dm^{-3}$ and during treatment process were raised to level about 8 mg $\cdot$ dm^{-3} into classical MBBR (with fine bubble aeration) inversely proportional to the concentration of COD and to level about 6 mg $\cdot dm^{-3}$ into Mammoth bioreactor. The reduction of COD values in MBBR reactors ranged from 52 to 82%, and significantly worse results obtained from the reactor type Mammoth 36 - 62%, (figure 2) which was reflected in changes in the dynamics of the oxygen profiles for both types of reactors during theprocess. The changes in thelevel of easily biodegradable organic compounds expressed as BOD_5 presented much better in MBBR reactors in the range 65 – 89% and for theMammoth reactor in therange 57 – 78% of total reduction (Fig. 3). Statistical analysis confirmed the statistically significant differences between results for the MBBR and Mammoth reactors, with exception of changes in theconcentration of NO_2^- ions. Many authors have confirmed the effectiveness of the reduction of organic pollutants in MBBR reactors [14, 17, 18]there are no articles devoted to acomparison of the type of aeration in the wastewater treatment efficiency for MBBR reactors, which differ in the types of aerators. The nutrients level analyses also showed statistically significant differences in changesof their concentrations PO_4^{3-}(11 – 80%) (Fig. 4a, 6) and NH_4^+ (4 – 67%) (Fig. 4a, 6) in relation to baseline levels.The results of NO^{3-} and NO^{2-} ions concentrations due to reactor operation mode were characterized by significant variability depending on the kinetics of process and the nitrification stage (Fig. 5) [19, 20, 21].The comparison of the wastewater treatment efficiency between two types of MBBR reactors (Fig. 1) also showed significantly better results for reactors equipped with disk diffusor.The final results clearly show that disk diffusers are better in the wastewater treatment process than air-lift reactor in Mammoth mode.

Based on the obtained results and literature reports [19] can be demonstrated that tested MBBR reactors are the appropriate solution for the reclamation of water reservoirs heavily polluted by organic compounds and with high nutrients concentration. The similar solutions are used in wastewater treatment of highly productive fish farms[22]. The conception project of floating reclamation platforms with MBBR reactors are still ongoing, with theapplication of new RES technologies, which allow them for the independent operation on the lakes. Despite the wastewater treatment process in laboratory bioreactors were conducted in quasi-flow mode, nevertheless, the obtained efficiencies allow us to conclude that the tested technologies are promising for the application in the reclamation of small water reservoirs [23]. Hydraulic and biotechnological studies allowed better estimation of the number of reactors necessary in the restoration of polluted water reservoirs, with respect to its volume. The proposed method of reclamation with MBBR reactors can be compared to the in-situ WWTP (wastewater treatment plant) technology. The high stability of the wastewater treatment process makes these types of reactors particularly valuable in temperate climate zones [24].

5 Conclusions

Tested laboratory models of MBBR reactors characterized by high efficiency of wastewater treatment, which proves their usefulness for the purposes of the degraded water reservoirs restoration.

Further studies are needed to determine their effectiveness of sewage treatment with low loads of organic pollutants.

Investigated bioreactors were designed on the semi-technical scale, which allows for extrapolation of laboratory results to field conditions, with no significant statistical error.

MBBR belongs to unique technologies, which can be easily adapted to the treatment processes of small water reservoirs.

References

1. EPA. Assessment, E. R. Guidelines for Ecological Risk Assessment.(1998)
2. G. W. Suter II, Ecological risk assessment. CRC press.(2016).
3. A.Hadipour, T.Rajaee, V.Hadipour&S.Seidirad, Multi-criteria decision-making model for wastewater reuse application: a case study from Iran. Desalination and Water Treatment, 57(30), 13857-13864.(2016).
4. E.Wąsik, K.Chmielowski&A. Operacz, PCA jako narzędzie eksploracji danych charakteryzujących pracę komór nitryfikacji oczyszczalni ścieków w Trepczy. ActaScientiarumPolonorumFormatioCircumiectus, **16**(1), 209–222. (2017).
5. E. Wąsik, P. Bugajski, K. Chmielowski, A. Nowak &R. Mazur. Crystallization of struvite and hydroxyapatite during removal of biogenic compounds on the filter bed. PrzemysłChemiczny, 96/8, 1739–1743.(2017).
6. P. Kaszycki, P. Petryszak&H. Kołaczek. "Biological treatment of wastewaters generated by furniture industry part 2. Construction of a specialized activated sludge and optimization of bioprocess parameters in semi-technical scale tests." Ecological Chemistry and Engineering. A 15.11 1257-1271.(2008)
7. G. T.Ooi, M. E.Casas, H. R.Andersen&K. Bester. Transformation products of clindamycin in moving bed biofilm reactor (MBBR). Water Research, 113, 139-148.(2017).

8. E.Wąsik&K. Chmielowski, Ammonia and indicator bacteria removal from domestic sewage in a vertical flow filter filled with plastic material. Ecological Engineering, Volume 106, Part A, Pages 378–384. (2017).
9. P. Bugajski, K. Chmielowski&G. Kaczor,Optimizing the Percentage of Sewage from Septic Tanks for Stable Operation of a Wastewater Treatment Plant. Polish Journal of Environmental Studies, 25(4), 1421–1425. (2016).
10. R. Mazur, D. Bedla, K. Chmielowski, A. Nowak&J. Mazurkiewicz, Wpływ warunków tlenowych na skuteczność oczyszczania ścieków bytowych w technologii zatapialnych filtrów włókninowych. Przemysł Chemiczny, 96(8), 1513–1517. (2016).
11. K. Chmielowski&R. Ślizowski, Effect of grain-size distribution of sand on the filtrate quality in vertical-flow filters. PrzemysłChemiczny, 87(5), 432-434. (2008).
12. P.Bugajski, K.Chmielowski&G. Kaczor,Wpływ wielkości dopływu wód opadowych na skład ścieków surowych w małym systemie kanalizacyjnym. Acta ScientiarumPolonorumFormatioCircumiectus, 15(2), 3–11.(2016).
13. T. Leiknes&H. Ødegaard,The development of a biofilm membrane bioreactor. Desalination, 202(1), 135-143.(2007).
14. H.Ødegaard, B.Rusten&F. Wessman, State of the art in Europe of the moving bed biofilm reactor (MBBR) process. In Proceedings of the 77th Annual Water Environment Federation Technical Exposition and Conference, New Orleans, Louisiana, (2-6).(2004).
15. K.Chmielowski&R. Ślizowski, Defining the optimal range of a filter bed's d(10) replacement diameter in vertical flow sand filters. Environment Protection Engineering, 34(3), 35-42.(2008).
16. R.Kollmann, G. Neugebauer, F.Kretschmer, B.Truger, H.Kindermann, G.Stoeglehner&M. Narodoslawsky, Renewable energy from wastewater-Practical aspects of integrating a wastewater treatment plant into local energy supply concepts. Journal of Cleaner Production, 155, 119-129.(2017).
17. G. Andreottola, P. Foladori, M. Ragazzi&F. Tatano,Experimental comparison between MBBR and activated sludge system for the treatment of municipal wastewater. Water Science & Technology, 41(4), 375-382.(2000).
18. K. Miksch&J. Sikora, Biotechnologia ścieków" Wydawnictwo Naukowe PWN(2010).
19. R.Delatolla, B.Young &A. Stintzi, Application of the MBBR Technology to Achieve Nitrification Below 1° C: Biofilm and Microbiome Analysis. In Frontiers International Conference on Wastewater Treatment and Modelling (pp. 513-517). Springer, Cham.(2017).
20. T.Abzazou, R., M.Araujo, M. Auset&H. Salvadó, Tracking and quantification of nitrifying bacteria in biofilm and mixed liquor of a partial nitrification MBBR pilot plant using fluorescence in situ hybridization. Science of the Total Environment, 541, 1115-1123.(2016).
21. B.Young, R.Delatolla, K.Kennedy, E.Laflamme&A. Stintzi, Low temperature MBBR nitrification: Microbiome analysis. Water Research.(2017).
22. A.Kamstra, E. Blom&B. F. Terjesen, Mixing and scale affect moving bed biofilm reactor (MBBR) performance. Aquacultural Engineering.(2017).
23. D. Młyński, K. Chmielowski & A. Młyńska. The analysis of the quantity variability of the sewage inflowing to the selected wastewater treatment plants of Sanok district ActaSci.Pol. Form. Cir. 16.1: 77-89. (2017)

24. B.Rusten, B.Eikebrokk, Y.Ulgenes&E. Lygren, Design and operations of the Kaldnes moving bed biofilm reactors. Aquaculturalengineering, 34(3), 322-331.(2006).

https://doi.org/10.1051/e3sconf/20183002005

Sewage Management Changes in the North-eastern Poland After Accession to the European Union

Szymon Skarżyński[1,*], and *Izabela* Bartkowska[2]

[1,2]Bialystok University of Technology, Faculty of Civil and Environmental Engineering, Department of Technology and Systems in Environmental Engineering, Wiejska 45A, 15-351 Bialystok, Poland

Abstract. Poland's accession to the European Union contributed to the infrastructure development of the whole country. One of the elements of the modernized infrastructure is the sewage network and facilities on this network, as well as facilities for waste water treatment and disposal of sludge. A wide stream of funds flowing to the country, and consequently also to the north-eastern polish voivodeships (Podlaskie, Warmian-Masurian, Lublin), allowed modernization, organize, and sometimes to build a new sewage management of this part of the country. The main factors and parameters that allow us to evaluate the development of the sewage management in north-eastern Poland are included: percentage of population using sewage treatment plants, number of municipal sewage plants with the division of their type, number of industrial plants, number of septic tanks, amount of sewage purified in a year, amount of sludge produced in the year, design capacity of sewage treatment plant, size of plant in population equivalent (PE). From a number of investments in the field of wastewater management carried out in the discussed area in the period after Poland's accession to the European Union, 9 investments were considered the most important, 3 from each of the voivodeships.

1 Introduction

Poland's accession to the European Union contributed to the infrastructure development of the whole country. One of the elements of the modernized infrastructure is the sewage network and facilities on this network, as well as facilities for sewage treatment and disposal of sludge. A wide stream of funds flowing to the country, and consequently also to the north-eastern Polish voivodeships, allowed modernization, organize, sometimes to build a new sewage management of this part of the country.

2 Characteristics of north-eastern Poland Voivodeships

The definition of north-eastern Poland defines three voivodeships: the Podlaskie Voivodeship (with the seat of the authorities in Białystok), the Warmian-Masurian

* Corresponding author: s.skarzynski@doktoranci.pb.edu.pl

Voivodeship (with the seat of the authorities in Olsztyn) and the Lublin Voivodeship (with the seat of the authorities in Lublin). As of 31 December 2016, the voivodeships covered by this area are inhabited by more than 4.76 million people on an area of nearly 70.000 km^2, resulting in a population density of 67.8 persons/km^2 [1]. Figure 1 present the location of the voivodeships on the Polish map and Table 1 presents the characteristics of each voivodeship.

Figure 1. Location of the voivodeships on the Polish map.

Table 1. Data characterizing chosen voivodeships and their comparison to national data [1].

Voivodeship	Population		Area		Population density
	person	% of total country	km^2	% of total country	person/km^2
Podlaskie	1,187,587	3.09	20,187.02	6.46	58.8
Warmian-Masurian	1,437,812	3.74	24,173.17	7.73	59.5
Lublin	2,135,715	5.56	25,122.46	8.03	85.0
Total	4,761,114	12.39	69,482.65	22.22	67.8
Poland	38,432,992	100.00	312,679.00	100.00	123.0

Analysing the above data, it can be noticed that the largest population, surface area and population density are Lublin Voivodeships, while the lowest values reach Podlaskie Voivodeship. The analysed voivodeships cover a total of nearly 12.5% of the country territory, and is inhabited by just over 22% of the population of the whole country. The population density in these voivodeships reaches 67.8 person/km^2 (123.0 person/km^2 in the whole country).

2.1 Podlaskie Voivodeship

The Podlaskie Voivodeship is located in the north-eastern part of Poland, in the geographical center of Europe. The seat of the province authorities is Białystok. Through its center flows the river Narew. The voivodeship occupy the area of 20,187.02 km^2 and is inhabited by 1,187,587 people, giving a population density of 58.8 persons/km^2. The Podlaskie Voivodeship is divided into 14 counties (consisting of 118 communes) and 3 cities with county status. The largest cities are Białystok (about 300 thousand inhabitants), Suwałki (about 70 thousand inhabitants) and Łomża (about 63 thousand inhabitants) [1].

The voivodeship is bordered by: Belarus, Lithuania, Lublin Voivodeships, Masovian Voivodeships and Warmian-Masurian Voivodeship. In the voivodeships there are 4 national parks, including the oldest in Poland, Białowieża National Park, as well as the largest area, the Biebrza National Park. The other two national parks are Wigry and Narew. Protection of nature and the environment in Podlaskie Voivodeship also includes 3 landscape parks, 85 nature reserves and 2051 nature monuments [1].

2.2 Warmian-Masurian Voivodeship

The Warmian-Masurian Voivodeship is located in the north-eastern part of Poland. The seat of the voivodeship authorities is Olsztyn. The voivodeship occupy the area of 24,173.17 km^2 and is inhabited by 1,437,812 people, giving a population density of 59.5 persons/km^2. Warmian-Masurian Voivodeship is divided into 19 counties (consisting of 116 communes) and 2 cities with county status. The largest cities are Olsztyn (about 173 thousand inhabitants), Elbląg (about 121 thousand inhabitants) and Ełk (about 60 thousand inhabitants) [1].

The voivodeship is bordered by: Russia, Kuyavian-Pomeranian Voivodeship, Masovian Voivodeship, Podlaskie Voivodeship, Pomeranian Voivodeship. From the north-west it borders the Vistula Lagoon. Protected areas in voivodeship occupy a total of 1,126,155.3 ha, which accounts for 46.6% of the voivodeship. In the region there are 8 landscape parks, 110 nature reserves, 71 protected landscape areas, 2,576 natural monuments, 14 special bird protection areas, 41 areas of Community importance, 297 ecological grounds, 18 natural-landscape teams and one documentation stand [1].

2.3 Lublin Voivodeship

Lublin Voivodeship is located in the northern part of Poland. The seat of the voivodeship authorities is Lublin. The voivodeship occupy the area of 25,122.46 km^2 and is inhabited by 2,133,340 people, giving a population density of 85.0 persons/km^2. Lublin Voivodeship is divided into 20 counties (consisting of 213 communes) and 4 cities with county status. The largest cities are Lublin (about 340 thousand inhabitants), Zamość (about 65 thousand inhabitants), Chełm (about 64 thousand inhabitants) and Biała Podlaska (about 57 thousand inhabitants) [1].

The voivodeship is located between the Vistula and the Bug River and borders with the Belarus, Ukraine, Subcarpathian Voivodeship, Podlaskie Voivodeship, Masovian Voivodeship and Świętokrzyskie Voivodeship. Lublin Voivodeship there are two national parks (Polesie National Park, Roztocze National Park), 17 landscape parks, 87 nature reserves, 1,513 nature monuments, 7 documentation stands, 267 ecological grounds and 7 natural-landscape teams [1].

3 Parameters of sewage management in the north-eastern Voivodeships of Poland

The main factors and parameters that allow us to evaluate the development of the sewage management in north-eastern Poland are presented in table 2. All data was obtained from the Local Data Bank provided by the Central Statistical Office.

Table 2. Parameters of sewage management in north-eastern Poland [1]

Parameter	Voivodeship								
	Podlaskie			Warmian-Masurian			Lublin		
	2003	2009	2015	2003	2009	2015	2003	2009	2015
population using sewage treatment plants [%]	60.20	62.49	67.40	67.53	72.72	75.90	49.93	52.84	57.20
total municipal sewage plants [units]	96	113	122	173	222	237	217	260	286
*mechanical	0	0	0	6	0	0	15	12	3
*biological	69	80	93	112	158	172	182	216	255
*with increased biogenic removal	27	33	29	55	64	65	20	32	28
industrial plants [units]	30	30	26	51	21	18	106	68	66
Onsite sewage facilities (OSSF) [units]	**nd**	3.90	13.79	**nd**	1.14	5.34	**nd**	7.68	22.33
septic tanks [units]	**nd**	80.20	74.33	**nd**	53.35	48.50	**nd**	184.30	171.25
sewage purified in a year [dam^3]	32.38	34.07	32.02	49.62	46.51	44.80	50.66	50.26	49.14
sludge produced [t/year]	13.31	17.41	14.45	24.58	27.87	18.74	20.94	23.00	19.73
total designed capacity of sewage treatment plant [m^3/d]	234.59	234.71	237.96	381.49	321.34	324.9	355.67	347.87	350.40
*mechanical	0	0	0	172	0	0	688	559	81
*biological	55.89	38.26	42.16	127.51	65.39	64.85	235.62	75.28	75.80
*with increased biogenic removal	178.70	196.45	195.80	253.81	255.95	260.05	119.37	272.04	274.52
size of municipal sewage plants [thousand PE]	1.23	1.26	1.36	2.19	1.94	2.03	1.90	2.20	2.32

nd – no data

The above table shows the three years data for the voivodeships. The year 2003 shows the situation before Poland's accession to the European Union, 2009 after several years of presence in the European Union and in 2015, the most up-to-date. Analysing this data can be seen more of the following regularities:

• percentage of the population with access to sewage treatment plants in the voivodeships north-eastern Polish has increased by about 7 percent compared with 2003, Warmian-Masurian Voivodeships are in 4th place, Podlaskie Voivodeship at 11th place and Lublin Voivodeship at 14th place in Poland [12],

• during the presence of Poland in the European Union the number of mechanical sewage treatment plants was reduced (and in some cases they were completely liquidated), however were noted more biological sewage treatment plants and sewage treatment plants with increased biogenic removal,

• in the analysed years, the number of industrial sewage treatment plants and septic tanks slightly decreased, while the number of industrial sewage treatment plants was greatly increased,

• the amount of sewage treated during the year remained relatively constant, but can be observed an increase (of 2009) and a slight decrease compared to 2003,

• the amount of sludge produced shows similar trends in the amount of sewage,

• the designed capacity of the sewage treatment plant has been slightly increasing in the Podlaskie Voivodeship, while in the remaining voivodeships, it has decreased in comparison with the base year 2003,

• noteworthy is the fact that the increased amount of sewage treated in sewage treatment plants with increased removal of biogenic.

4 Characteristics of the most important investments in sludge management

From a number of investments in the field of wastewater management carried out in the discussed area in the period after Poland's accession to the European Union, 9 investments were considered the most important, 3 from each of the voivodeships. These investments have the highest social and environmental importance and their cost to be significant [2].

4.1 Podlaskie Voivodeship

The most important investment in the field of sewage management in Podlaskie Voivodeship was the development and modernization of the water and sewage system in Białystok and the community of Wasilków. The aim of the project was to ensure the compliance of the Bialystok agglomeration requirements contained in the relevant EU Directives and Polish law regarding the conduct of water and sewage management. The scope of the project was [3]:

• construction of water supply and sanitary sewage systems in areas not covered by the sanitary sewage system,

• separation of sewerage channels through the construction of rain channels and the installation of equipment for the purification of rainwater from petroleum substances and suspensions,

• modernization and renovation of the existing sanitary sewage system to improve its functioning,

• modernization and renovation of the existing water supply network and the replacement of the network made from asbestos and steel pipes in order to improve the reliability of water supply and improve the quality of water intended for consumption,
• improve water distribution by supplementing the metering of water supply network.

This investment was co-financed by the Cohesion Fund under the Infrastructure and Environment Operational Program with the amount of almost 97 million PLN, with a total value of almost 144 million PLN, which constitutes approximately 67% of the co-financing. The project was realized in the years 2008 – 2013 [3, 4].

The second important investment in the voivodeship was "Improvement of water quality in Białystok". This four-step project assumed [3]:
• development of technological processes at the Water Treatment Station "Wasilków - Pietrasze",
• construction of a preliminary ozonation facility for the underground water treatment in the Water Treatment Station "Jurowce",
• modernization of fermenting chambers in Białystok sewage treatment plant,
• modernization of the biological part in the sewage treatment plant in Białystok.

This investment was co-financed by the Cohesion Fund under the Infrastructure and Environment Operational Program with an amount of over 41.6 million PLN, with a total project value of over 76.4 million PLN, which constitutes approximately 56% of the co-financing. The project was realized in the years 2004 – 2006 [3, 4].

The last investment to look out for is the modernization of the sewage treatment plant and the development of water and sewage infrastructure in Suwałki. This project has been divided into two phases. The general objectives of the project was the comprehensive modernization of the sewage treatment plant and extension of water supply and sewerage in Suwałki to improve infrastructure conditions necessary to ensure the proper socio-economic development of the area covered investment. Both stages of the described investment were co-financed from the Cohesion Fund under the Infrastructure and Environment Operational Program with an amount of over 34.6 million PLN, with a total project value of over 60.5 million PLN, which constitutes approximately 57% of the co-financing. The project was realized in the years 2007 – 2013 [4, 5].

4.2 Warmian-Masurian Voivodeship

One of the most important investments realized in the Warmian-Masurian Voivodeship was a project called "Water and sewage management in Olsztyn". The aim of the project was to introduce a modern three-stage sewage pollutant removal method, which contributed to the protection of Łyna River waters of the, and finally the Baltic Sea or the achievement of European standards for municipal sewage treatment. In addition, the aim was to improve the quality of drinking water and supply all residents with drinking water, improving the efficiency of the sewage system and rainwater, reduce flooding with rainwater and reduce surface water pollution resulting from discharges of rainwater, as well as the long-term solution to the problem of final disposal of sludge. The investment was co-financed by the Cohesion Fund under the Infrastructure and Environment Operational Program with the amount of almost 92.8 million PLN, with a total value of nearly 296.9 million PLN, which constitutes over 31% of the co-financing. The project was realized in the years 2004 – 2006 [4, 6].

Another important investment in the sewage management of the Warmian-Masurian Voivodeship was a project called "Improvement of water and sewage management in the agglomeration of Ełk". The scope of the project included comprehensive rebuilding of sewage pumping stations, sanitary sewers, sewage wells, and water supply with replacement of equipment. Reconstruction of canals and water pipeline was implemented by trenchless methods. The investment was co-financed by the Cohesion Fund under the Infrastructure and Environment Operational Program with the amount of almost 33.2 million PLN, with the total value of the project equalling nearly 72.7 million PLN, which constitutes over 46% of the co-financing. The project was realized in the years 2007 – 2013 [4, 7].

The last investment to look out for is "Construction of industrial sewage treatment plant for Tymbark Sp. z o.o. Sp. k. in Olsztynek". The purpose of the project was to expand and modernize the sewage treatment plant installation to fruit and vegetable processing. Industrial sewage is purified in an anaerobic-aerobic process, using the resulting biogas combustion in the local boiler room. As part of the investment were constructed facilities and technological equipment: the mechanical and biological part of the sewage treatment plant, as well as sewage sludge processing elements. The investment was co-financed by the European Regional Development Fund under the Infrastructure and Environment Operational Program with an amount of almost 10.3 million PLN, with a total project value of almost 43.0 million PLN, which constitutes over 24% of the co-financing. The project was realized in the years 2007 – 2013 [4, 8].

4.3 Lublin Voivodeship

A significant investment in the Lublin Voivodeship was "Development and modernization of water supply and sewage disposal in Lublin". The main objective of the project was to organize the water and sewage systems in the city of Lublin. For the purposes of side include [9]:

- maintaining the quality and reliability of drinking water supply in Lublin, including the reduction of water losses in the network,
- elimination of ground sources pollution and surface waters in the form of individual septic tanks,
- reducing the negative impact of the sewerage system on groundwater,
- ensuring the reliability of the operation of the "Hajda" Sewage Treatment Plant,
- improving the comfort of the inhabitants,
- improving the efficiency of the water system in Lublin,
- improving the economic development of Lublin and its surroundings,
- development of the sewerage system in the currently unarmed area.

This investment was co-financed by the European Cohesion Fund under the Infrastructure and Environment Operational Program with the amount of nearly 147.4 million PLN, with a total project value of over 311.8 million PLN, which constitutes over 47% of the co-financing. The project was realized in the years 2007 – 2013 [4, 9].

Another project of the Lublin Voivodeship was called "Inter-municipal water and sewage system within the agglomeration of Puławy". The main aim of the project was to adjust the Puławy agglomeration area to the current national and European standards by including [10]:

- increase channelling agglomeration Puławy from 88.3% to 95.2%,
- providing access to water of adequate quality and quantity for the inhabitants of the agglomeration of Puławy,
- allowing the connection to the sanitary sewage system of inhabitants from the agglomeration,
- elimination of local sources of pollution of soil, groundwater and surface by liquidations of individual septic tanks,
- reducing the negative impact of the sanitary sewage system on the environment by modernizing emergency and exploited sections of the sewer system in Puławy,

• increasing the economic development potential of Puławy, Końskowola and Żyrzyn communes,
• increasing the aesthetic appeal of the countryside,
• to create a single, common system of sewage management,
• connection to sanitary sewage system over 4 thousand people,
• connection to the water supply network over 2 thousand people.

This investment was co-financed by the European Cohesion Fund under the Infrastructure and Environment Operational Program with an amount of nearly 44.2 million PLN, with a total project value of over 89.7 million PLN, which constitutes almost 50% of the co-financing. The project was realized in the years 2007 – 2013 [4, 10].

The last investment of the Lublin Voivodeship was a project called "Modernization of the Chełm Wastewater Treatment Plant with the expansion of the water and sewerage system". The aim of the project was to equip the Chełm agglomeration with technical infrastructure enabling [11]:
• reception and cleaning of municipal sewage in the Chełm agglomerations in accordance with the requirements of Directive 91/271/EEC and the Ordinance of the Minister of the Environment of 24 July 2006,
• contribute to fulfilling the requirements of the National Program of Municipal Sewage Treatment,
• ensuring continuity of drinking water supply, corresponding to the standards laid down in Directive 80/778/EEC of 15 July 1980 and 98/83/EC of 3 November 1988 and the Ordinance of the Minister of Health of 29 March 2007.

This investment was co-financed by the European Cohesion Fund under the Infrastructure and Environment Operational Program with the amount of almost 40.5 million PLN, with the total value of the project equal to almost 84.5 million PLN, which constitutes almost 48% of the co-financing. The project was realized in the years 2007 – 2013 [4, 11].

5 Summary

Poland's accession to the European Union has significantly contributed to the modernization and expansion of the North-East Poland sewage infrastructure. Most of the funds came from the European Cohesion Fund under the Infrastructure and Environment Operational Program, contributing to improving the state of environment in Poland. In conclusion, it can be seen that:
• total amount of investments described in this article was 1,179.5 million PLN (280.9 million PLN in Podlaskie Voivodeship, 412.6 million PLN in Warmian-Masurian Voivodeship and 486.0 million PLN in Lublin Voivodeship),
• total amount of co-financing from European Union programs was 541.6 million PLN (173.2 million PLN in Podlaskie Voivodeship, 136.3 million PLN in Warmian-Masurian Voivodeship and 232.1 million PLN in Lublin Voivodeship),
• value of investment co-financing was on average 47.3% (60.0% in Podlaskie Voivodeship, 33.7% in Warmian-Masurian Voivodeship and 48.3% in Lublin Voivodeship).
• amount spent on investment per capita in the region was 248 PLN/person of which 114 PLN/person came from co-financing,
• amount spent on investment per capita in Podlaskie Voivodeship was 237 PLN/person of which 145 PLN/person coming from co-financing,
• amount spent on investment per capita in Warmian-Masurian Voivodeship was 287 PLN/person of which 95 PLN/person coming from co-financing,

• amount spent on investment per capita in Lublin Voivodeship was 228 PLN/person of which 109 PLN/person coming from co-financing.

The study was carried out within MB/WBiIŚ/10/2017 work and financed by the Ministry of Science and Higher Education.

References

1. Central Statistical Office Department of Demography and Labor Market Research (2017) Population and natural movement in 2016 in Poland, Warsaw.
2. EU subsidy map (www.mapadotacji.gov.pl) [access 27.03.2017].
3. Wodociągi Białostockie Sp. z o.o. (www.wobi.pl) [access 11.06.2017].
4. Infrastructure and Environment Operational Program (www.pois.gov.pl) [access 27.03.2017].
5. Przedsiębiorstwo Wodociągów i Kanalizacji w Suwałkach Sp. z o.o. (www.pwik.suwalki.pl) [access 11.06.2017].
6. Przedsiębiorstwo Wodociągów i Kanalizacji w Olsztynie Sp. z o.o. (www.pwik.olsztyn.pl) [access 11.06.2017].
7. Przedsiębiorstwo Wodociągów i Kanalizacji w Ełku Sp. z o.o. (www.pwik.elk.com.pl) [access 11.06.2017].
8. Tymbark - MWS Sp. z o.o. Sp.k. (www.tymbark.com) [access 11.06.2017].
9. Miejskie Przedsiębiorstwo Wodociągów i Kanalizacji w Lublinie Sp. z o.o. (www.mpwik.lublin.pl) [access 11.06.2017].
10. Miejskie Przedsiębiorstwo Wodociągów i Kanalizacji "Wodociągi Puławskie" Spółka z o.o. (www.mpwik.pulawy.pl) [access 11.06.2017].
11. Miejskie Przedsiębiorstwo Gospodarki Komunalnej w Chełmie Sp. z o.o. (www.mpgk.chelm.pl) [access 11.06.2017].
12. Strategia Rozwoju Województwa Podlaskiego do roku 2020, Urząd Marszałkowski Województwa Podlaskiego

Use of a water treatment sludge in a sewage sludge dewatering process

Justyna Górka[1,*], *Małgorzata* Cimochowicz-Rybicka[1] , and *Małgorzata* Kryłów[1]

[1] University of Technology, Department of Environmental Engineering, 24 Warszawska, Cracow 31-155, Poland

Abstract. The objective of the research study was to determine whether a sewage sludge conditioning had any impact on sludge dewaterability. As a conditioning agent a water treatment sludge was used, which was mixed with a sewage sludge before a digestion process. The capillary suction time (CST) and the specific filtration resistance (SRF) were the measures used to determine the effects of a water sludge addition on a dewatering process. Based on the CST curves the water sludge dose of 0.3 g total volatile solids (TVS) per 1.0 g TVS of a sewage sludge was selected. Once the water treatment sludge dose was accepted, disintegration of the water treatment sludge was performed and its dewaterability was determined. The studies have shown that sludge dewaterability was much better after its conditioning with a water sludge as well as after disintegration and conditioning, if comparing to sludge with no conditioning. Nevertheless, these findings are of preliminary nature and future studies will be needed to investigate this topic.

1 Introduction

Dewatering of sludge is a very important stage of the sludge processing. It reduces both sludge mass and volume, which consequently reduces costs of a further sludge processing, facilitates its transport and allows for its thermal processing. Generally, a sewage sludge has a very high water content (92-99.5%), which contributes to high costs of its further processing. So even a slight decrease of a water content may result in a significant reduction of a sludge volume, facilitating its subsequent disposal [1]. It is also worth noting that a growing acceptance of a sludge incineration, as the final sewage sludge disposal, stimulates a search for methods that would intensify dewatering processes.

At sewage treatment plants, it is mostly mixed sludge that is dewatered after thickening and anaerobic or aerobic digestion. Processing of a sewage sludge is difficult because the sludge is rich in organic compounds that form gel-like webs and may slow down dewatering [2]. Therefore, different methods of sludge conditioning are applied during its processing and dewatering. Sludge conditioning modifies a structure and properties of the processed sludge and enables a more efficient removal of water captured within sludge flocks [3]. Nevertheless, the selection of optimal and most efficient measures used in this process remains still an open question [4]. Sludge conditioning can be carried out with physical or chemical methods.
Physical methods of sludge conditioning include:

* Corresponding author: justynagrka@gmail.com

- High temperature,
- Freezing,
- Addition of conditioners that modify a sludge structure, such as: coal dust, cement dust and ash.

Chemical conditioning, on the other hand, involves supplementing sludge with some organic (polymerized cationic, anionic or neutral polyelectrolytes) or inorganic (iron and aluminum salts) compounds. In addition to the commonly used sludge conditioning techniques, also sludge from water treatment processes may be used as such supplementing agent that facilitates a dewatering process; it may contain aluminum or iron precipitates, depending on the water treatment methods used.

Currently, there is a number of studies underway focusing on this topic and investigating the issues on management and utilization of a water treatment sludge [5-7] . The authors have shown that conditioning of a sewage sludge with a water sludge improves its dewaterablity. The addition of sludge from water treatment plants, especially the ones employing coagulation with aluminum salts, improved dewaterablity of excess sludge. Moreover, sludge filtration properties improved along with a larger fraction of a water treatment sludge added to a sewage sludge [8]. Dewaterablity of digested sludge mixed with aluminium precipitates was also investigated and it was confirmed that they improved the sewage sludge dewaterblility [9].

The objective of the research was to determine whether sewage sludge conditioning had any impact on sludge dewaterability. As a conditioning agent a water treatment sludge was used, which was mixed with the digested sludge before another digestion. Such an interference in anaerobic digestion (modification of sludge before the process) affects the sludge final dewaterability. In the next step, the authors tried to assess how disintegration of a water sludge affects sewage sludge dewaterablity. The findings were of preliminary nature and future studies would be needed to thoroughly investigate this topic.

2 Material and methods

A sewage sludge (SS) from a municipal wastewater treatment plant with a capacity of 150 000 $m^3 \cdot d^{-1}$ was used for the study. The sludge was a mixture of primary and excess sludge. The water treatment sludge (WTS) came from a municipal water treatment plant with a capacity of 100 000 $m^3 \cdot d^{-1}$. The sludge was produced during coagulation of water with aluminum coagulant (PAX) and a sand filtration. In order to compare the samples for filtration properties tests were used:

- Sewage sludge from wastewater treatment plant mixed with water treatment sludge from water treatment plant,
- Sewage sludge from wastewater treatment plant mixed with water treatment sludge from water treatment plant – the mixture was digested.

The research included respirometric batch tests, which had been employed for energy research; enabled to analyze both quantity and quality of a fermentation gas. The digestion process was carried out under mesophilic conditions at 35 °C for 21 days in three separate runs. The sludge samples with of 1 liter volume had concentrations of 4.0-5.0 g TVS$\cdot dm^{-3}$. To ensure the appropriate process conditions, pH of the samples was adjusted to pH=7.0 with NaOH. The sample was purged with technical nitrogen for three minutes, before measurements were taken. The analysis during anaerobic batch tests also included: dry solids (TS), total volatile solids (TVS), chemical oxygen demand (COD), alkalinity, pH, ammonia nitrogen and total phosphorus, measured according to the current EU standards.

In the first stage of the study, the optimum dose of a WTS was selected; a selection was done while testing different proportions of a water treatment sludge to a sewage sludge 0.1-1.0 g TVS of a water treatment sludge per 1.0 g TVS of a sewage sludge. The criterion for selecting the optimum water treatment sludge dose included both an improvement in sludge dewaterability and a biogas production (higher efficiency of the digestion process – not included in the article).

In the next stage of the study, to further improve sludge dewaterability as well as the digestion efficiency, the water treatment sludge was disintegrated and mixed with sewage sludge. Disintegration was carried out at the most favorable disintegration parameters. According to the previous research (not included in the article), it was found that the optimum disintegration parameters were:

- Ultrasonic disintegration: intensity 24 $kW \cdot m^{-2}$, time 10 min,
- Thermal disintegration: temperature 55°C, time 60 min.

The tests series were run on the following materials:

- SS,
- SS+WTS in a ratio of 1:0.3,
- SS+WTS (WTS after thermal disintegration) in a ratio of 1:0.3,
- SS+WTS (WTS after ultrasonic disintegration) in a ratio of 1:0.3.

Sludge dewaterability was measured by a capillary suction time (CST) and a specific filtration resistance (SRF). The capillary suction time measurement was based on the Baskerville and Galle standard PN-EN 1407-1: 2007, which measures the passage of the head filtrate layer due to a suction force of the Whatman filter paper.

The specific filtration resistance was determined based on the PN-EN 14701-2 standard. This method follows the flow of a fluid through a porous medium in accordance with the Darcy's law. The specific resistance to filtration is determined by pouring a suitable volume of sludge into a filtering device, allowing the liquid to be filtered under constant vacuum or pressure (50 kPa), while recording the amount of filtrate over time. The specific resistance to filtration is calculated as follows:

$$SRF = \frac{2 \cdot \Delta p \cdot A^2 \cdot b}{\mu \cdot m} \; [m \cdot kg^{-1}] \qquad (1)$$

where:

Δp – pressure drop across the filter,

A – filtration area,

b – the slope of the linear part of the curve obtained by plotting t/V vs. V (Fig. 1),

μ – the viscosity of filtrate at the temperature of the sludge,

m – the mass of solids deposited on the filtering medium per unit volume of filtrate.

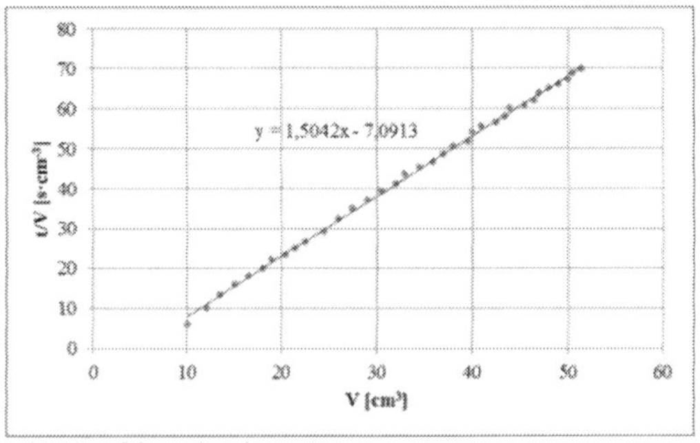

Fig. 1. Linear part of the curve obtained by plotting t/V vs. V.

Based on the f(V) = t/V graph (Fig. 1) the water content can be estimated in the individual phases of sludge filtering and the water volume in the sludge according to the method of attachment to a solid structure.

3 Results

The physic-chemical characteristics of the WTS and the SS prior to the anaerobic digestion are shown in Table 1. The water treatment sludge had lower values of such parameters as; TVS (by about 48%), COD (by 66%), phosphorus (by 70%), total nitrogen (by 73%) and ammonium nitrogen (by 88%) than the sewage sludge.

Table 1. The physic-chemical characteristics of the sludge.

Parameter		Water treatment sludge	Sewage sludge
pH	[-]	6.65	7.02
TS	[$gTS \cdot dm^{-3}$]	35.13	32.17
TVS	[$gTVS \cdot dm^{-3}$]	9.87	19.11
COD	[$mgO_2 \cdot dm^{-3}$]	8000	24000
sCOD	[$mgO_2 \cdot dm^{-3}$]	74	320
TP	[$mgP \cdot dm^{-3}$]	174	581
TN	[$mgN \cdot dm^{-3}$]	448	1646
N_{NH4}	[$mgN_{NH4} \cdot dm^{-3}$]	78	638

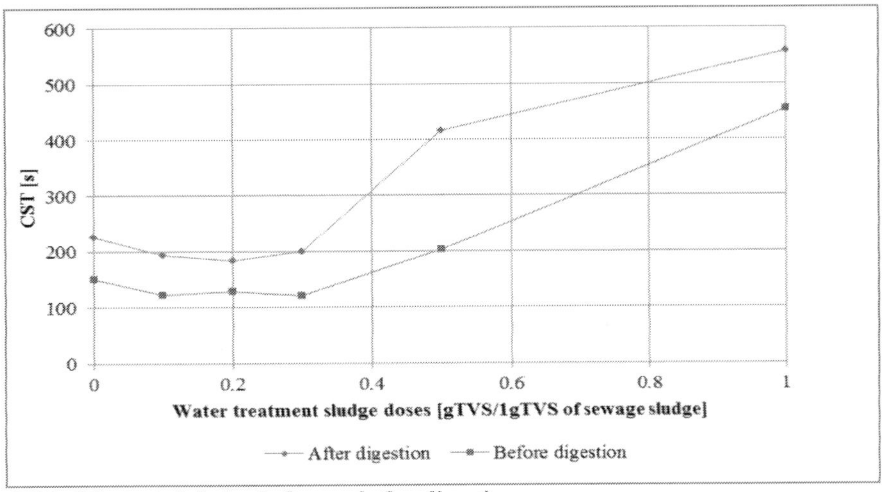

Fig. 2. CST of the tested sludge before and after digestion.

In order to determine the optimum ratio of a water tratment sludge to a sewage sludge, the changes of the CST were investigated for the tested sludge, before and after the digestion process. The tests were run for different doses of the water treatment sludge and their results are shown in Figure 2. It is worth noting that the CST values observed before the digestion process were lower than the ones after digestion. The average CST values for sludge that had not been conditioned with water treatment sludge and was not digested were 150s, and they were 66% lower than the ones after digestion. It has also been observed that above the dose of 0.3 g TVS of water treatment sludge/g TVS of sewage sludge the CST values increased, which indicates that the use of a water treatment sludge as a sewage sludge conditioner may adversely affect sludge dewaterability. The optimal ratio of water to sewage sludge proved to be 1:0.3 (1.0g TVS of sewage sludge to 0.3g TVS of water treatment sludge). The average capillary suction time for the sample (before digestion) was 120 s and it was 60% lower than the ones observed for digested samples. The optimal ratio of 1:0.3 was selected for the further studies.

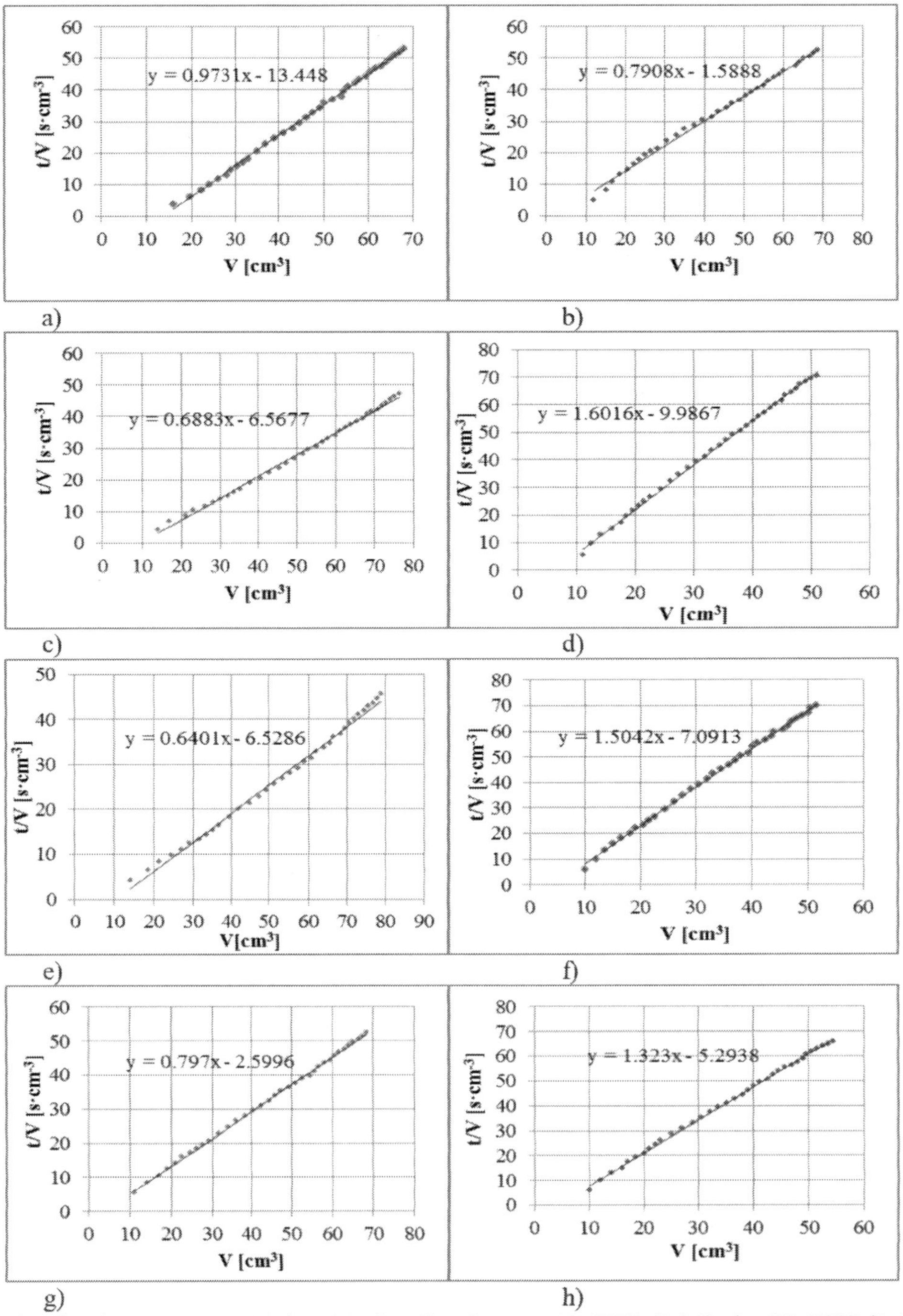

Fig. 3. Filtration curves: a – SS, b – SS after digestion, c – SS+WTS (1:0.3), d – SS+WTS (1 :0.3) after digestion, e – SS+WTS thermal (WTS after thermal disintegration) in a ratio of 1:0.3, f –

SS+WTS thermal (WTS after thermal disintegration) in a ratio of 1:0.3 after digestion, g – SS+WTS ultrasonic (WTS after ultrasonic disintegration) in a ratio of 1:0.3, h – SS+WTS ultrasonic (WTS after ultrasonic disintegration) in a ratio of 1:0.3 after digestion.

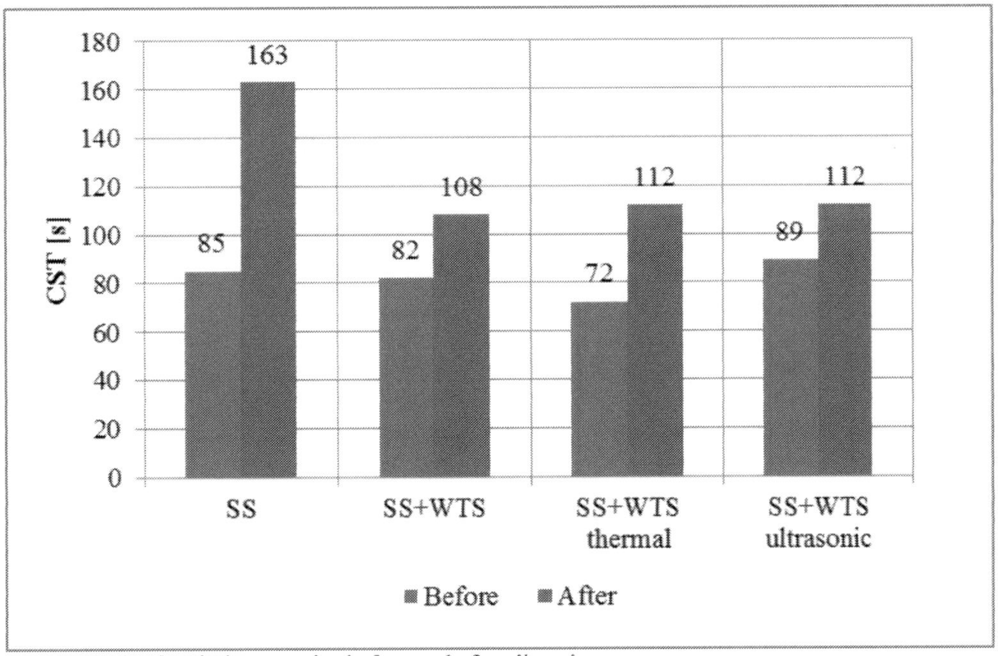

Fig. 4. CST for the sludge samples before and after digestion.

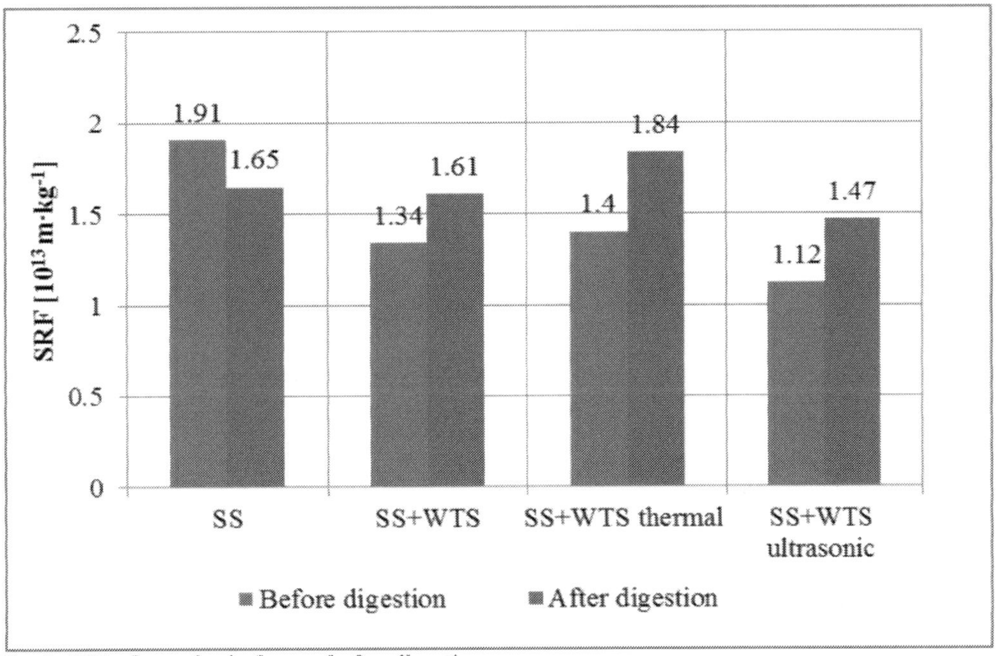

Fig. 5. SRF of samples before and after digestion.

Before sludge digestion, a slight difference in capillary suction times (a change from 72 to 89 s) was observed (Fig. 4). This indicates that both conditioning with a water treatment sludge and using disintegration methods hardly effect sewage sludge filtration properties. In turn, an anaerobic digestion resulted in a major increase of the CST in all samples - ranging from 108 to 163 s. The highest CST values were observed in the sewage sludge (163 s). The sewage samples conditioned with a water treatment sludge (regardless of disintegration) had the CST about 66% lower than the sewage sludge samples. A thermal disintegration, like an ultrasonic one, did not significantly affect a sludge dewatering ability.

Analizing the Figure 3, it can be concluded that with increasing dry solids in the sludge cake, the SRF increased. Dewatering of sewage sludge in a vacuum filter, during the SRF measurement, increased filtration resistance (Figure 5) for samples conditioned with water treatment sludge after anaerobic digestion: for sludge conditioned with a water treatment sludge by 17% and for sludge conditioned with a water treatment sludge after ultrasonic and thermal degradation by 24 %. In turn, SRF for only sewage sludge was lower by 15%.

4 Conclusions

Based on the research, the following conclusions were made:

1. Conditioning of a sewage sludge with sludge from water treatment processes improved a sewage sludge dewatering characteristic. The studies have shown that conditioning of sludge with sludge from water treatment processes after digestion resulted in higher sludge dewaterability i.e. lower CST values.
2. The changes of the CST values depended on the dose of a water treatment sludge; the optimum dose of a water treatment sludge was 0.3 g TVS per 1 g TVS of sewage sludge.
3. The study clearly stated that anaerobic digestion increased the CST and SRF values. However, for sludge conditioned with a water treatment sludge as well as for a disintegrated water sludge, the CST values were lower than for a sewage sludge.
4. Both thermal and ultrasonic disintegration of sludge from water treatment processes did not significantly affect sewage sludge dewaterability.

References

1. M. Christensen, K. Keiding, P.H. Nielsen, M.K. Jorgensen. *Wat. Res.*, **82** (2015).
2. S. Skinner, L. Studer, D. Dixon , P. Hillis, C. Rees, R. Wall, R. Cavalida, S. Usher, A. Stickland, P. Scales. *Wat. Res.*, **82** (2015).
3. P. Wolski, L. Wolny. *Rocz. Ochr. Środ.*, **13** (2011).
4. M. Dębowski, M. Zieliński, M. Krzemieniewski. *Ochr. Środ.*, **30** (2008).
5. T. Ahmad, K. Ahmad, M. Alam. *Journ. Clean. Prod.*, **124** (2016).
6. K. Dassanayake, G. Jayasinghe, A. Surapaneni, C. Hetherington. *Was. Man.*, **38** (2015).
7. M. Cimochowicz-Rybicka, J. Górka. *Przem. Chem.*, 96/**8** (2017).
8. J. Lai, J. Liu. *Wat. Sc. & Tech.*, **9** (2004).
9. Y. Yang, Y. Zhao, A. Babatunde, P. Kearney. *Wat. Envir.*, **13** (2007).

Effect of organic nitrogen concentration on the efficiency of trickling filters

Łukasz Kopeć[1,*], *Jakub* Drewnowski[2], *F. J.* Fernandez-Morales[3]

[1] Ekofinn-Pol Ltd, ul. Leśna 12, 80-297 Banino, POLAND
[2] Gdansk University of Technology, Faculty of Civil and Environmental Engineering, Narutowicza 11/12, 80-233 Gdansk, POLAND
[3] Departamento de Ingeniería Química. Instituto de Tecnologías Química y Medioambiental, Universidad de Castilla-La Mancha, Avda. Camilo José Cela s/n. 13004 Ciudad Real.

Abstract. The study was conducted in Poland at six selected wastewater treatment plants (WWTP) based on the trickling filters Bioclere® technology. The aim of the study was to find the relationship between the influent organic nitrogen concentration and the purification efficiency expressed as effluent COD concentration. In the tests performed, the COD to BOD_5 relationship was close to 2 and the ratio of BOD_5 to TN was lower than 4. The research indicated that this specific chemical composition of raw wastewater causes appearance of filamentous bacteria on the surface of trickling filter filling and strongly affect the effluent quality.

Introduction

In Poland, the trickling filter technologies are widely used for the treatment of domestic sewage in non-urbanized areas. One of the example of this technology are Bioclere® trickling filters. The main advantages of these devices are the high resistance to biomass wash-out and easy operation under variable wastewater flow-rates. In the trickling filters, the biomass forming the biofilm develops as a result of the scattering of wastewater over the surface of the filling and consists of different microbial species. The species distribution of the biofilm depends primarily on the influent wastewater characteristics. Especially, when the ratio of COD to BOD is greater than 2, undesirable microorganisms grow, such as filamentous bacteria, resulting in deterioration of the effluent quality. When operating trickling filters, as well as in other facilities of this type, little attention is paid to nitrogen compounds, mainly because it is not stablished as one of the main effluent characteristic required to discharge the effluent into the receiving water body. Preliminary studies and observations indicate that the concentration of organic nitrogen, including urea (carbonic acid diamide), has a significant impact on the performance of the trickling filters. The occurrence of very high loading rates of organic nitrogen into the biological treatment could be explained due to the continuous water-saving by the people using the sewage system [1].

* Corresponding author: lk@ekofinn.pl

In this context, the aim of the study was to study the relationship between the influent organic nitrogen concentration and the purification efficiency.

Bioclere® treatment plants are based on low-rate trickling filter technology. There is Hufo® media inside the reactors. Media is made of polypropylene (PP) and present a great specific surface about 120 m^2/m^3. Biofilm, consisting of a mixed culture, grows on media surface. This biofilm preforms the specific reactions removing the pollutants from the wastewater.

In recent years, the number of new Bioclere® treatment plants systematically rise and average about 700 units in Poland. The Biocleres® are good solution for rural areas and work with applications like: scattered households, schools, hotels, restaurants and mountain refugees. It can work in following configurations (the loading is refer to volume of bed as kg $BOD_5/m^3 \cdot d$):

1. one-stage unit (basic unit) – run with 0.4 kg $BOD_5/m^3 \cdot d$ and enable to reduce 90% organic loading (Fig. 1a).
2. two-stage unit without nitrification – I° trickling filter works with 0.8 kg $BOD_5/m^3 \cdot d$ and II° trickling filter works with 0.4 kg $BOD_5/m^3 \cdot d$. This solution enable to reduce 95% organic loading (Fig. 1b).
3. two-stage unit with nitrification – I° trickling filter works with 0.4 kg $BOD_5/m^3 \cdot d$ and II° trickling filter with 0.1 kg $BOD_5/m^3 \cdot d$ and 0,025 kgTN/$m^3 \cdot d$ (nitrification bed). This system can secures organic concentration below 15 mg O_2/L, nitrogen concentration below 6 mgTN/L and phosphorus loading below 1 mgP/L (with chemical precipitation).

Bioclere® treatment plants are appropriated for the treatment of domestic wastewater. The chemical constitution of sewage should be: BOD_5: 200-600 mgO_2/L; COD: 700-1200 mgO_2/L; SS: 200-400 mg/L. According to Heidrich et al. [2] efficient removal is possible when $COD/BOD_5 \leq 2$; $BOD_5/TN \geq 4$ and $BOD_5/TP \geq 25$.

Material and methods

The study was conducted in Poland at six selected wastewater treatment plants (WWTP) based on the trickling filters Bioclere® technology. The research was conducted in the villages of Kwitajny, Protowo (Warmia and Mazury District), Pronie, Gisiel (Pomeranian District), Rządza and Gąsiorowo (Masovian District) for 6 months, from November 2016 to April 2017. The WWTP were configured as two-stage units, with and without secondary sedimentation tank (Tab. 1). Every installation is assigned for service small number of inhabitants (from 75 to 210 people).

Table 1. Characteristic of selected Bioclere® plants (ST – septic tank, TF1 – trickling filter I°, TF2 – trickling filter II°, SST – secondary sedimentation tank

No. of object	Name of plant	pe	$Q_{av.}$ [m^3/d]	Unit System
1.	Kwitajny	150	12	ST-TF1-TF2-SST
2.	Protowo	210	14	ST-TF1-TF2-SST
3.	Pronie	115	10	ST-TF1-TF2
4.	Gisiel	115	11	ST-TF1-TF2
5.	Rządza	75	6	ST-TF1-TF2-SST
6.	Gąsiorowo	150	20	ST-TF1-TF2-SST

The sewage treatment facilities were characterized by similar capacities and were located in non-urbanized areas. Only wastewater from dwellings and households came to these WWTPs. Discharges of other sewage than household were excluded. BOD, COD, ammonia, organic nitrogen (as a measure of urea concentration) and pH were determined in raw wastewater. Standard methods were used according to [4]. In order to calculate the effectiveness of the purification process for each WWTP, the effluent COD concentration was determined. 12 measurement series were carried out (two per month), and the average concentration values were used for further calculations.

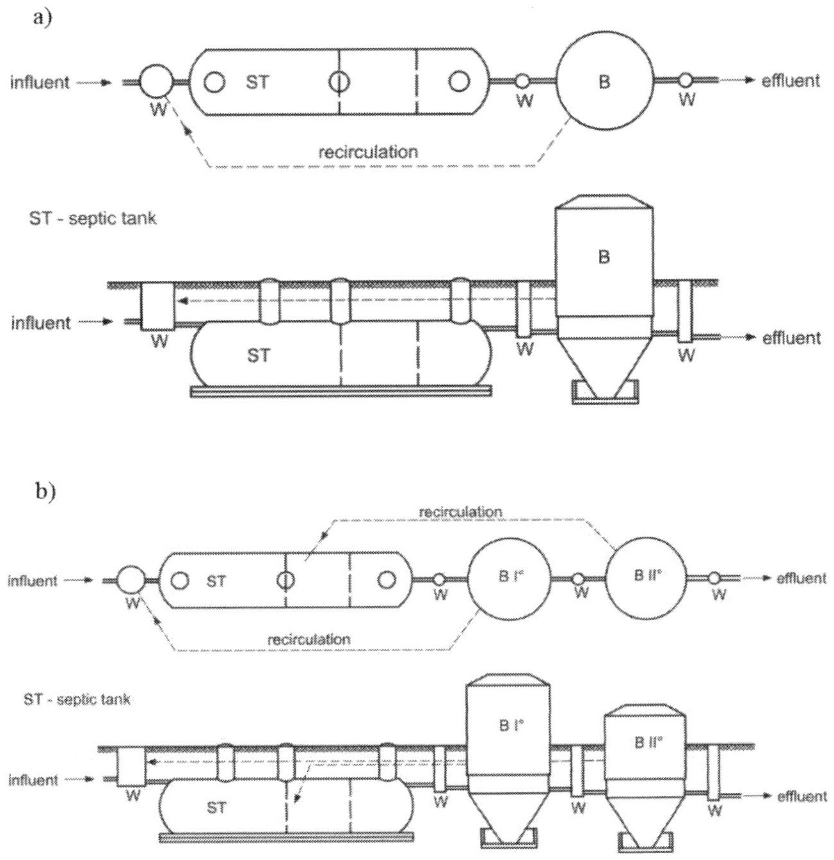

Fig 1. One-stage (a) and two-stages (b) unit applied in Bioclere® trickling filters, W – inspection chambers; B I°, B II° - trickling filters; ST – septic tank [3]

Results and discussion

Table 2 shows the wastewater characteristics studied in this work. It can be seen that the chemical composition of raw sewage is not the conventional one. As can be seen in Table 2, the concentration of Kjeldahl nitrogen was too high in all trickling filters to allow its complete removal and therefore to reach a good effluent quality.

Table 2. Values of selected parameters in studied objects.

Name of plant	BOD_5 [mg/L]	COD [mg/L]	Ammonia nitrogen [mgN/L]	Kjeldahl nitrogen [mgN/L]	pH	COD [mg/L] (effluent)	Q_d [m^3/d]
Kwitajny	210	488	102	112	7.2	96	12
Protowo	360	754	105	120	6.8	106	14
Pronie	330	612	86	110	6.9	150	10
Gisiel	290	687	107	133	7.1	171	11
Rządza	280	564	77	92	7.0	112	6
Gąsiorowo	210	341	60	90	6.7	170	20

In all the tests performed, the COD to BOD_5 relationship was close to 2 (Fig. 2), which indicates that the wastewater fed to the trickling filters was hardly biodegradable [5]. The ratio of BOD_5 to TN was much lower than 4 as shown in Figure 3.

Additional, it must be highlighted that the concentrations of organic nitrogen had a significant effect on the effluent COD concentration. WWTP presenting organic nitrogen concentrations above 20 mg/l at the beginning of the biological treatment were characterized by a nearly doubled COD effluent concentration (Fig. 4). These results indicated that the influent nitrogen concentration clearly influence the effluent quality.

The combination of the low biodegradability and the high concentration of nitrogen compounds leads to changes in the appearance of the biofilm and reduce the quality of the effluent obtained [6]. This chemical composition of raw wastewater causes appearance of filamentous bacteria on the surface of trickling filter filling (Fig. 5a and 5b). Specific chemical composition is the main factor affecting the occurrence of filamentous bacteria [7].

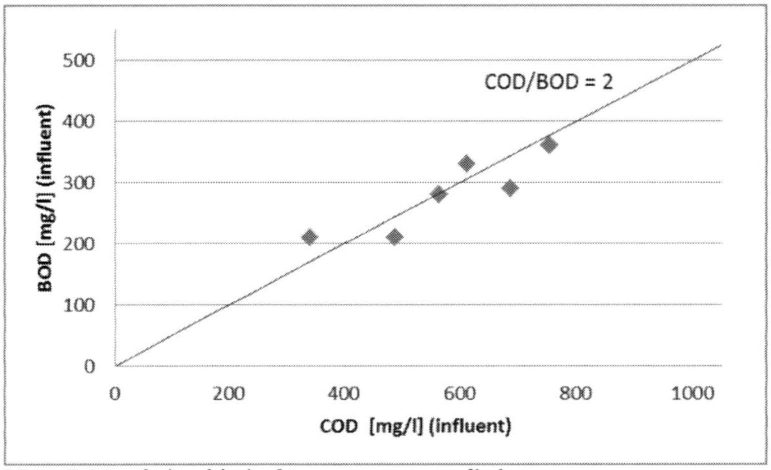

Fig. 2. COD to BOD relationship in the wastewaters studied

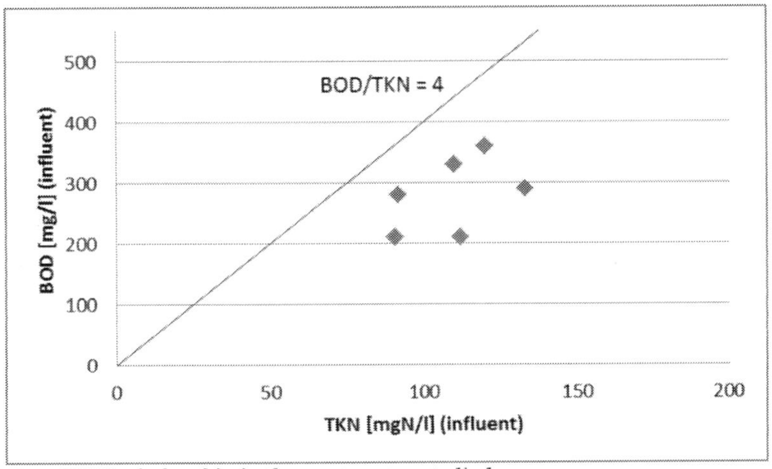

Fig. 3. BOD to TKN relationship in the wastewater studied

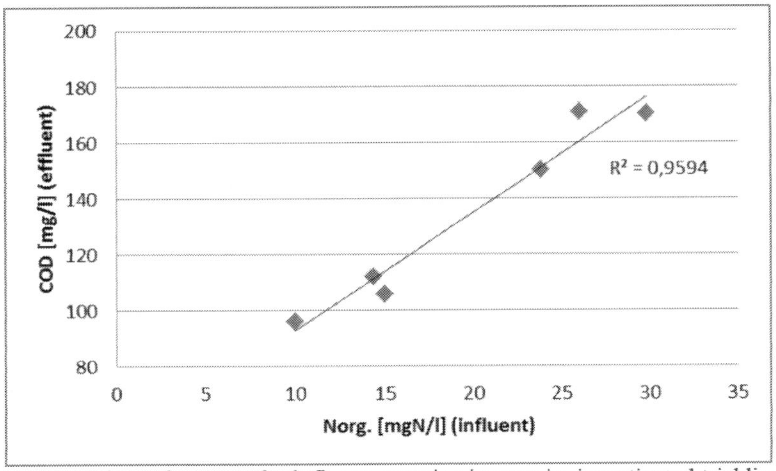

Fig. 4. Effluent COD dependency on the influent organic nitrogen in investigated trickling filters.

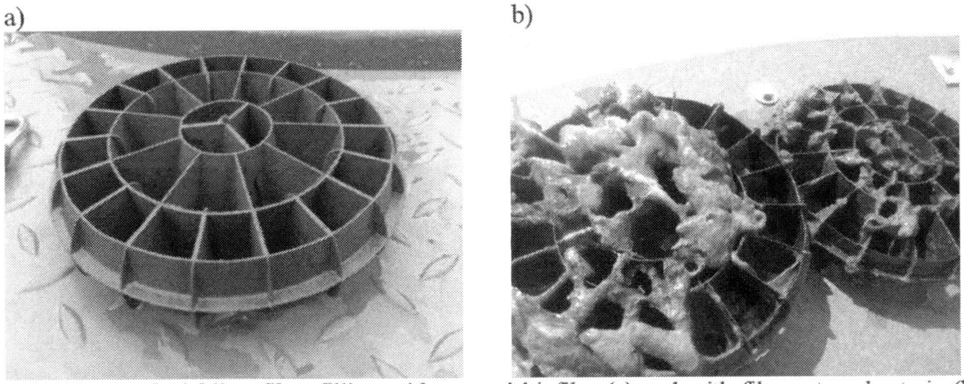

Fig. 5. Shape of trickling filter filling with normal biofilm (a) and with filamentous bacteria (b), (Kopeć 2013)

The investigated research show that the influent sewage contains hardly biodegradable and toxic to most bacteria and protozoa forms of nitrogen. According to the latest sources, compounds such as: urea, sulfoamides, antibiotics, analgesics, and other amine-containing pharmaceuticals are present in wastewater [8,9] and would lead to lower qualities in the microbial population degrading these pollutants and therefore in the effluent quality.

Conclusions

Achieved results show that Bioclere® trickling filters are able to effective removal of pollutants from wastewater to the allowable effluent values. The exceptions to the properly functioning trickling filter plants are related to the unbalanced proportions in the chemical composition of the influent wastewater. Especially disproportionate is the BOD_5 to total nitrogen compounds, including organic nitrogen. The presence of high organic nitrogen levels in the influent wastewater leads to a process where some pollutants are not removed, getting the effluent and degrading its quality.

The causes of high amount of nitrogen must lie in the behavior of people and how their use of the sewerage network. People in rural areas in Poland have low water consumption, so raw sewage is more concentrated and easier to rotting.

The results of the article confirm the possibility of large quantities of harmful substances for bacteria in sewage in non-urbanized areas. However, the determination of impact of these compounds on the effectiveness of purification based only on organic nitrogen concentrations is insufficient. Further research should be extended by chemical analyzes of specific substances which are present in wastewater and slow down biochemical processes.

Acnowledgements: Ekofinn-Pol Ltd., Poland, for accessing Bioclere® trickling filters.

References

1. M. Gajewska, Ł. Kopeć , H. Obarska-Pempkowiak. The operation of a small wastewater treatment facilities in a scattered settlement. Rocz. Ochr. Środ., **13**, 207-225 (2011)
2. Z. Heidrich, M. Kalenik, J. Podedworna, G. Stańko Sanitacja wsi. Wyd. „Seidel-Przywecki. Warszawa. (2008).
3. Ekofinn-Pol Ltd., Information materials, 36, (2015).
4. APHA. Standard Methods for the Examination of Water and Wastewater, 18th edn. American Public Health Association/ American Water Works Association/Water Environment Federation, Washington, DC, USA, (1992).
5. Z. Dymaczewski Poradnik eksploatatora oczyszczalni ścieków – praca zbiorowa. Polskie Zrzeszenie Inżynierów i Techników, Poznań, 1152, (2011).
6. Ł. Kopeć, Mikroorganizmy błony biologicznej w złożach zraszanych. *Gaz, Woda i Technika Sanitarna*, **2**, 53-58. (2013).
7. S. Rossetti, C. Maria, M.C. Tomei, P.H. Nielsen, V. Tandoi ''Microthrix parvicella'', a filamentous bacterium causing bulking and foaming in activated sludge systems: a review of current knowledge. FEMS Microbiology Reviews **29**, 49–64. (2005)
8. B. Kraigher, T. Kosjek, E. Heath, B. Kompare, I. Mandic-Mulec. Influence of pharmaceutical residues on the structure of activated sludge bacterial communities in wastewater treatment bioreactors. *Water Research* **42**, (2008).
9. T. Heberer. Occurrence, fate and removal of pharmaceutical residues in the aquatic environment: a review of recent research data. *Toxicology Letters*, **131**, (2002).

Ciliated protozoa in the impact zone of the Uzhgorod treatment plant

Volodimir Pliashechnyk[1], *Yaroslav* Danko[2], *Grzegorz* Łagód[3], *Jakub* Drewnowski[4], *Tatiana* Kuzmina[5], and *Roman* Babko[6]

[1]State institution of higher education "UzhNU", Narodna Square 3, 88000 Uzhhorod, Ukraine
[2]Sumy Makarenko State Pedagogical University, Romenska 87, 40002 Sumy, Ukraine
[3]Lublin University of Technology, Environmental Engineering Faculty, Nadbystrzycka 40B, 20-618 Lublin, Poland
[4]Gdansk University of Technology, Civil and Environmental Engineering Faculty, Narutowicza 11/12, 80-233 Gdańsk, Poland
[5]Sumy State University, N. Rimsky-Korsakov 2, 40007 Sumy, Ukraine
[6]Schmalhausen Institute of Zoology National Academy of Sciences of Ukraine, B. Khmelnitsky 15, 01601 Kiev, Ukraine

Abstract. This paper presents the results of studies on the Uzh River (Ukraine, Zakarpattia Oblast) near the effluent point of a sewage treatment plant in Uzhgorod. The samples were taken at various sites of the treatment plant along the stages of purification process, as well as in the river, at a number of different points above and below the wastewater discharge. At each of these objects, the temperature and O_2 were measured. The structure of ciliate assemblage was analyzed along the stages of the treatment process in the WWTP and in the river before and after the sewage discharge. A total of 26 ciliate taxa were observed and included in the analysis. All the studied stations were considered as a continuum in which populations of protozoa spread freely according to their ecological preferences. The majority of ciliate species were encountered in each of the examined stations, but their quantitative development differed significantly, reflecting their response to the environmental conditions at the stations. The analysis of the qualitative and quantitative distribution of ciliate populations by the stations enabled to group them in respect to the peculiarities of the local conditions. The study showed that the majority of the ciliate species, typical of bioreactors, are equally common at the stations of the Uzh River below wastewater discharges. The ciliate assemblage in the oxygen gradient demonstrated a wide spectrum of ecological tolerance at the species level. These findings confirm that ciliates are very good indicators of the environmental quality, provided that detailed information about their environmental priorities is available.

1 Introduction

In the 19th century, many rivers in Europe which flow through settlements, received huge amounts of domestic and industrial wastewater. As a result of the industrial development and the growth of cities in the 19th century, even large rivers such as the Rhine, the Oder, the Thames, the Vistula, the Seine and other European rivers were critically polluted. Creation of sewage treatment plants, enabled to prevent the final degradation of European rivers. The development of technology over the past 100 years has significantly reduced the impact of sewage. However, so far it has not been possible to eliminate their impact on hydrobionts completely. In most cases, the influent sewage locally changes the trophic status of the reservoirs [1, 2], provoking the growth of bacteria [3], protozoa and thereby providing nutritional requirements for many invertebrate species [4], which entails changes in the structure of aquatic biocenosis [5-10]. A significant negative aspect of wastewater discharge points is that they induce the effect of pollutant accumulation and decrease the resistance of the hydroecosystem [4]. For example, Madoni discovered that the Parma River at the moment of study was more polluted than ten years earlier, and its ciliate community changed dramatically [11]. Often, the sewage discharged to a river induces processes leading to a decrease of dissolved oxygen content or even local anoxia (Spänhoff, 2007, Wakelin, 2008)[3, 12]. In the context of water bodies protection, apart from the improvement of wastewater treatment quality, an important role is played by the assessment of treated water impact on natural biocenoses. The complexity of such assessments and uncertainty of their results is obvious, since systemic disturbances depend not only on the content (chemical composition) and the amount of wastewater, but also on the self-cleaning potential of natural reservoirs, their ecological plasticity and resistance to contamination. Nowadays, there are many methods for assessing the state of hydrobiocenoses, starting from an organoleptic assessment of water quality to high-precision physicochemical methods and biotesting. However, as far as assessing the quality of aquatic ecosystems is concerned, the current preferences of researchers are increasingly shifting towards the use of systemic environmental indicators, which enable to evaluate the quality of the aquatic environment as a habitat for the biocenosis. In this respect, the ciliated protozoa are used most frequently. In reservoirs, they are represented by a significant variety of species [13, 14] the quantitative and qualitative distribution of which reflects the reaction of biocenoses to the external influences [11, 15-18]. Although there is a considerable amount of data on the occurrence of species under different conditions, objectively, for many species, the available data concern only narrow ranges of factors. Obtaining objective data on the realization of a population in the continuum of a particular limiting factor within a natural body of water is a very difficult task. However, it is very relevant in the context of improving the methods for assessing the quality of the environment.

2 Materials and methods

The studies were carried out on the Uzh River (Ukraine, Zakarpattia Oblast) near the effluent point of a sewage treatment plant in Uzhgorod in January-April 2017. The samples were taken at various sites of the treatment plant along the stages of purification process, as well as in the river at a number of different points above and below the wastewater discharge. Thus, the investigated objects are of the two types. The first includes the main elements of Uzhgorod WWTP's: primary clarifier, aeration tank, secondary clarifier, mixing channel, effluent channel. On the other hand, the second comprises a fragment of Uzh river: 50 m upstream from wastewater discharge (uzh0), at the WWTP's discharge site, 50, 100, 250 and 300 m below the discharge.

At each of those objects, the measurements of temperature and O_2 were conducted by means of the HACH HQ40d Portable Multi-Parameter Meter. In all of the investigated sites, samples of water were taken with a 300 mL syringe. On the river, the samples were taken using a syringe with a plastic tip up to 0.5 m long, which enabled to take a bulk sample of bottom sediments. After the sampling, the samples were immediately delivered to the laboratory of Uzhgorod University and processed. In order to preserve the unprocessed samples, they were kept in the refrigerator at a temperature of +5°C, which slowed down the degradation processes. Species identification was based on [19-23]. When necessary, cells were stained with the use of 1% methyl green or silver nitrate [30]. Estimation of population density was performed on the basis of counting organisms in sub-samples of 25 μL each, using a micropipette, as described by [24]. Counting was performed 5 times for each active sludge sample and 11 times for bottom sediments. Determination of the species composition and calculation of the abundance of ciliates was performed with an Olympus CX41 microscope in transmitted light. Data were processed using R version 3.4.0 [25]. Hierarchical clustering was performed with hclust from core R package stats, fuzzy clustering with fanny from cluster [26], Principle Components Analysis (PCA) with fviz_cluster from factoextra [27], Non-metric multidimensional scaling (NMDS) with metaMDS from vegan [28]. Plots were produced using R packages ggplot2 [29], factoextra [27], ggrepel [30], directlabels [31].

3 Results and discussion

The structure of ciliate assemblage was analyzed along the stages of treatment process in the WWTP and in the river before and after sewage discharge. The treated sewage from WWTP entered the river Uzh from the right bank. The speed of the current in the river at the discharge point varies from 1 to 2 m per second. The station that is not affected by effluents and can be considered as a control is located 50 m upstream of the wastewater discharge. Changes in the structure of ciliate assemblage in the river were monitored at distances of 50, 100, 250 and 300 m downstream the WWTP discharge. At all the stations on the river and in the WWTP, the water temperature ranged from 2 to 8°C, and the concentration of dissolved oxygen was between 2 and 14.5 mg L^{-1}. A total of 26 ciliate taxa were observed, 24 out of which were identified to the species level and included in the analysis. All the studied stations can be considered as a continuum in which populations of protozoa spread freely according to their ecological preferences. Considering that the ranges of organic matter content, oxygen concentration and temperature amplitude were quite wide, the development of the protozoan populations, *a priori*, should reflect their environmental priorities with respect to these factors. Indeed, the majority of ciliate species were encountered at all examined stations, but their quantitative development differed significantly, reflecting their response to the environmental conditions occurring at the stations. The analysis of the qualitative and quantitative distribution of ciliates populations by the stations enabled to group the stations in accordance with the peculiarities of the local conditions. The results of the cluster analysis are shown in Fig. 1.

The cluster analysis showed that the stations studied is divided into two major groups of clusters, which indirectly confirms the presence of significant structural rearrangements in the protozoan assemblages. The first group combines an aeration tank with stations on the river, located at a distance of 50, 100 and 150 meters downstream from the wastewater inlet. The second group combines primary and secondary clarifiers and derivation channel of the treatment plant with stations on the river, where the impact of the WWTP is minimal – 300 m downstream from wastewater discharge – and maximal – at the point of wastewater discharge. Obviously, proximity in space is not the only factor that forms these two clusters. A more significant factor than spatial proximity is the presence or absence of the activated sludge at the considered stations. The first cluster is composed of stations where the activated sludge is present; in contrast, the second one is where the sludge absent. Stations at 50-150 m downstream (belonging to the first group of clusters) are under the conditions resembling those in the aeration tank: energy subsidies in the form of organic matter from treatment plants, vigorous mixing and sufficient concentrations of dissolved oxygen. For obvious reasons, activated sludge is absent at most stations from the second group of clusters.

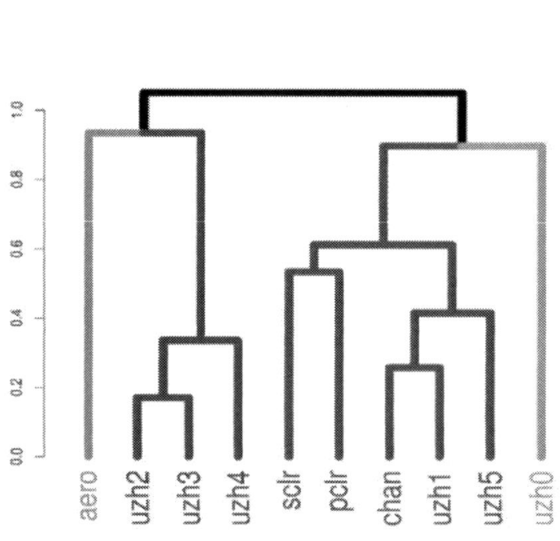

Fig. 1. Ward clustering of a matrix of Kulczynski distances among stations. Data were Hellinger transformed. Stations: aero — aeration tank, chan — derivation channel, pclr — primary clarifer, sclr — secondary clarifer, uzh0 — river 50 m upstream the treated wastewater discharge, uzh1 — river at the discharge site, uzh2, uzh3, uzh4, uzh5 — river 50, 100, 250 and 300m.

The absence of the activated sludge in the river at the wastewater inlet is explained by the specific conditions at this point, i.e. where sewage and river water meet. It can be said that under the conditions of flow, space plays the role of time and the activated sludge in the river requires a certain amount of time to form, which in this case means that it forms only at a certain distance downstream.

The considerable amplitude of the conditions realized at the considered stations enabled to observe a quantitative realization of the ciliate populations in the gradient of two important environmental factors. Our research was focused on the spatial distribution of the ciliates populations along oxygen and temperature gradients. Nonmetric multidimensional scaling was used for visualization of similarities in ciliate species composition and abundance of the studied sites.

Relative positions of the considered stations and ciliate species with oxygen gradient superimposed obtained by NMDS are presented in Fig. 2. The impact of WWTP is evident. Only site upstream of WWTP (uzh0) is clearly outside of impact. All other sites in the NMDS space are located quite close to secondary clarifier or derivation channel. The location of sites in those space correlates ($r^2 = 0.58$, $p < 0.05$) with the dissolved oxygen concentration. The results show a highly differentiated response of the infusoria species to the oxygen gradient. At the same time, some of the species, common for activated sludge, reached a maximum under the river conditions. The presence of the species under the conditions of the bioreactor (activated sludge) is not the basis for the assertion that its ecological optimum is located exactly under these conditions. However, it should be noted that in the space of the nonmetric multidimensional scaling, the species encountered in the treatment facilities most frequently — *Aspididsca cicada*, *Plagiocampa rouxi*, *Acineria uncinata*, *Vorticella aquadulcis*, etc. are closest to the station of aeration tank. On the other hand, the species that are practically not registered under the conditions of treatment plants are clearly located in this coordinate system in the area of the station not affected by effluents.

The temperature is obviously an important ecological factor; however since the studies were conducted in the winter-spring period, a sufficient amplitude for the temperature gradient was not obtained. The temperature gradient was directed against the oxygen gradient, but its impact was less pronounced ($r^2 = 0.45$, $p < 0.1$) Fig. 3.

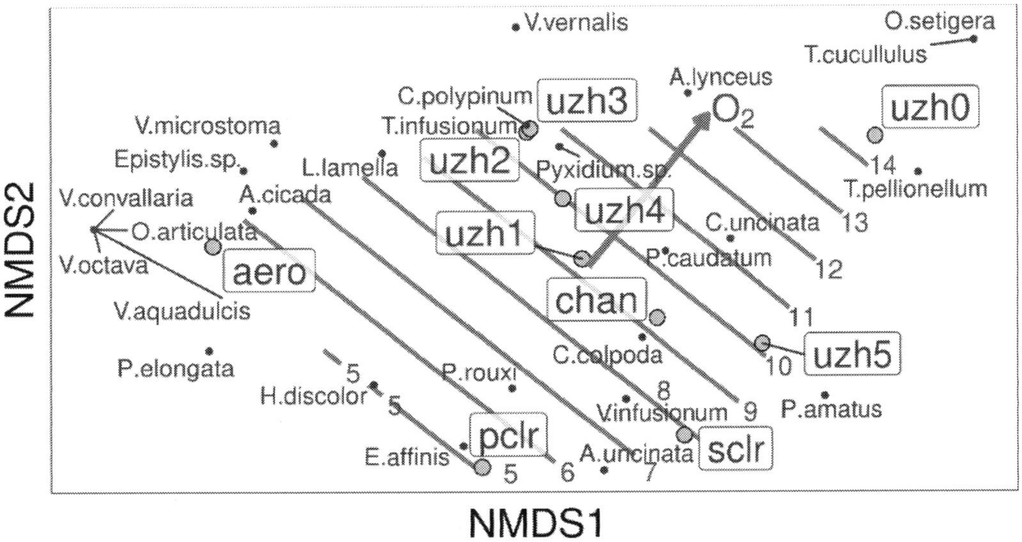

Fig. 2. Non-metric multidimensional scaling biplot of a Kulczynski distance matrix of Hellinger-transformed data on the ciliates abundance with the gradient of dissolved oxygen superimposed. The designations of the stations are the same as in Fig. 1.

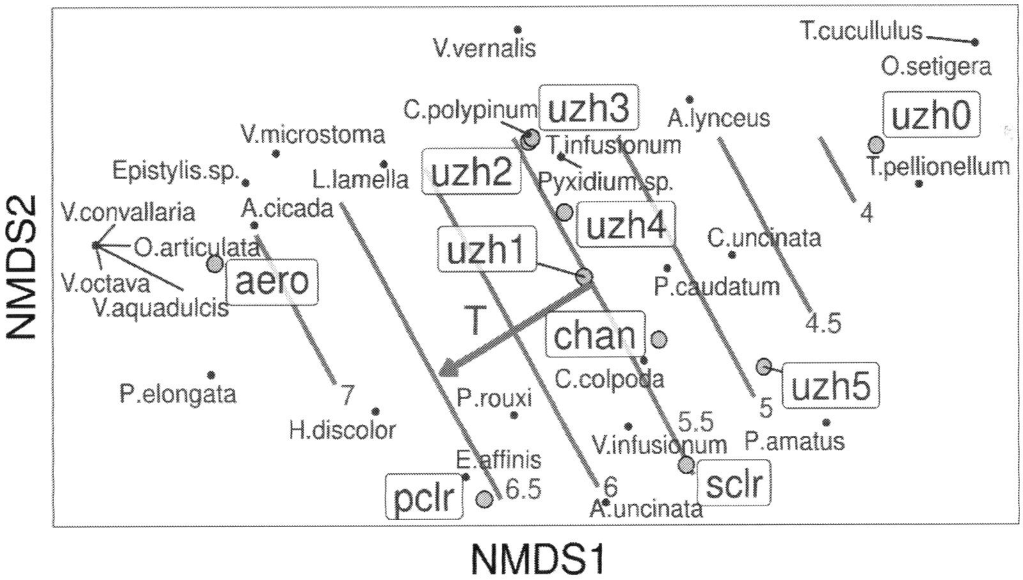

Fig. 3. Non-metric multidimensional scaling biplot of a Kulczynski distance matrix of Hellinger-transformed data on the ciliates abundance with the gradient of dissolved temperature superimposed. The designations of the stations are the same as in Fig. 1.

However, the obtained data show that species oriented toward low concentrations of dissolved oxygen are better realized at higher temperatures. Nevertheless, in order to formulate more reliable conclusions regarding the environmental priorities of ciliates associated with a temperature gradient, more materials are needed, primarily collected at higher temperatures. On the basis of the acquired data, several statements can be made which are presented in the conclusions.

4 Conclusions

1. The study showed that the majority of the ciliate species , typical for bioreactors, are equally common at the stations of the Uzh River below wastewater discharges.
2. Moreover, contrary to the expectations, many species of infusoria, characteristic of the activated sludge, have reached maximum abundance precisely under the river conditions.
3. The ciliate assemblage in the oxygen gradient demonstrated a wide spectrum of ecological tolerance at the species level.
4. All these facts confirm that ciliates are very good indicators of environmental quality, provided that detailed information about their environmental priorities is available.

References

1. V. H. Smith, G. D. Tilman, J. C. Nekola, Environ. Pollut., **100**, 1-3 (1999)
2. P. F. Strom, V.A. Matulewich, M. S. Finstein, Appl. Environ. Microbiol., **31** (1976)
3. S. A. Wakelin, M. J. Colloff, R. S. Kookana, Appl. Environ. Microbiol., **74**, 9 (2008)
4. B. Gücker, M. Brauns, M. T. Pusch, J. N. Am. Benthol. Soc., **25** (2006)
5. J. L. Carter, V. H. Resh, M. J. Hannaford, M. J. *Myers Macroinvertebrates as biotic indicators of environmental quality* (Methods in Stream Ecology. Amsterdam: Academic Press, 2006)
6. I. A. Wright, B. C. Chessman, P. G. Fairweather, L. J. Benson, Aust. J. Ecol., **20** (1995)
7. EU Water Framework Directive 2000/60/EC. (2000)
8. D. Kominkova, D. Stransky, G. St'astna, J. Caletkova, J. Nabelkova, Z. Handova, Water Sci. Technol., **51**, 2 (2005)
9. M. Gorzel, R. Kornijow, Ecohydrol. Hydrobiol., **7**, 1 (2007)
10. M. A. Lewis, Ohio J. Sci., **86**, 1 (1986)
11. P. Madoni, Hydrobiologia, **264**, 3 (1993)
12. B. Spänhoff, R. Bischof, A. Böhme, S. Lorenz, K. Neumeister, A. Nöthlich, et al., Water Air Soil Pollut., **180** (2007)
13. J. O. Corliss, Acta Protozool., **41** (2002)
14. B. J. Finlay, G. F. Esteban, Biodivers Conserv., **7** (1998)
15. S. Yun-Fen, A. L. Buikema, jr., W. H. Yongue, jr., J. R. Pratt, J. Cairns jr., J. Protozool., **33** (1986)
16. D. H. Lynn, G. L. Gilron, J. Aquat. Ecosyst. Health, **1**, 4 (1992)
17. J. R. Lawrence, G. D. W. Swerhone, L. I. Wassenaar, T.R. Neu, Can. J. Microbiol., **51** (2005)
18. J. D. Stout, Ecology, **37**, 1 (1956)
19. H. Berger, W. Foissner *Illustrated guide and ecological notes to ciliate indicator species (Protozoa, Ciliophora) in running waters, lakes, and sewage plants* (Steinberg C, Editor. Handbuch Angewandte Limnologie. Ecomed Verlag, 2003)
20. S. Serrano, L. Arregui, B. Perez-Uz, P. Calvo, A. *Guinea, Guidelines for the Identification of Ciliates in Wastewater Treatment Plants* (IWA Publishing. wastewater-treatment-plants, London, 2008)
21. W. Foissner, H. Berger, Freshwater Biol., **35** (1996)
22. A. Warren, Bull. Br. Mus. Nat. Hist. (Zool.), **50**, 1 (1986)

23. A. Warren, Bull. Br. Mus. Nat. Hist. (Zool.), **52**, 1 (1987)
24. P. Madoni, Water Res., **28** (1994)
25. R Core Team, R: A language and environment for statistical computing. R Foundation for Statistical Computing, Vienna, Austria (2017)
26. M. Maechler, P. Rousseeuw, A. Struyf, M. Hubert, K. Hornik, cluster: Cluster Analysis Basics and Extensions. R package version 2.0.6, (2017)
27. A. Kassambara, F. Mundt, factoextra: Extract and Visualize the Results of Multivariate Data Analyses. R package version 1.0.4. (2017)
28. J. Oksanen, F. G. Blanchet, M. Friendly, R. Kindt, P. Legendre, D. McGlinn, P. R. Minchin, R. B. O'Hara, G. L. Simpson, P. Solymos, M. Henry, H. Stevens, E. Szoecs, H. Wagner, vegan: Community Ecology Package. R package version 2.4-3 (2017)
29. H. Wickham, *ggplot2: Elegant Graphics for Data Analysis* (Springer-Verlag, New York, 2009)
30. Slowikowski, K. ggrepel: Repulsive Text and Label Geoms for 'ggplot2'. R package version 0.6.5 (2016)
31. T. D. Hocking, directlabels: Direct Labels for Multicolor Plots. R package version 2017.03.31. (2017)

Photo-oxidation of PAHs with calcium peroxide as a source of the hydroxyl radicals

Jolanta Kozak[1,*], *Maria* Włodarczyk-Makuła[2]

[1] Czestochowa University of Technology, Department of Chemistry Water and Wastewater Technology, Dąbrowskiego st 69, 42-200 Czestochowa, Poland
[2] Czestochowa University of Technology, Department of Chemistry Water and Wastewater Technology, Dąbrowskiego st 69, 42-200 Czestochowa, Poland

Abstract The efficiency of the removal of selected PAHs from the pretreated coking wastewater with usage of CaO_2, Fenton reagent ($FeSO_4$) and UV rays are presented in this article. The investigations were carried out using coking wastewater originating from biological, industrial wastewater treatment plant. At the beginning of the experiment, the calcium peroxide (CaO_2) powder as a source of hydroxyl radicals (OH•) and Fenton reagent were added to the samples of wastewater. Then, the samples were exposed to UV rays for 360 s. The process was carried out at pH 3.5-3.8. After photo-oxidation process a decrease in the PAHs concentration was observed. The removal efficiency of selected hydrocarbons was in the ranged of 89-98%. The effectiveness of PAHs degradation was directly proportional to the calcium peroxide dose.

1. Introduction

PAHs, along with other environmental pollutants, such as: pesticides, endocrine disruptor contaminants (EDCs), PPCPs- Pharmaceuticals and Personal Care Products, retardants and others belongs to a group of compounds known as persistent environmental pollutants. These compounds are characterized by high toxicity, persistence and bioaccumulation potential [1-4] Sources of PAHs are mainly anthropogenic activities, especially during incomplete combustion of organic matter in various industries and other human activities like tobacco smoking, burning coal, petrol, gas and other fossil fuels [5-7] The main sources of pollution aquatic environment are untreated industrial wastewater, in particularly coking wastewater (coking wastewater, petrochemical). The above mentioned wastewater is the most polluted with PAHs, thus they have been choosing for investigations [8-9] The presence of PAHs in pretreated coking wastewater is confirmed by literature sources [10-11] In Poland there are not available regulations of the permissible PAHs concentration in the treated wastewater discharging to the receiver. It is only pointed out that carcinogenic compounds should be monitored in the surface water.

Advanced Oxidation Processes (AOPs) have been increasingly applied in the treatment of industrial wastewater. In AOPs methods the formation of a hydroxyl radical is formed.

[*] Corresponding author: j.kozak@is.pcz.pl ;mwm@is.pcz.czest.pl

Hydroxyl radicals are able to decompose organic compounds into CO_2 and H_2O. In AOPs processes are used: H_2O_2, potassium manganese, ozone, UV radiation, ultrasound, or methods involving the combined action of a chemical reagent with irradiation or ultrasound [12-15] The literary investigations indicate that simultaneous usage of several reagents results in the most effective degradation of pollutants. The former studies have suggested that CaO_2 is more efficient source of hydroxyl radicals (OH·) than the H_2O_2 [16-17] Among other reagents able to release OH (such as Na_2CO_3, MgO_2) CaO_2 has been selected for these studies due to its relatively low cost. The release of OH·from CaO_2 is gradual, and appears only under acidic, aquatic environments [18] The process of dissolving CaO_2 to H_2O_2 and $Ca(OH)_2$ takes place according to the reaction (1) [14].

$$CaO_2 + 2H_2O \rightarrow H_2O_2 + Ca(OH)_2 \qquad (1)$$

Therefore, it seems reasonable to carry out the experiments using the afore mentioned reagent.
The main objective of the study is to determine the effectiveness of PAHs removal with CaO_2 as an alternative source of hydroxyl radicals in the photo-Fenton process. The focus of the study was on PAHs analysis with 4-6 benzene rings selected from the US EPA list due to their high carcinogenicity [19] Table. 1. Characteristics of selected PAHs.

Table. 1. Characteristics of study PAHs [19]

Name	Chemical formula	Molar mass [g/mol]	Numbers of benzene rings	Carcinogenicity index
Pyrene	$C_{16}H_{10}$	202,25	4	Not classifiable for human
Benzo(k)fluoranthene	$C_{20}H_{12}$	252,32	4	0,1
Benzo(b)fluoranthene	$C_{20}H_{12}$	252,32	4	0,1
Benzo(a)anthracene	$C_{18}H_{12}$	228.29	4	0,1
Chrysene	$C_{18}H_{12}$	228,29	4	0,01
(Indeno(1,2,3-cd)pyrene	$C_{22} H_{12}$	276,34	5	0,1
Dibenzo(a,h)anthracene	$C_{22}H_{14}$	278,35	5	5
Benzo(a)pyrene	$C_{20}H_{12}$	252,31	5	1
Benzo(g,h,i)perylene	$C_{22}H_{12}$	276,33	6	0,01

2. Experimental procedure

2.1 Materials

In the study coking wastewater from the industrial wastewater treatment plants was used. Wastewater was biologically treated with an activated sludge methods. In the coking wastewater organic compounds (COD) and PAHs were determined.

2.2 Experimental procedure

The studies were carried out using pretreated coking wastewater. In the first step was characterized COD according to the US shortcut method. Samples of wastewater were taken as temporary, then 9 PAHs, listed by US Environmental Protection Agency (EPA) were determined. Hydrocarbons containing 4-6 benzene rings are the most carcinogenic. At the beginning of the experiment CaO_2 was added to the wastewater samples in the following amounts: 1 g; 2 g; 3 g per/L respectively. Samples were acidified at pH 3,5-3,8.Samples were acidified at pH 3,5-3,8. Then Fenton reagent was added to the samples and samples were exposed to UV rays for 360 s in a plastic cuvette (the thickness of the exposed layer was equal to 2 mm). Then, qualitative and quantitative analysis of PAH and COD was carried out.

2.3 PAHs analysis

PAHs analysis included extraction organic phase from coking wastewater. Cyclohexane and dichloromethane were added to the wastewater (5:1 v/v). The mixture was shaken for 60 minutes maintaining constant amplitude. Extracts were separated from the wastewater sample on the laboratory separator. The extracts were purified on silica gel column (densification of the mixture was carried out with a mixture of dichloromethane and cyclohexane (v/v 1:5) three times). The obtaining extract was concentrated to the volume of 2 mL under a nitrogen stream. The final step was the quantitative and qualitative determination of 16 PAHs compounds using a GC-MS gas chromatography. The assays were carried out using a GC-MS. The temperature program was 40 ° C/min, heating 5 to 120 ° C, and the final temperature was 280 ° C for 60 minutes.

3. Results

Preliminary investigations showed that the COD concentration was equal 558 mg O_2/L in initial wastewater sample. Addition of reagents and applications of UV light resulted in the removal of COD by 30-35%. The total concentration of the analyzed hydrocarbons was 5603 µg /L in pretreated coking wastewater. After the oxidation process total concentration of the analyzed hydrocarbons was 203 µg /L (after adding 3 g/L CaO_2). Degree of removal was 97%. Figure 1 presented decrease in concentration of pyrene depending on doses of calcium peroxide.

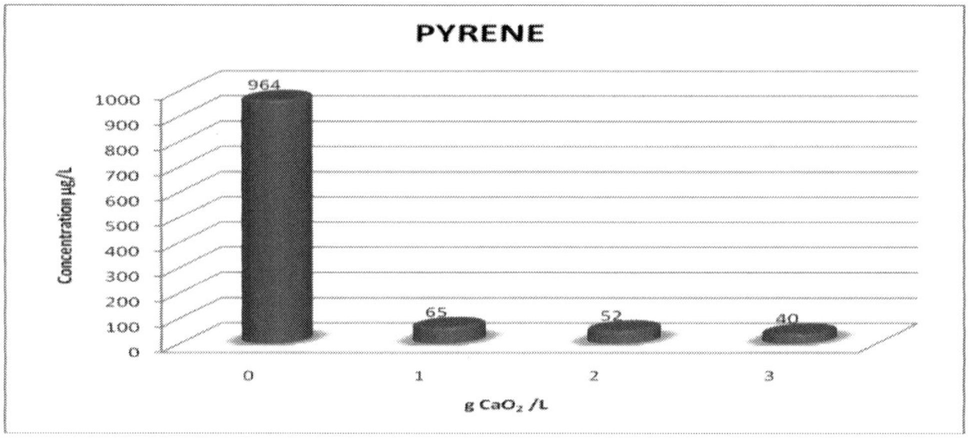

Fig. 1. Changes in the concentration of pyrene depending on dose of CaO_2 added to samples.

The highest decrease of pyrene concentration was observed for dose CaO_2 3 g/L. In this case efficiency of removal was 96 %. The degree of removal of pyrene increased gradually as the dose increased.

The highest decrease of concentration was observed for benzo(b)fluoranthene (98%) and for its isomer benzo(k)fluoranthene(98%) after adding 3 g/L CaO_2.

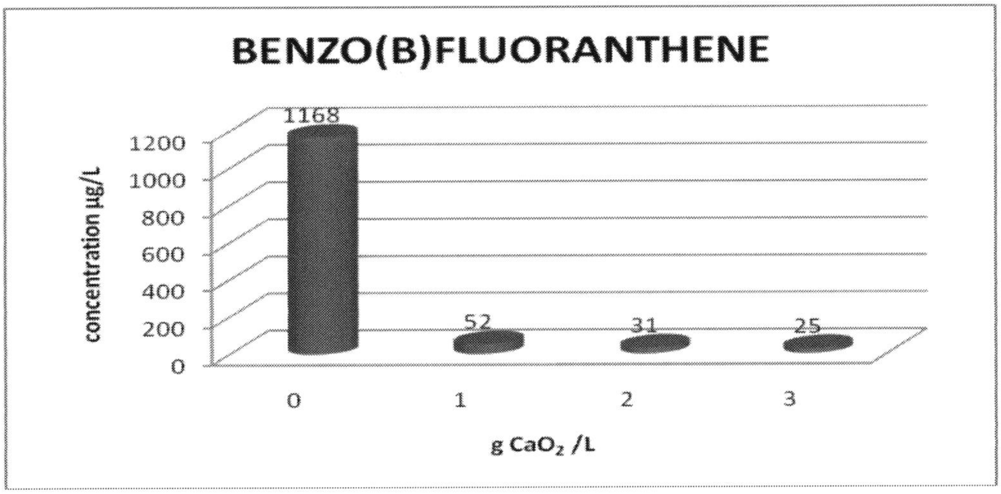

Fig. 2. Changes in the concentration of benzo(b)fluoranthene depending on dose of CaO_2 added to samples.

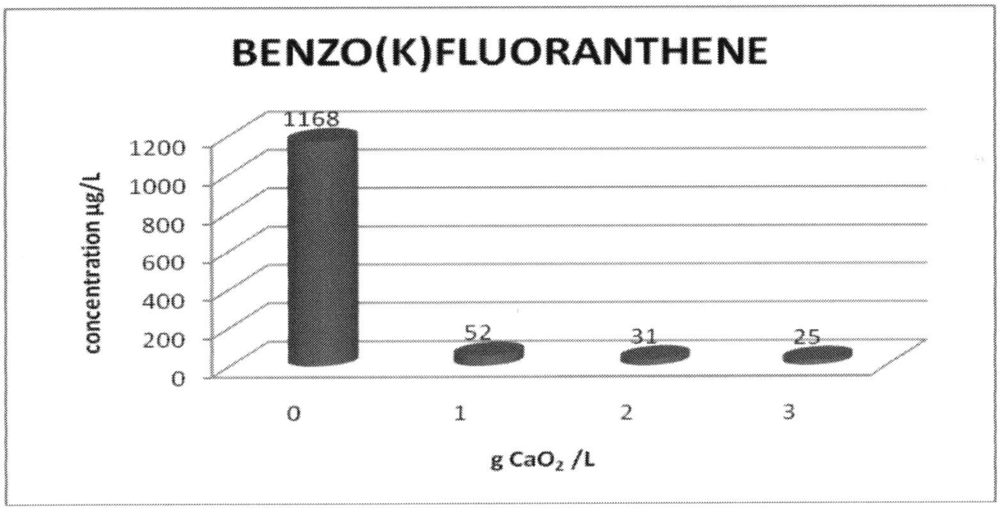

Fig. 3. Changes in the concentration of benzo(k)fluoranthene depending on dose of CaO_2 added to samples.

Concentration of benzo(b)fluoranthene and its isomer gradually decreased with the increase of oxidizer dosage (Figure 2; Figure 3).

For benzo(a)anthracene and chrysene, the situation was similar. Also after adding 3 g / L removal efficiency was greatest and was 93% and 97%. (Figure 4;Figure 5). Total concentration of hydrocarbons with 4 benzene rings was 4593 µg/L in pretreated coking wastewater. After oxidation process for dose 3g/L CaO_2 total concentration was 157 µg/L. Degree of removal was 96%

 https://doi.org/10.1051/e3sconf/20183002009

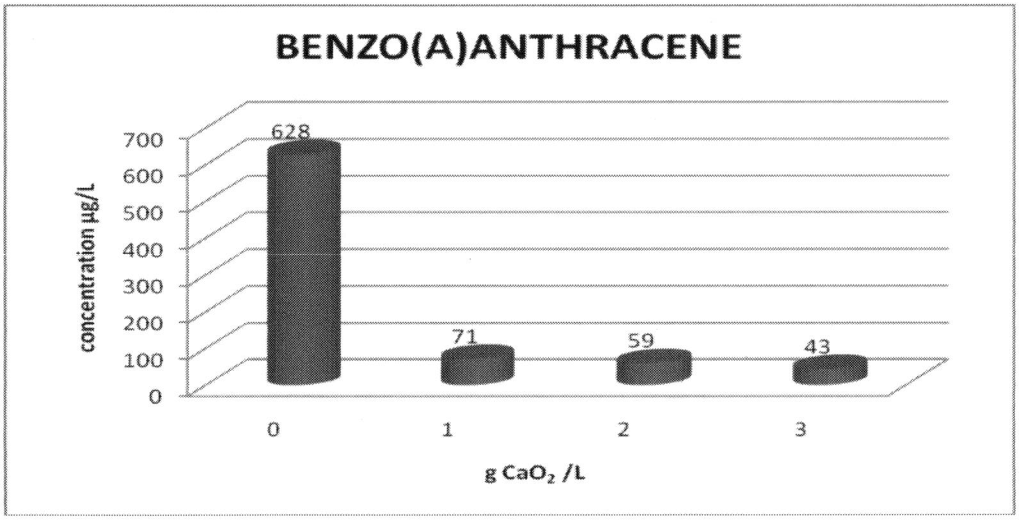

Fig.4. Changes in the concentration of benzo(a)anthracene depending on dose of CaO_2 added to sample.

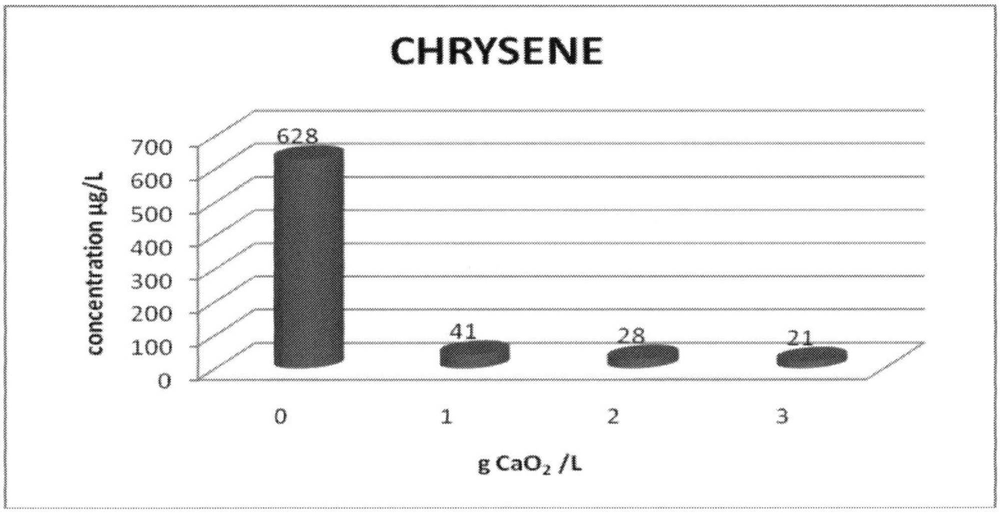

Fig. 5. Changes in the concentration of chrysene depending on dose of CaO_2 added to sample.

Figure.6 shows the changes in the concentration of benzo(a)pyrene depending on the CaO_2 dose. Benzo(a)pyrene has 5 benzene rings. Total concentration of hydrocarbons with 5 benzene rings was 1716µg/L in pretreated coking wastewater. The highest decrease in the concentration of benzo(a)pyrene was for the dose of 3 g/L CaO_2. The efficiency in the removal of benzo(a)pyrene was equal to 98 %

The highest decrease in the concentration of indeno(1,2,3-cd)pyrene was for the dose of 3 g/L. CaO_2. Degree of removal was 98% (Figure7)

The highest decrease in the concentration of dibenzo(a,h)anthracene was for the dose of 3 g/L CaO_2.. The efficiency of the removal was equal to 96 % After oxidaton process total concentration of hydrocarbons with 5 benzen rings was 38 µg/L. For hydrocarbons with 5 benzene ring also the highest decrease in the concentration was for the dose of 3 g/L CaO_2.Removal efficiency form hydrocarbons with 5 benzene rings was 98 % (Figure 8).

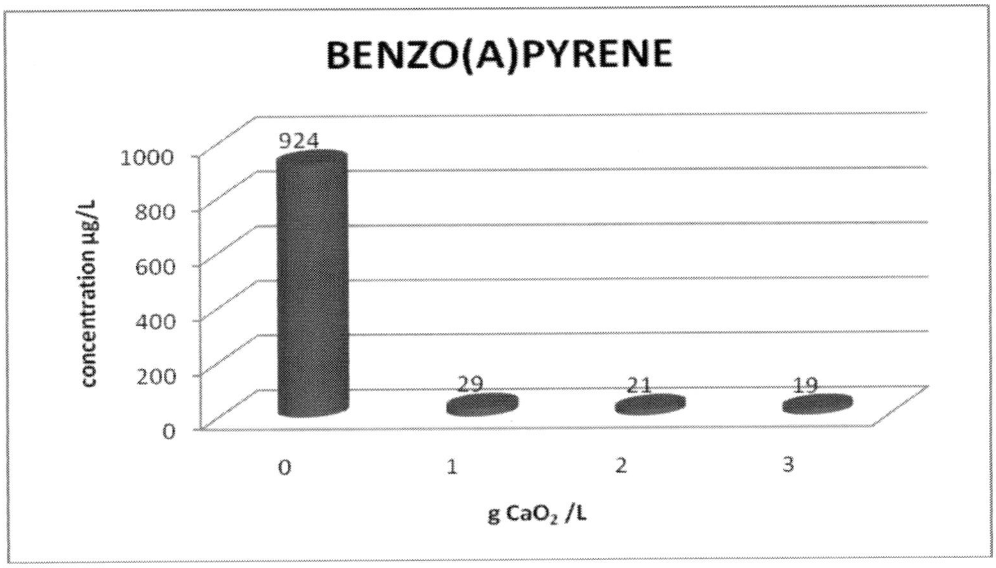

Fig. 6. Changes in the concentration of benzo(a)pyrene depending on dose of CaO_2 added to sample.

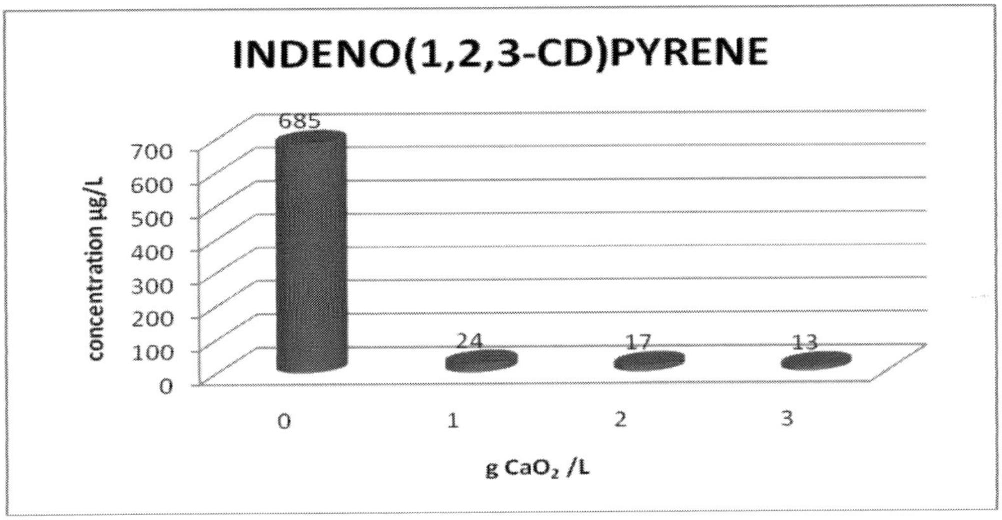

Fig. 7. Changes in the concentration of indeno(1,2,3-cd)pyrene depending on dose of CaO_2 added to sample.

Benzo(g,h,i)perylene has 6 benzene rings. Figure 9 shows the changes in the concentration of benzo(g,h,i)perylene depending on the dose of calcium peroxide Removal efficiency of benzo(g,h,i)perylene was 97% and occurred after adding 3 g/L of CaO_2. As shown the highest removal efficiency of all tested hydrocarbons was for 3 g /L calcium peroxide.

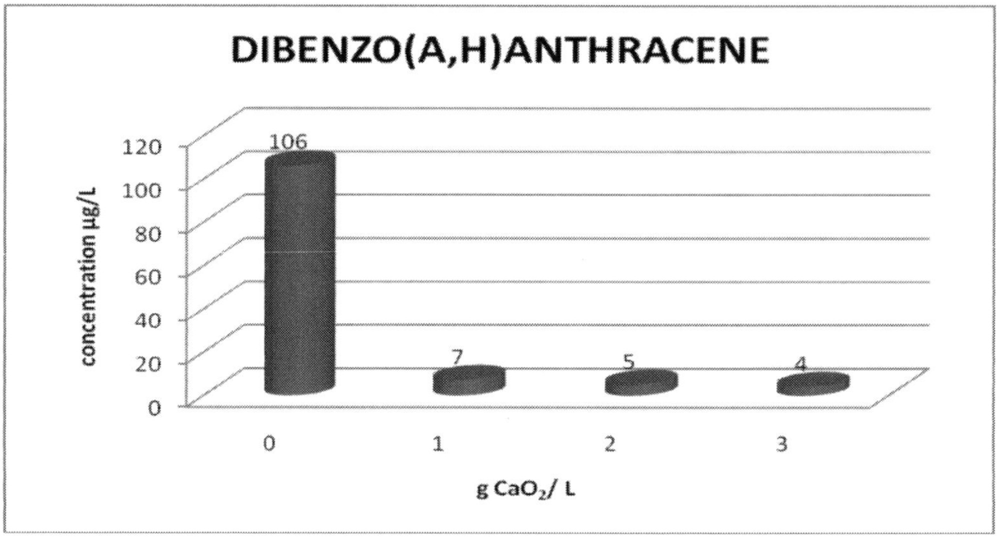

Fig. 8. Changes in the concentration of dibenzo(a,h)anthracene depending on dose of CaO_2 added to sample.

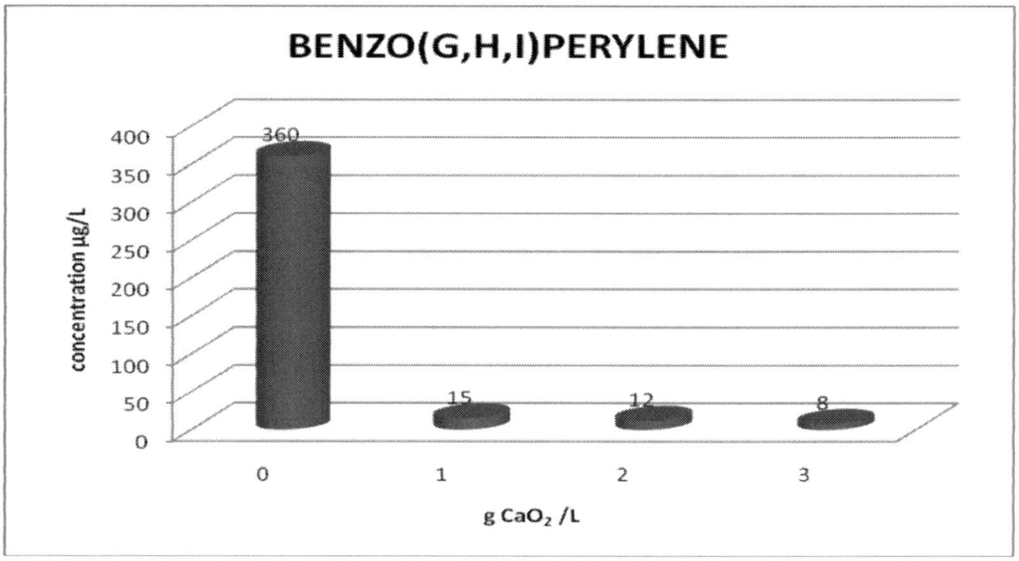

Fig. 9. Changes in the concentration of benzo(g,h,i)perylene depending on dose of CaO_2 added to sample

Effectiveness of decrease the concentration pyrene was 96 %. It has four benzene rings in which there are 8 delocalized π bonds. For both hydrocarbons with 4, 5 and 6 benzene rings the removal efficiency was similar and was 96-98%. This suggests that the amount of rings had no effect on the oxidation efficiency. Only in case benzo(a)anthracene removal efficiency was smaller. Benzo(a)anthracene has 4 benzene rings like pyrene but its benzene rings are connected in a different way. This can cause more difficult oxidation.

4 Conclusion

Based on the results, the following conclusions can be drawn.

1. Removal of organic compounds expressed by COD were the range of 30-35%.

2. The effectiveness of PAH degradation was directly proportional to the calcium peroxide dose
3. The efficiency in the removal of 4-ring of PAHs were as follows : pyrene 93%-96%, benzo(a)anthracene 89-94%, chrysene 93-97%, benzo(b)fluoranthene 96-98%, benzo(k)fluoranthene 96-98%. Removal efficiency was 97% .
4. The efficiency in the removal of 5-ring of PAHs were as follows benzo(a)pyrene 97-98%, dibenzo(a,h)anthracene 95-98%, indeno(1,2,3-cpyrene 96-98%. After oxidaton process total concentration of 5-ring hydrocarbons was 98% lower than initial concentration.
5. The efficiency in the removal of benzo(g,h,i)perylene, were in the range of 96-98%.
6. The results of the studies shown that CaO_2 can be applied as a source of hydroxyl radicals to the degradation of organic pollutants, such as PAHs being present in coking wastewater.
7. Photo-Fenton process can be used for wastewater treatment highly polluted with organic pollutants such as PAHs, and primarily purified in conventional processes.

References

1. A. Kot-Wasik, D. Dąbrowska, Pub. House of Faculty of Chem Gdańsk. **700** (in Polish) (2012)
2. F. Busetti., A. Heitz., M. Cuomo., S. Badoer , P., Traverso , J.Chromatogr., **1102**, 104–115 (2006)
3. G.S. Brown, L.L Barton, B. Thomson, Waste Manag., **23**, 737-740 (2003)
4. J. Wąsowski, A. Piotrowska, Environment. Protect., **85,** 27-32 (2002)
5. D. Gonzalez, L.M. Ruiz, G. Garralon., F. Plaza, J. Arevalo, J. Parada et al., Desalin Water Treat.**,** **42**, 94-9 (2012)
6. I.D. Manariotis, H.K Karapanagioti et al., Water Res, **45**, 2587-94 (2011)
7. K. Ravinda, R. Sokhi van Grieken., Atmos. Environ., **42**, 2895-921 (2008)
8. X. Zhao, Y. Wang, Z. Ye, A.G.L. Borthwick, J. Ni, Process Biochem., **41**, s. 1475-1483 (2006)
9.E. Bartulewicz, J. Bartulewicz, J. Gałkowski, Pub. House State. Depart. Hygien., 95-109 (in Polish) (1997)
10. M Włodarczyk-Makuła, Pub. House of Częstochowa University of Tech., **126**, (in Polish) (2007)
11. M Włodarczyk-Makuła, E. Wiśniowska, A. Turek, Desalin Water Treat., **57 (3)**, 1262 - 1272 (2016)
12. A. Dugay, C. Herrenknecht, M. Czok, F. Guyon, Pages N, J. Chromatogr., **958**, 1-7 (2002)
13. K. Marttinnens, R.H. Kettuner, J.A. Rintala, Sci.Total Environ., **301**, 1-3 (2003)
14. A. Rubio-Clemente, R.A. Torres-Palma, G. Panuela, Sci. Total Environ**, 478,** 201-225 (2014)
15. A. Northup, D.J. Cassidy, Hazard. Mater, **152**, 1164–1170.(2008)
16. M. Arienzo, Chemosphere **40**,331–337 (2000)
17 K. Barbusiński , Pub House Silesian University of Technology, (in Polish) (2013)
18. I. Skoczko, Annual set the Environ. Protec. **15,** 1460-1473 (in Polish) (2013)
19. T. Petry, P. Schmid, C. Schlatter, Chemosphere, **32**, 639-648 (1996)

Analysis and comparison of methods for the preparation of domestic hot water from district heating system, selected renewable and non-renewable sources in low-energy buildings

Maciej Knapik[1,*]

[1] Institute of Thermal Engineering and Air Protection, Faculty of Environmental Engineering, Cracow University of Technology

Abstract. The article presents an economic analysis and comparison of selected (district heating, natural gas, heat pump with renewable energy sources) methods for the preparation of domestic hot water in a building with low energy demand. In buildings of this type increased demand of energy for domestic hot water preparation in relation to the total energy demand can be observed. As a result, the proposed solutions allow to further lower energy demand by using the renewable energy sources. This article presents the results of numerical analysis and calculations performed mainly in MATLAB software, based on typical meteorological years. The results showed that system with heat pump and renewable energy sources Is comparable with district heating system.

1 Introduction

Low-energy and passive constructions are a result of the introduction of new ideas in building design process. Their main objective is to achieve a significant reduction in demand for non renewable primary energy, necessary to cover the needs of these buildings, mostly related to their heating, ventilation and domestic hot water. [1, 2]

Buildings of this type are characterized by increased demand of energy for domestic hot water preparation in relation to the total energy demand compared to existing buildings.

Research [3] shows that, the use of renewable sources such as wind and solar energy significantly reduces the primary energy demand in the building.

Electricity generated by wind turbines with vertical axis of rotation and photovoltaic cells can be used to heat the hot water in the tank through the warmer or as a source of electrical energy for the motive of a heat pump to prepare domestic hot water. Combining in one building, installations using these two types of renewable energy sources will provide benefits of mutual complementarity, when the energy generation in one of these installations will disappear or will be reduced. In the autumn and winter period, when the possibilities of solar energy are significantly reduced, wind turbines will produce much more energy than photovoltaic cells due to usually occurring windy weather in this time.

* Corresponding author: maciek.knapik@gmail.com

This situation is changing in the summer period, when the photovoltaic cells produce more energy, due to the longer day and much greater intensity of solar radiation. In the summer months, while the air masses are not moving too quickly, in result, it is observed lower wind speeds and therefore wind turbines produce limited amounts of power.

Application of the above described installations, which use renewable energy sources, can allow to cover a significant demand for electricity, that is required to power the heat pump [4, 5] working for purposes of heating and domestic hot water. In case of excess electricity and the lack of heating needs, this energy can be used to heat the hot water in the tank. The use of heat pump in the above-mentioned system, will reduce the energy demand for primary fuels used to generate electricity.

In case of disadvantageous wind and solar conditions, the installation of renewable energy sources may be uneconomical, then the preparation of domestic hot water from district heating system [6] or natural gas installation [7] should be analyzed.

The basis for the analysis is the installation with the priority of hot water preparation. Building heat losses exist, but for the purposes of this analysis they are omitted.

The main purpose of the article is to present an economic analysis and comparison of selected methods for the preparation of domestic hot water in a building with low energy demand. This article also presents the results of numerical analysis and calculations performed in MATLAB software, based on typical meteorological years and the results, which presents that the system with heat pump and renewable energy can be economically justified.

2 The assumptions for the analysis and simulation (heat pump system)

Table 1. The assumptions for the analysis and simulation. (study of Author)

Description	Assumption
Building type	Multifamily, modernized, low energy building
Location	Legnica, Poland (Polish third climatic zone) [8]
Inhabitants	30 people
Hot water tank	2*0.75 m^3
Wind turbines	Selected 1 wind turbine with power 1 kW
Photovoltaic cells	Selected 4 photovoltaic cells with power 300 W each (orientation S45^0)
Heat pumps	Selected 2 heat pumps with power 11 kW each

On the basis of the typical demand for hot water for 30 people in a multi-family building (Fig. 1), the typical heat demand for domestic hot water purposes has been calculated in the following months of the year. (Fig. 2). Heat demand for hot water purposes in installation with water tank is based on calculation procedure given in Polish standard [9]

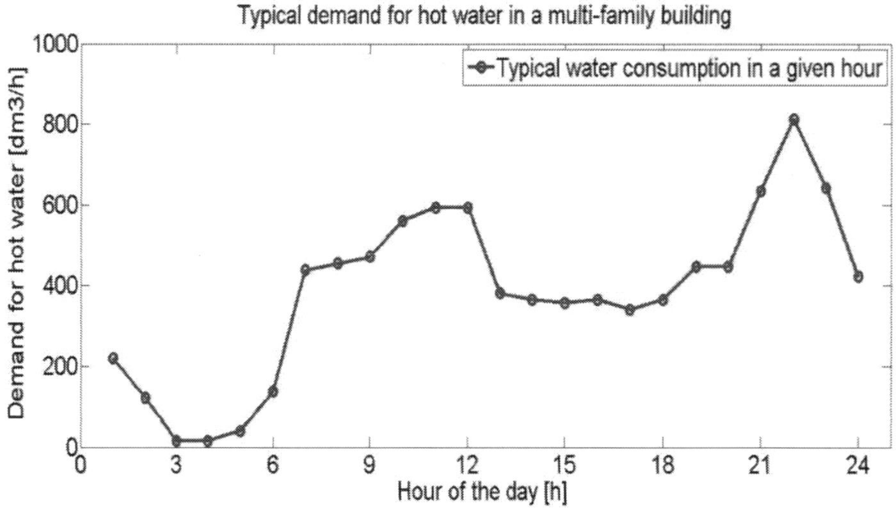

Fig 1. Typical demand for hot water for 30 people in a multi-family building. (study of Author)

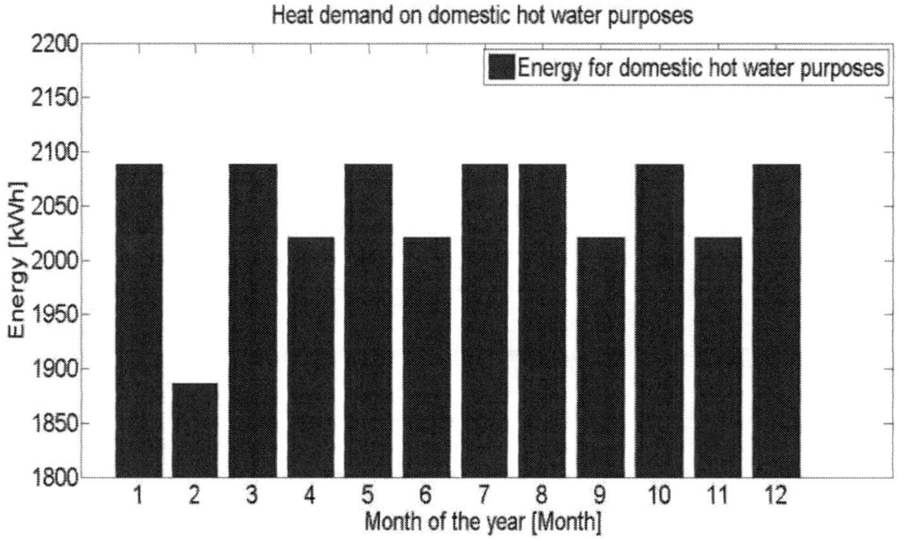

Fig 2. Typical heat demand on domestic hot water purposes during the year. (study of Author)

3 Domestic hot water from district heating system

One of the alternatives of supplying buildings with heat is district heating system [3]. In this case, heat is generated outside of the heated building and it supplied to the building via high-parameter pipe network, and then by a heat exchanger directed to the specific receivers. Particular advantage of district heating is the lack of a heat source at the customer, therefore, are not emitted to the environment any contaminants associated with combustion of non-renewable sources.

District heating substation is located in the building of the user, converts the high parameter of the heating medium, on the primary side, through a heat exchanger to the required parameter of hot water on the secondary side. District heating substation completely covers the energy demand for heating purposes. The heat demand is variable and adjusted by using the weather controller. Weather controller is changing the primary power supply parameters. Fees for the heat from the district heating network are calculated according to the current tariff and an indication of the heat meter is in a district heating substation.

Table 2. Tariff for consumers, which are supplied with heat in the form of hot water from district heating network in Legnica - B1-Lg. [10]

Tariff group	Heat production		Transmission and distribution	
	Annual fee for ordered thermal power	Cost of heat	Constant annual fee for heat transmission services	Variable fee rate for heat transmission services
	PLN/MW/year	PLN/GJ	PLN/MW/year	PLN/GJ
B1-Lg.	65486.01	25.25	19125.58	14.79

Thermal power of system with a tank of water, based on the maximum hourly water flow, multifamily, modernized, low energy building (30 people) amounts 13 kW.

The ordered thermal power amounts 0.013 MW. Based on typical consumption of hot water in the building (Fig. 2), annual demand of energy for domestic hot water in low energy building will amount 74.2 GJ per year.

On the basis of the district heat supplier tariff data (Tab. 2), annual operating costs for domestic hot water purposes are shown in the Tab. 3.

Table 3. The annual cost of hot water preparation according to the current WPEC Legnica tariff for heat from district heating network. (study of Author)

Description	PLN/year
Annual fee for ordered thermal power	851.32
Cost of heat	1873.63
Constant annual fee for heat transmission services	248.63
Variable fee rate for heat transmission services	1097.47
Summary (EC)	4071.05

4 Domestic hot water from renewable sources – heat pump powered by wind turbines and photovoltaic cells

Heat pump [4, 5] can be powered by electricity from wind turbines [11] and photovoltaic cells and it will reduce the energy demand for primary fuels used to generate electricity. In case of insufficient amounts of energy from renewable sources, the heat pump will be supplied from electrical network.

The use of the heat pump significantly reduces the need for energy for domestic hot water.

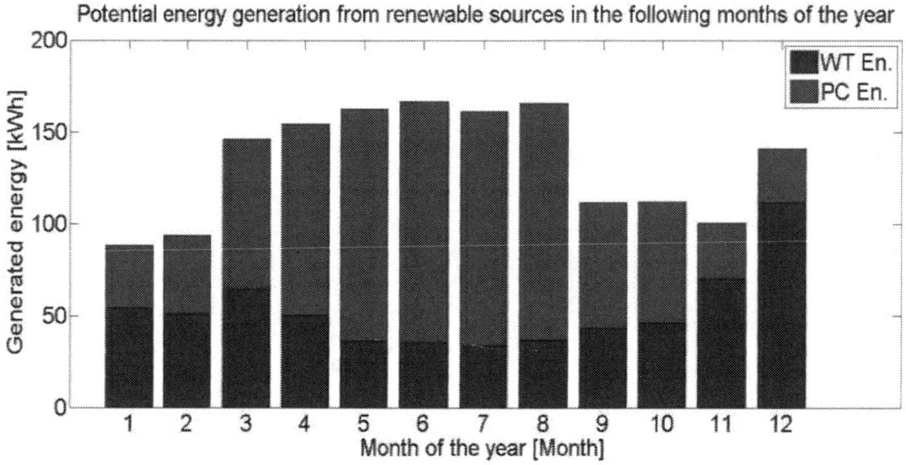

Fig 3. Potential energy generation from wind turbines and photovoltaic cells in the following months of the year. (study of Author)

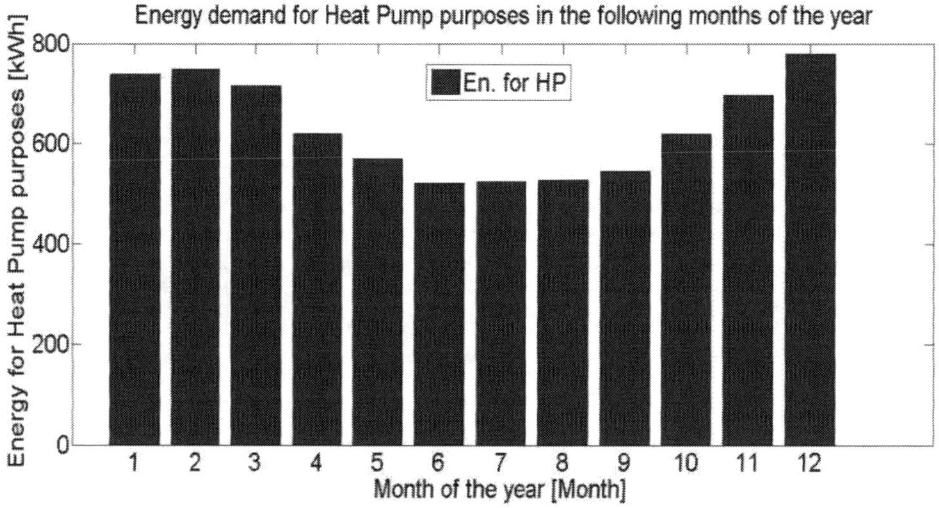

Fig 4. Electrical energy demand for heat pump purposes (motive power) in the following month of the year. (study of Author) , [12]

Fig. 3 shows the potential energy produced by renewable energy sources [12] in the analyzed system, the real value is weather dependent. Compared to Fig. 2, there can be observed a large energy deficit that must be covered by the electrical network.

The system with the heat pump [13] is able to fully cover the demand for domestic hot water. Energy demand for hot water purposes has been reduced by using an air-to-water heat pump [14] Economically, energy from renewable sources is not able to provide the total heat pump needs (Fig. 4), so the remaining energy must be provided from electrical network (Tab. 4).

Table 4. Electrical energy tariff [15] and the annual cost of hot water preparation in system with heat pump, powered by wind turbines and photovoltaic cells. (study of Author)

Description	Results
Generated energy from renewable sources	1604 kWh

Annual energy demand for heat pump, preparing domestic hot water	7601 kWh
Cost of energy daily tariff	0.316 PLN/kWh
Cost of energy night tariff	0.164 PLN/kWh
Annual summary costs (EC)	1350.29 PLN/year

5 Domestic hot water from non-renewable sources – natural gas

In this study is described a heating system powered by condensing gas boiler. Natural gas is the kind of ecological source of energy. As consequence of combustion of natural, it is being created a carbon dioxide and water vapour. The heat generated by the combustion of natural gas allows quick preparation of domestic hot water with minimal maintenance of the system. [16]

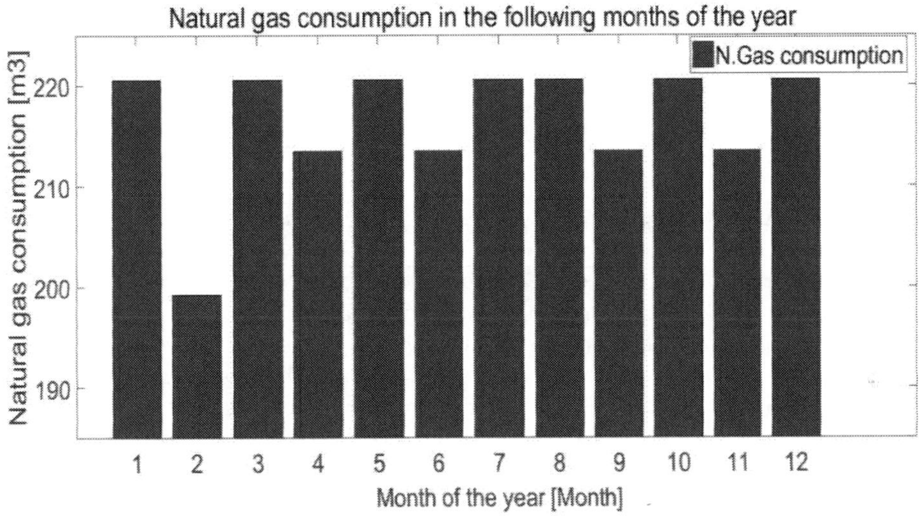

Fig 5. Natural gas consumption in the following months of the year. (study of Author)

Natural gas is one of the most expensive non-renewable energy sources. The amount of gas consumed during combustion depends on its caloric value and the efficiency of the gas boiler. The gas condensing boilers are characterized by the highest efficiency [7].

Natural gas consumption (Fig. 5) in the following months of the year can be calculated as follows:

$$V_{\mathrm{NGpM}} = \frac{E_{DHWP} * 3{,}6}{C_{VNG}} * \frac{1}{\eta} \qquad (1)$$

where:

V_{NGpM} - natural gas consumption in the following month of the year, m3/month

E_{DHWP} - energy for domestic hot water purposes in the following month of the year, kWh/month
C_{VNG} - caloric value of natural gas, MJ/m3
η - total efficiency of the domestic hot water preparation system

Gas installation is able to quickly prepare the proper amount of hot water, but this carries an economic consequences. (Tab. 5)

Table 5. Natural gas parameters [17] and the annual cost of hot water preparation in system with gas condensing boiler. (study of Author)

Description	Results
Caloric value of natural gas	35.5 MJ/m3
Efficiency of the system	96%
Annual natural gas consumption	2597 m3
Natural gas cost	2.40 PLN/m3
Annual summary costs (EC)	6233.27 PLN/year

6 Investment costs of analyzed systems

Investment costs are included without any subsidies (especially for renewable sources and heat pumps) and with discounts from individual manufacturers. A summarized costs of the analyzed systems are comparised on Fig. 6.

The analysis includes the cost of purchasing individual components such as wind turbines and photovoltaic cells, domestic hot water tanks, gas condensing boilers and heat pumps. Assembly costs for individual installations are also included.

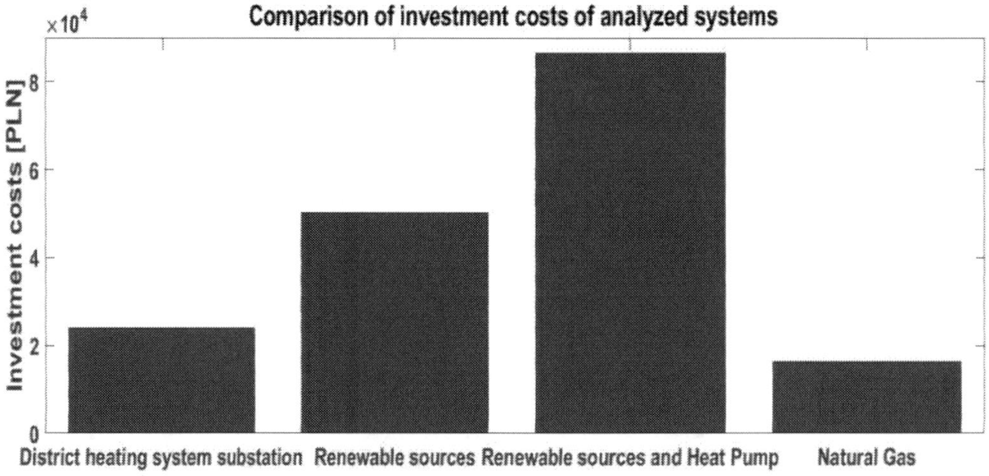

Fig 6. Comparison of investment costs of analyzed systems (IC_{SYS}). (study of Author)

7 Summary - calculations

The aim of the study was an analysis and comparison of the results of domestic hot water preparation from district heating system, renewable sources, non-renewable sources and heat pump powered by electricity from renewable sources in low-energy buildings. The economic effect of analysis can be observed comparing the mentioned systems together. In an article, for the analysis and comparison, it has been deliberately disregarded coal due to significant emissions to the atmosphere and considerable participation in the formation of smog in Polish cities.

The duration of the analysis is 15 years (photovoltaic cells warranty + 5 years). It has been taken into account the annual increase in the price of fuel (natural gas) and electricity by 1%.

Investment and operating costs (Fig. 7) in the analyzed period of time of the year can be calculated as follows:

$$EC_{(i+1)} = EC_{(i)} * 1{,}01 \tag{2}$$

where:
$EC_{(i)}$ - operating costs in the following year, PLN/year
$EC_{(i+1)}$ - operating costs in the next year, PLN/year
i - following year of the analysis, i ∈ <1,15>

$$C_{SYS(i)} = -IC_{SYS} - EC_{(1)} - \sum_{i=2}^{n=15}(-EC_{(i)}) \tag{3}$$

where:
C_{SYS} - investment and operating costs in the following year of the analysis, PLN
IC_{SYS} - investment costs of the analyzed system, PLN
$EC_{(1)}$ - operating costs in the first year of analysis, PLN
$EC_{(i)}$ - operating costs in the following year, PLN
i - following year of the analysis, i ∈ <1,15>

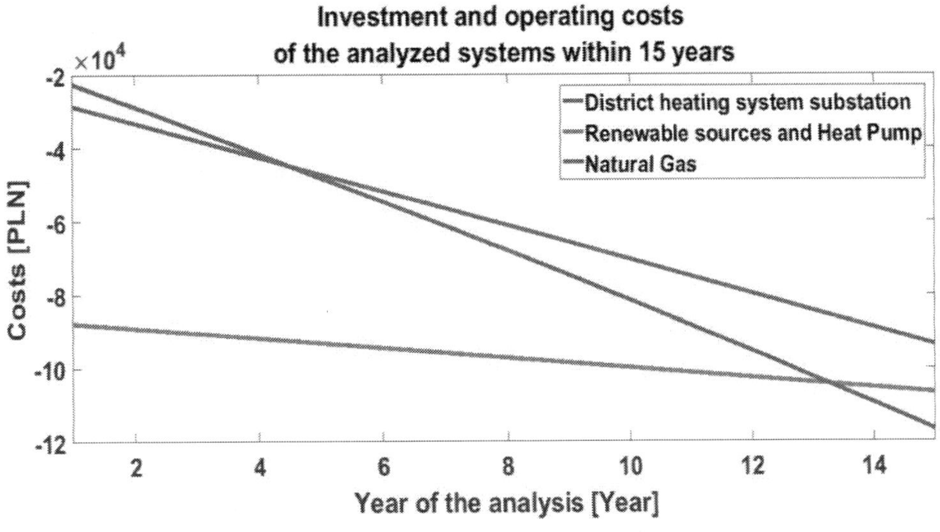

Fig 7. Investment and operating costs of the analyzed systems within 15 years. (study of Author)

The simulation (electrical energy generation by wind turbines and photovoltaic cells, based on typical meteorological years) and analysis in MATLAB software [12] allow to conclude that, it is preferable to use renewable energy sources which produce electricity to motive heat pumps or district heating system, if it is possible, because not all city units have a district heating network. The economic effect of analysis can be observed comparing the mentioned systems together (Fig. 7).

Considering the investment and operating costs, a better economic effect is achieved by district heating system compared to the gas system . A better result is achieved after 4 years of operation (Fig. 7).

Analyzing the system with renewable energy sources and comparing it with the gas system, better result is achieved after 13 years of operation (Fig. 7).

8 Summary - discussion

Research [18, 19] shows that, individual gas or oil boilers today and could be substituted by district heating or a more efficient individual heat sources as heat pumps powered by renewable sources. In such overall perspective, the best solution will be to combine a gradual expansion of district heating with individual heat pumps in the buildings with obsolete heating systems.

In the analyzed period, economically and ecologically as the research [20, 21] shows, the best choice is district heating system. The second system, based on the heat pump does not take into account the subsidy and is weather dependent. In the analyzed case (Fig. 3), can be observed that the wind turbines have a poor performance in comparison to the price.

Studies [22] shows that each different arrangement of Savonius rotor affects its performance.

In the case of using photovoltaic cells, instead of a wind turbine the results could be more energy efficient. Other studies [23] shows that the usage of photovoltaic cells can also be advantageous from the energetic point of view.

This article presents results that renewable energy sources, can allow to cover a significant demand for electricity, that is required to power the heat pump and it is economically justified, in comparison to natural gas system and in comparison to district heating system it is not significantly different.

Article draws attention to the process of proper selection of the domestic hot water preparation system. Proper selection should be preceded by a thorough analysis of the prevailing weather conditions and the recognition of the use of district heating network.

References

1. W. Feist, U. Münzenberg, J. Thumulla, *Podstawy budownictwa pasywnego*, (2009, ISBN 83-923807-0-3)
2. E. Rylewski, *Energia własna*, (2002, ISBN 83-917314-2-1)
3. M. Knapik, *Analiza i wybór źródła grzewczego przygotowującego ciepłą wodę z wykorzystaniem energii odnawialnej*, RI, **9** (2016), 36-38
4. K. Wojtas, *Sprężarkowa pompa ciepła jako alternatywne źródło ciepła w budynku (cz. 1)*, Polski Instalator, **3** (2011), 40-44
5. K. Wojtas, *Sprężarkowa pompa ciepła jako alternatywne źródło ciepła w budynku (cz. 2)*, Polski Instalator, **4** (2011), 40-44
6. E. Szczechowiak, *Energooszczędne układy zaopatrzenia budynków w ciepło*, (Envirotech, Poznań 1994)
7. R. Śnieżyk, *Dostawa ciepłej wody zasilanej gazowym kotłem kondensacyjnym*, RI, **5** (2014), 76-80

8. PN–EN 12831 – 2006: *Instalacje ogrzewcze w budynkach. Metoda obliczania projektowego obciążenia cieplnego*
9. Polish standard, PN-90/B-01706 - 1992: *Instalacje wodociągowe -- Wymagania w projektowaniu*
10. Taryfa dla ciepła sieciowego na 2017 rok, WPEC, Legnica 2017
11. Blitzmann, technical materials and catalogs
12. M. Knapik, *Analysis of the possibility to cover energy demand from renewable sources on the motive power of the heat pump in low-energy building*, E3S Web of Conferences **17**, 00039 (2017)
13. DAIKIN, technical materials and catalogs
14. PN-EN 15316-4-2 – 2008: *Heating systems in buildings – Method for calculation of system energy requirements and system efficiencies – Part 4-2: Space heating generation systems, heat pump systems*
15. Taryfa dla energii elektrycznej na 2017 rok, PGE, 2017
16. M. Kowalik, J. Boryca, B. Hałusiak, *Analiza wpływu cen gazu ziemnego i energii elektrycznej na koszty ogrzewania za pomocą 2-funkcyjnego kotła gazowego w budownictwie wielorodzinnym*, COW, **3** (2013), 96-100
17. Taryfa dla gazu ziemnego GZ-50 na 2017 rok, PGNiG, 2017
18. H. Lund, B. Möller, B. V. Mathiesen, A. Dyrelundb, *The role of district heating in future renewable energy systems,* Energy, **3** (2010), 1381-1390
19. M. Münster, P. E. Morthorst, H. V. Larsen, L. Bregnbæk, J. Werling, H. H. Lindboe, H. Ravn, *The role of district heating in the future Danish energy system,* Energy, **1** (2012), 47-55
20. M. Thyholt, A. G. Hestnes, *Heat supply to low-energy buildings in district heating areas: Analyses of CO_2 emissions and electricity supply security,* Energy and Buildings, **2** (2008), 131-139
21. X. Chen, L. Wang, L. Tong, S. Sun, X. Yue, S. Yin, L. Zheng, *Energy saving and emission reduction of China's urban district heating,* Energy Policy, **Vol. 55** (2013), 677-682
22. J. V. Akwa, H. A. Vielmo, A.P. Petry, *A review on the performance of Savonius wind turbines,* Renewable and Sustainable Energy Reviews, **5** (2012), 3054-3064
23. J. Ji, G. Pei, T. Chow, K. Liu, H. He, J. Lu, C. Han, *Experimental study of photovoltaic solar assisted heat pump system,* Solar Energy, **1** (2008), 43-52

Energy from Landfill Gas as an Example of Circular Economy

Józef Ciuła[1], *Krzysztof* Gaska[2], *Agnieszka* Generowicz[3] and *Gabriela* Hajduga[3,*]

[1]Nowy Sącz Water Networks Company, 22 Wincentego Pola Street, 33-300 Nowy Sącz, Poland
[2]Chair of Technologies and installations for Waste Management, Silesian University of Technology, 18 Konarskiego Street, 44-100 Gliwice, Poland
[3]Institute of Water Supply and Environmental Protection, Cracow University of Technology, 24 Warszawska Street, 31-155 Cracow, Poland

Abstract. Landfill biogas becomes an important factor in elimination of fossil fuels as a result of fast- growing use of renewable energy sources. The article presents an analysis of operation of the plant where landfill biogas was utilized for energy production. The average annually (gross) productions of electric energy and heat at the plant were 1217 MWh and 1,789 MW, respectively. The average calorific value of biogas was 17 MJ/m^3, which corresponds to 4,8 kW/m^3. According to the measurements and actual readings acquired during operation of a cogeneration unit, it can be stated that the CHP system has been working within its average operation limits and still has some power reserves to utilize. Therefore, the authors concluded that a landfill can be operated both as a producer and a supplier of prosumer energy.

1 Introduction

According to the Central Statistical Office (GUS) [1, 2] there have been 394 municipal landfills in operation in Poland at the end of 2014. They occupied a total area of over 4762 acres. At the same time, 66 landfills were closed in 2014 with the total area of almost 390 acres. Also, about 10 330.4 thousand tons of municipal waste (an increase by 8.3%, if compared to 2013.) have been collected and 5 978.7 thousand tons (52.6%) were disposed at landfills in 2014. The data show that landfills are still the leading method of solid waste disposition in Poland.

Municipal waste stored in landfills has a high content of organic materials of different humidity. Methane fermentation produces biogas. The composition of biogas is mainly methane (35 - 70%), carbon dioxide (20 - 50%) and other pollutants, among others. nitrogen, hydrogen, hydrogen sulphide, oxygen, carbon monoxide. On average, 150-400 m3 of waste can be recovered from the 1 tonne mass of waste [3 – 5, 6, 7].

Biogas must be cleaned and used for energy purposes or burned in a torch. The most common way of managing landfill gas is to burn it in specially prepared torches or heat-generating plants that supply heat to adjacent plants or housing estates. However, it is best to use landfill gas as a fuel for generating sets equipped with heat exchangers. Not only does it have "clean" ecological electricity as well as warmth that can be used freely.

A general problem with biogas resources is their local nature. The EU policies concerning renewable energy systems (RES) have set forward a fixed goal of supplying 20% of the European energy demands from RES by year 2020. A major part of the renewable energy will originate from European farming and forestry. At least 25% of all bioenergy in the future can originate from biogas, produced from wet organic materials such as: animal manure, whole crop silages, wet food and feed wastes, etc. [6 – 9].

The worldwide biogas production is unknown, but the production of biogas in the European Union was estimated to be around 69 TWh in 2007. To arrive at a rough estimate of global landfilling, we started with the known rate of landfilling in the US (220 million tonnes). The European Union (EU), and the rest of the "golden billion" who enjoy a high standard of living generate an estimated 420 million tonnes of MSW of which at least 210 million tonnes (50%) are landfilled. Waste management studies in developing nations, including some in Africa, have shown that the MSW generation is always higher than 0.2 tonnes per capita, most of which is food and yard wastes and is landfilled. This results in the estimate of 1080 million tonnes for the 5.4 billion people in the developing world. Adding up these estimates indicates that the global MSW landfilled is somewhere close to 1.5 billion tonnes of MSW[6, 7, 10].

Optimization of energy consumption at solid waste disposal facilities is a major challenge while managing these objects. Rising operation costs (in particular energy costs) are forcing policymakers to search for new solutions, including new energy sources. Landfill biogas can be considered as such a potential energy source.

2 Waste disposal installations – Waste-to-Energy Units

The landfill, which collects municipal waste with a biodegradable fraction becomes a "bioreactor", where organic matter is transformed into landfill biogas. According to data provided by the GUS, 342 Polish landfills had degassing installations in 2014; they accounted for 86.8% of all active landfills where municipal waste were deposited. The detailed summary of the Polish landfills with degassing installations in –service in 2014 is presented in the table below.

Table 1. Degasification installations and energy recovery systems at the Polish landfills

Year	Landfills with degasification	Landfills with a gas discharge to atmosphere	Landfills with individual burners	Landfills with a flare	Landfills with heat recovery	Landfills with electricity generators
2014	342	159	59	82	16	62
2013	362	199	61	78	15	59
2012	430	244	63	81	13	58
2011	322	225	30	28	5	31

More than 42% of the degassing installations were units that discharge gas directly into the atmosphere (a drop by 6.2%, comparing to 2013) while at 4.2% of the installations the captured landfill gas was neutralized and used for recovery of thermal energy; 16.4% of units use landfill gas for electricity production. As a result of landfill gas burning about 81 415 thousand MJ of thermal energy and about 148 348 thousand kWh of electricity were recovered [6 - 8].

The optimum parameters of landfill biogas, used for energy purposes, reported in the literature should remain within the following ranges: methane 45 ÷ 58%, carbon dioxide 32 - 45% nitrogen, 0 ÷ 5%, hydrogen, 1 ÷ 2%, oxygen 2% and trace amounts of other compounds such as hydrogen chloride, alkaline compounds, hydrogen sulfide 10-200 ppm, ammonia, etc.In fact, every landfill is different in terms of a chemical composition of biogas, its performance, humidity and energy potential. Therefore, every degassing installation at landfills should be treated individually, especially while looking into use of biogas for energy.

Currently, cogeneration (CHP) i.e. landfill operation with energy recovery is becoming more popular in Poland. This process can be defined as a thermodynamic process in which chemical energy from landfill biogas is converted into useful energy forms such as electricity and heat (for plant's technological/maintenance needs). The main advantage of cogeneration is overall conversion efficiency achieved in the cogeneration process, which is higher than when electricity and heat are produced in a separate way. The overall process efficiency exceeds 85% [9 - 12].

3 Landfill as a source of renewable energy

Energy from renewable sources it is energy obtained from regular natural processes and extracted from renewable non-fossil energy sources including e.g. landfill biogas. Renewable Energy Sources (RES) are an alternative to traditional, non-renewable primary energy carriers [12 – 15, 16, 17].

The amount of primary energy, exerted from renewable sources, and its share in total amount of energy obtained in individual EU countries varies. In most EU countries, renewable energy and its share in total primary energy acquisition is growing. In Poland, a significant increase in the amount of energy coming from renewable sources was observed in the last 10 years (Table 2).

Table 2. Energy from renewable sources as a percentage of primary energy in UE-27 and Poland [18]

Categories	2008	2009	2010	2011	2012	2013	2014
Energy from renewable sources as a part of primary energy in the EU (total)	16,2	17,9	19,6	20,2	22,3	24,3.	25,1
Energy from renewable sources as a part of primary energy in Poland	7,6	9,0	10,2	10,9	11,7	11,9	11,8

Consequently, a recent increase of renewable energy source utilization is accompanied by a growing role of landfill biogas in elimination of fossil fuels. According to data published by the GUS [18], concerning the amount of electricity produced from biogas in Poland in 2014 (Figure 1), biogas comprises 3,66% of all renewable sources. The data include landfill biogas, biogas from wastewater treatment plants and remaining biogas (from agriculture and fermentation of waste from brewery, food processing and slaughter houses).

Electricity production from biogas can be further divided, as follows [6, 7, 8]:

1) Landfill biogas 225 GWh,
2) Biogas from wastewater treatment plants: 253 GWh,
3) Remaining biogas: 338 GWh

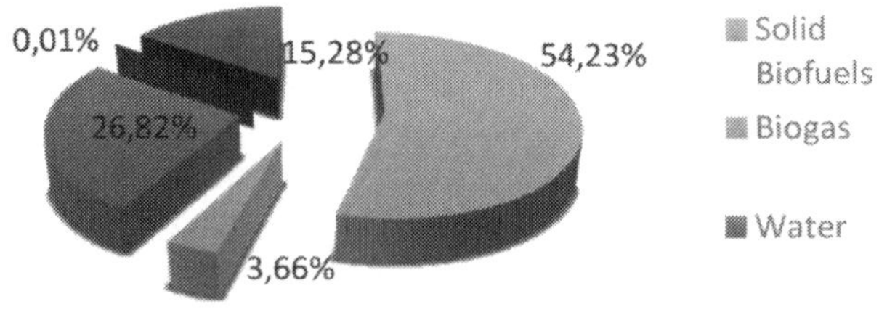

Fig. 1. Renewable energy carriers and their share in electricity production in 2014 [18]

Accordingly, heat production from biogas in 2014 was as follows:
1) Landfill biogas: 69 TJ,
2) Biogas from wastewater treatment plants: 86 TJ,
3) Remaining biogas: 144 TJ.

The lowest heat production attributed to landfill biogas results from a low heat demand at a landfill, compared to other producers[17- 20]. There are some advantages of use of biogas as an energy source. They include: environmental effect coming from discontinued emissions (methane, carbon dioxide), economic benefits due to use of heat and electricity to satisfy user's own needs and the added value of so called "green certificates" [21 - 25].

4 Waste disposal plant – energy prosumer

At most waste disposal plants a landfill is just one of units that comprise a waste disposal system and require a constant supply of energy. Electricity and heat produced from landfill biogas can be primarily spent to satisfy facility's own needs while the surplus can be transferred (sold) to a power grid as "green energy". This is the new role of the facility as an energy prosumer.

While looking into the way electricity is managed in prosumer systems three types of installations can be distinguished:

- directly connected to the network (on-grid) – that transfer the produced energy directly to the grid. The owner of installation benefits from the energy sale. At present, such a solution requires a special license from the President of the Energy Regulatory Office.
- with no connection to the external grid (i.e. off-grid or island systems) - that operate independently; the plant powers a single facility with all its machinery and equipment (e.g. a landfill). In such cases a detailed balance of the energy demand is required for each day and time of year.
- mixed systems - most commonly used systems, combining features of on- and off-grid systems. During peak energy demands and "green energy" scarcity a renewable energy is supplemented by energy from a grid. When more energy is produced than can be utilized its excess can be transmitted to the grid using intelligent energy meters (bidirectional system)[20 – 22, 25].

Electricity and heat generated at landfills make up the external electric power system. Prosumer energy, produced on site, can significantly affect the overall energy balance at the landfill and consequently reduce costs of its operation[23 - 29] as well as its harmful environmental effects.

5 Waste disposal plant and prosumer energy use

The facility (landfill) is located in the Malopolska province in the area of valuable natural assets. The facility comprises 4 installations used for: waste disposal (landfill), mechanical-biological waste treatment, landfill degassing and disposal of landfill biogas and its use for energy.

The landfill, that has been collecting non-hazardous and inert wastes, was put into operation in 1999 and currently it covers the area of about 10 acres; its operational capacity is about 600 000 m^3. Currently, the landfill is filled up to about 70% of its volume. The landfill operates an installation for landfill degassing, which consists of vertical and horizontal degassing wells, biogas transporting pipelines and a station for biogas acquisition and processing with a flare mounted at its roof. An installation for active degassing of the landfill was put into operation in 2009, and in January 2010 the Company

 https://doi.org/10.1051/e3sconf/20183003002

managing the unit received the licensed to produce electricity from landfill biogas. The energy is produced by burning biogas extracted from the waste in a gas engine or the flare.

A package cogeneration unit (in container) is a part of the installation for a landfill biogas disposal and energy use. The unit produces 365 kW of electric power and 455 kW of heat. The biogas produced from landfill waste deposits is used as fuel to power a piston engine, which drives a synchronous generator and converts mechanical power into electricity. The unit works as a biogas CHP in a CHP (Combined Heat and Power) system, where both electricity and heat are produced. The volumes of landfill biogas collected and disposed at the landfill in the years 2010 - 2015 are presented in the Table 3. Utilization of landfill biogas in a cogeneration system is the optimal way of a chemical energy use in a thermodynamic process. Currently methane, the main component of landfill gas, represents 48 to 50% of its volume. In 2010, when the production of electricity was initiated, a methane content was about 56%. Changes in the methane content in the landfill biogas in the years 2010 - 2015 are shown in the figure below.

Since according to the new regulations non-biodegradable waste cannot be deposited at landfills, it may be assumed that volumes of both biogas and methane produced at landfills will steadily decline in the coming years. Therefore, the installation's managers will face a problem how to optimize production of electricity and heat to power the engines.

Currently, 100% of electricity produced in the installation is distributed through a mixed type system of on - and off-grid. Electricity is mostly used to satisfy plant's own needs and power an electricity generator (process maintenance); its surplus is transmitted to the external power grid for further distribution. The balance of electricity generated from a renewable source (landfill biogas) was presented in the Table 3.

Table 3. Landfill biogas disposal (Authors' elaboration)

Year	Landfill biogas collected and disposed	Biogas burned in an engine	Biogas burned in a flare
	m^3	m^3	m^3
2010	832398	778498	53900
2011	1120998	1072998	48000
2012	1037215	991295	45920
2013	773748	697499	76249
2014	822110	815390	6720
2015	816364	808464	7900
Total	**5402833**	**5164144**	**238689**

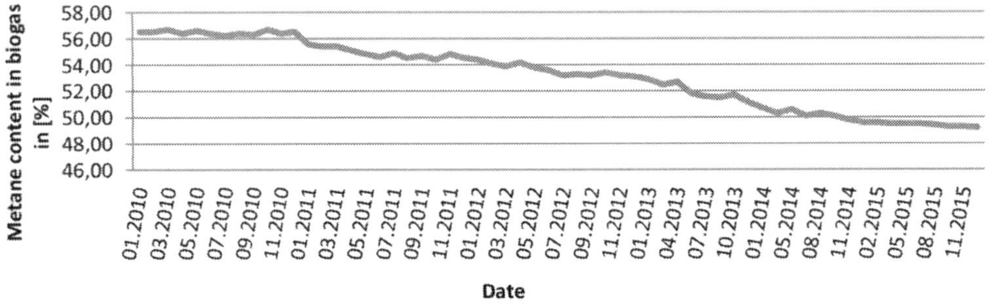

Fig. 2. Methane in the landfill biogas at the installation in years 2010–2015 (Authors' data)

Table 4. Balance of electricity generated and used at the location (Authors' own data)

Year	Electricity generated (gross)	Electricity transmitted to the external grid (net)	Electricity utilized on- site (total)	Electricity utilized for plant's own needs	Electricity utilized for installation maintenance
	MWh	MWh	MWh	MWh	MWh
2010	1223,5	1040,7	182,8	121,4	61,4
2011	1670,3	1417,6	252,7	143,5	109,2
2012	1436,4	1196,0	240,4	143,0	97,4
2013	971,4	775,8	195,6	124,9	70,7
2014	1012,5	802,3	210,2	133,5	76,7
2015	991,7	751,4	240,2	167,4	72,9
Total	**7305,8**	**5983,7**	**1322,0**	**833,7**	**488,4**

Electricity is transmitted to the external power grid, under the terms of the contract with the distributor, at the price per 1 MWh set each year by the President of the Energy Regulatory Office. Production of energy from renewable sources has also the added value. This is a support system that transfers property rights to the produced electricity (green certificates) and enables its sale within the actual financial mechanisms.

Heat generated in a biogas generator equipped with a heat recovery unit is used for every-day operation (heating of a sorting cabin - heat exchanger: glycol/air) and for technological processes (drying of waste in a rotary screen - heat exchanger). Currently, about 25% of the heat is used during a day-to-day operation. The amount of heat generated in a cogeneration system is shown in the Table 5.

Table 5. Heat generated from landfill biogas at the location in years 2010 – 2015

Year	Heat generated (gross)	Heat utilized at the waste disposal plant (net)	
	GJ	GJ	%
2010	5861	1008	17,2
2011	8440	1806	21,4
2012	7464	1649	22,1
2013	5252	1229	23,4
2014	6139	1523	24,8
2015	5500	1408	25,6
Total	**38653**	**8623**	

The average quantity of heat generated over 6 years was 6443 GJ, which corresponded to 789,789 kW of energy. To increase a heat consumption at waste disposal plant, an internal heating system will soon be constructed and put in operation to heat up a working area and produce hot water on site.

Biogas is not the only renewable energy source at waste disposal plant. The plant has also in liquid solar collectors to convers solar energy and use it for hot water production and as a back-up for a gas boiler. These methods of energy production may serve as good examples of implementation of the idea of prosumer energy in spread systems, using renewable energy sources.

Operation of installation using landfill biogas as an energy source was analysed in the years 2010 – 2015. It showed that about 1217 MWh of electricity and about 1,789 MW of heat was annually produced there (average gross values).

Cogeneration systems should meet the efficiency requirements for electricity and heat cogeneration if they are to work properly and provide a basis for the prosumer energy.

There are several most commonly used indicators describing efficiency of energy conversion in a CHP system with piston engines, which include:

1) Efficiency of electricity generation in a combined cycle: 25÷40% (1),

2) Efficiency of heat generation in a combined cycle: 30÷50% (2),

3) Overall CHP system efficiency (chemical energy utilization factor - EUF): 70÷90% (3),

4) Association index: 0,5÷1,0 (4).

Based on the average annual operating parameters of a cogeneration system with landfill biogas (6-year operation) the above indicators have been calculated. Their values are as follows:

$$el_{EC} = \frac{Eel}{P \cdot Wd} = \frac{1217625kW}{860691m^3 \cdot 4{,}8kW/m^3} = 0{,}29 \quad (1)$$

$$\eta q_{EC} = \frac{Q}{P \cdot Wd} = \frac{1789789kW}{860691m^3 \cdot 4{,}8kW/m^3} = 0{,}43 \quad (2)$$

$$\eta c_{EC} = EUF = \frac{Eel+Q}{P \cdot Wd} = \frac{1217625kW+1789789kW}{860691m^3 \cdot 4{,}8kW/m^3} = 0{,}72 \quad (3)$$

where$\eta el$is a electricpower generated in a combined cycle, $Q$is a heatgenerated in a combined cycle, $P$is a fuel stream and$Wd$is a fuel caloric value.

The average caloric value of biogas was 17 MJ/m^3, which is equivalent to 4,8 kW/m^3.

The input data for the above calculations came from actual measurements and readings acquired during cogeneration unit operation. It can be concluded that they fall within the range of average values so the CHP has been working within the operation limits and still has some reserves to utilize.

Based on the available data on electricity produced from landfill biogas in 2015 and the way it was utilized a prosumer energy balance may be prepared. It shows to which extend the installation serves as an electricity consumer or a producer or a distributor.

The balance of electricity generated at the plant shows that the most electricity (75,78%) was sold to the external grid (distribution) while the plant owns overall needs accounted for 24,83% of electricity: 16,89% of this value represents electricity used to power facilities, machinery and equipment at the landfill and the facility itself, while 7.94% is used by the degassing installation (maintenance works).

Electricity from the external power grid is purchased only in case of planned maintenance services/repairs and failures of the cogeneration unit. In 2015, the waste disposal plant and the installation for degassing and disposal of landfill biogas consumed a total of 240.2 MWh of electricity, while only 7.4 MWh were purchased from an external supplier. The balance of electricity used at the plant in 2015 shows that 3% was bought from the external power grid, while 97% was a prosumer energy

Considering the above data, it should be stated that the plant tries to implement the concept of prosumer energy in a manner consistent with the idea. Since both electricity and heat are produced from waste it can be firmly stated that the process operates in a closed waste-energy cycle where waste becomes a product (energy carrier).

The landfill as a prosumer meets its own energy needs and at the same time acts as a "green energy" provider for households, offices, cities, etc. This way, it contributes to the development of a prosumer energy.

6 Summary

The use of landfill biogas as a renewable energy source enables to utilize local energy resources. Prosumer energy serves as an example of a comprehensive approach, where transition from local issues to global issues in energy conservation is observed. It can be noticed throughout the course of its obtaining, efficient processing, transmission and finally proper utilization.

While looking into basic functions of the landfill as: facility, bioreactor, renewable energy source, consumer and ultimately an energy prosumer one can conclude that in this case the landfill becomes both a producer and a supplier of electricity needed for landfill operation.

It seems appropriate to extend the concept of prosumer energy utilization to waste management companies, which often, as regional installations, utilize energy generated during anaerobic digestion of biodegradable waste. Such examples can be already found in Poland and they should be encouraged to take a comprehensive approach to energy and waste issues, authorized by clear legislations.

References

1. I. Adamczyk, B. Różańska, M. Sobczyk, *Główny Urząd Statystyczny*, 24-33 (2015)
2. K. Gaska, K. Pikon,Japca.J. Air Waste. Ma.**60(7)**, 782-788 (2010)
3. D. D. Dillah, Heating Landfill Facilities Using Infrared Heaters
4. United States Environmental Protection Agency,*Emerging technologies for the management and utilization of landfill gas*, EPA-600/R-98-021(1998)
5. D. Deublein, A. Steinhauser*Biogas from Waste and Renewable Resources: An Introduction*(Viley – VChVerlag GmbH & Co KGaA, 2011)
6. J.B. Holm-Nielsenab, T.Al Seadib, P.Oleskowicz-Popielc, Bioresource Technol.**100 (22),**5478 – 5484 (2009)
7. N. J. Themelis, P. A. Ulloa, Renew. Energ.**32 (7)**, 1243 – 1257 (2007).
8. K. Ribeiro Salomon, E. Eduardo, S. Lora,Biomass Bioenerg. **33(9)**, 1101–1107 (2009)
9. B. Ozkaya, A. Demir, M. S. Bilgili, Environ. Modell. Softw. **22**, 815 – 822, (2007)
10. J.D. Murphy, E. McKeogh, Renew.Energ.**29(7)**, 1043–1057 (2004)
11. G. Zappini, P. Cocca, D. Rossi, Energy, **35(12)**, 5063-5069 (2016)
12. J. Ciuła, *Aura Ochrona Środowiska*, **5**, 8-9 (2009)
13. J. Mikosz, Desalin. Water. Treat.**51(10-12)**, 2461–246 (2010).
14. W. M. Budzianowski, RenewSust.Energ. Rev.**16 (1)**, 342–349 (2012)
15. S. Rasi, J. Läntelä, J. Rintala - Trace compounds affecting biogas energy utilisation – A review, Energ. Convers. Manage.**52 (12)**, 3369–3375 (2011)
16. M.Pöschla, S.Warda, P.Owendea, Appl.Energ.**87(11)**, 3305–3321 (2010)
17. J. Mikosz,Int. J. Environ. Sci.Te.**12(3)**, 827–836 (2013)
18. J.D. Nixon, P.K.Dey, S.K. Ghosh, P.A. Davies, Energy**59(15)**, 215-223 (2013)
19. M.D. Rey, R. Font, I. Aracil, Energy, **63(15)**, 161-167 (2013)
20. N. Kythreotou, S.A. Tassou, G.Florides, Energy, **47(1)**, 253-261 (2012)
21. Z. Kowalski, A. Generowicz, A. Makara, J. Kulczycka,Environ. Prot.Eng.**41(4)**, 167-179 (2015)
22. H. Emkes, F. Coulon, S. Waglan, Waste. Manage.**43**, 307-318 (2015)
23. R. Slezak,L. Krzystek, S. Ledakowicz, Waste. Manage.**43**, 293-299 (2015)

24. W. Balcerzak, A. Generowicz, Z. Mucha, Pol. J.Environ.Stud.**23(3)**, 983-987 (2014)
25. W. Kostkowski, S. Uson, W. Stanek, Energy **76**, 10-18 (2014)
26. D. Chwieduk, *Gospodarka Komunalna w Miastach*, 336 – 345 (2009)
27. Q. Tu, Ch.Zhu, D.C.McAvoy, Waste Manage.**39**, 258-265 (2015)
28. L. Lombardi, E.Carnevale, A. Corti, Energy **39(15)**, 3208–3219 (2006)
29. G. Hajduga, A. Generowicz, E3S Web Conf.**17** (2017)

Possibilities of heat energy recovery from greywater systems

Kaja Niewitecka[1,*]

[1]Faculty of Building Services, Hydro and Environmental Engineering, Warsaw University of Technology, 20 Nowowiejska Street, 00-653 Warsaw, Poland

Abstract. Waste water contains a large amount of heat energy which is irretrievably lost, so it is worth thinking about the possibilities of its recovery. It is estimated that in a residential building with full sanitary fittings, about 70% of the total tap water supplied is discharged as greywater and could be reused. The subject of the work is the opportunity to reuse waste water as an alternative source of heat for buildings. For this purpose, the design of heat exchangers used in the process of greywater heat recovery in indoor sewage systems, public buildings as well as in industrial plants has been reviewed. The possibility of recovering heat from waste water transported in outdoor sewage systems was also taken into consideration. An exemplary waste water heat recovery system was proposed, and the amount of heat that could be obtained using a greywater heat recovery system in a residential building was presented. The work shows that greywater heat recovery systems allow for significant savings in preheating hot tap water, and the rate of cost reimbursement depends on the purpose of the building and the type of installation. At the same time, the work shows that one should adjust the construction solutions of heat exchangers and indoor installations in buildings to the quality of the medium flowing, which is greywater.

1 Introduction

Increasing demand for energy, observed all over the world, forces to take action aiming at managing it more effectively. In Poland, a new approach to energy management results from imposed by the European Union country's commitment concerning the improvement of energy efficiency by increasing the share of renewable energy sources and reducing the negative impact of energy economy on the environment. One of the practical ways allowing to achieve these goals is to use waste heat from waste water.

Currently, many European countries are implementing technical solutions allowing to restore a portion of the lost heat and manage the recovered energy as an alternative source of energy for a building, for instance, to preheat water, heat the buildings or as a source of chill for air conditioning during summer. All this can be achieved using suitable devices such as heat exchangers, which can be installed both in buildings and in sewage systems.

* Corresponding author: kaja.niewitecka@pw.edu.pl

Recently, what is gaining popularity in Poland are so-called "nearly zero-energy" buildings, where one of many technical solutions to reduce energy consumption could be the heat recovery from the underrated energy which is greywater. Yet, this is still an innovative issue, rarely applied in Poland. Literature indicates, however, that this form of waste energy use has its own energy, environmental and economic justification.

2 Legal regulations of greywater use

The use of greywater as an alternative energy source in a building is not subject to any specific legal regulations in Poland. Thinking about waste water heat recovery, we have to distinguish two concepts: grey sewage and black sewage. These concepts are defined in the Polish standard PN-EN 12056-1: 2002P. What is meant by the term "black sewage water" is cold waste water discharged from toilet bowls and urinals, at the temperature of about 10°C. The "grey sewage water", however, is the outflows from bathtubs, showers, washbasins, automatic washing machines or dishwashers. They contain less pollutant load and their temperature is higher (about 30°C). In the regulations an important reference for the possibility and purpose of the use of grey sewage is the record in the standard PN-EN 12056-1: 2002, which classifies a sewer system discharging grey and black sewage by separate sewage systems (so-called system IV). Therefore, in order to use the energy potential of grey sewage discharged from the building, it is necessary to start with separating the sewage system. This will allow waste water heat recovery system to be mounted on the grey sewage drain tube [1].

Modern technologies allow for the collection of heat from waste water also at the stage of their transportation to the sewage treatment plant, i.e. in the collector pipe. This idea, however, is met with negative opinions, and this is related to the reduction of the temperature of sewage before their inflow to the sewage treatment plant, which can adversely affect the biological processes ongoing there [2].

Waste water heat energy can be obtained in several places in the sewage system: near the place of their formation and their outlet to the sewage system, at the stage of their transportation, i.e. in the collector pipe, and at the outlet of the collector from the sewage treatment plant [3].

3 Energy potential of sewage

The parameters that characterize waste water in terms of their energy potential are, first of all, their temperature and quantity. The next criteria assessing the waste water energy potential are their quality, technical condition of the sewers, the distance between the heat recovery system and the customers, as well as the type of customers and their energy needs [4].

The actual amount of greywater discharged from residential buildings depends on many factors such as the price of water, the quantity and quality of sanitary utensils, the habits of the residents, and the devices used in buildings to reduce water consumption. The amount of greywater discharged from the average household is about 60 dm^3/Md. Their average temperature is about 30°C [1].

On the other hand, the average temperature of domestic waste already in the sewage system is about 15°C. Over the summer it reaches 20°C and can be successfully used as an alternative energy source. Compared to other low temperature heat sources used as the lower energy source for heat pumps (soil, soil water with a temperature 7-13°C), waste water is distinguished by relatively high temperature throughout the year [4].

Apart from residential buildings, industrial plants have a high energy potential in terms of the amount and the temperature of waste water discharged. In many of them, technology is based on the processes in which high-temperature water is used. In the case of the textile industry there are so-called hot baths, and in the laundry facilities there is a lot of washing and rinsing [5]. Industrial waste is characterized by significantly higher temperatures in comparison to domestic waste. In addition, in accordance with the Regulation of the Minister of Construction on the realisation of the obligations of industrial waste water providers and the conditions of discharging sewage into sewerage devices [6], the temperature of sewage from industrial plants discharged to sewerage devices must not be more than 35°C. This is another argument that encourages to manage the heat from industrial waste water.

Other facilities where heat recovery may be a good solution are public buildings such as indoor swimming pools or hotels. In such facilities, large amounts of warm waste water are discharged into the sewer system and replaced by fresh warm water. An excellent source of energy to manage is the used water from swimming pools. Warm sewage comes mainly from showers, filter rinsing, and pool drainage. Their temperature is respectively [7]:

- 39-41 °C for waste water from showers,
- 25-26 °C for outflows from swimming pools,
- 28-30 °C for outflows from recreational pools,
- 30-32 °C for outflows from children's pools,
- 25-35 °C for rinse water from filters.

The average swimming pool produces about 30,000 liters of "grey" sewage per day. Such an amount causes a reflection on the use of a heat recovery system. However, a significant problem arising here is considerable pollution of used swimming pool water with slurry, fats, detergents, hair.

4 Possibilities of heat recovery from sewage

4.1 Heat recovery from internal installations

Much of the heat lost with sewage can be recovered and managed at the place where it was produced by installing a special exchanger. Exchangers used in sewage systems making heat energy recovery possible are called Drain Water Heat Recovery (DWHR). Of all types of DWHR, the devices in vertical systems are the most effective and most frequently used. They work as countercurrent exchangers (of opposite flow directions). Among them there are spiral heat exchangers and "pipe-in-pipe" exchangers. Vertical systems are manufactured by numerous manufacturers, mainly in the United States, Canada, Norway and the United Kingdom [8].

Spiral heat exchangers are anything but a drain pipe made of copper with a larger diameter, which discharges the used water from sanitary utensils, around which a copper tube is wound through which cold water flows in the opposite direction. The spiral is smaller in diameter than the drain pipe and, additionally, at the point of contact it is flattened, which increases the contact surface of both liquids and enables a more efficient transfer of heat from the effluent into the water. Spiral exchangers have been applied by many companies, but they vary in diameter, length of drains (70-200 cm) and the degree of flattening the pipes and the efficiency (30-65%). The diagram of the spiral heat exchanger is presented in the figure 1. This is a single copper pipe with a diameter of 18mm spirally wrapped around a downpipe with a diameter of 110 mm. The length of the device is 100 cm. According to the producent, thanks to the use of such an exchanger, we can recover 40-60% of the energy lost with sewage, and thus reduce the cost of preheating water by up to 40% [9].

https://doi.org/10.1051/e3sconf/20183003003

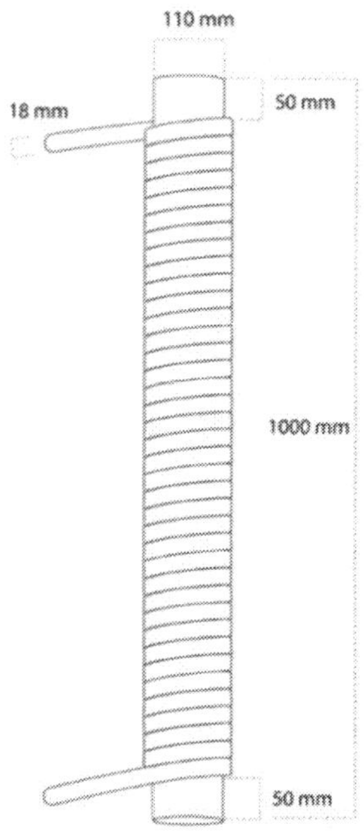

Fig. 1. A spiral heat exchanger - Ecoteb 110-100-01011.

The alternative for spiral exchangers are "pipe-in-pipe" exchangers. The components constituting this device are two pipes, the inner one of a smaller diameter and the outer one which forms a casing of the device. Heat is transferred between warm sewage flowing through the inner tube and tap water flowing in the space between the sewage transfer pipe and the outer pipe. Sewage flows down the inside of the drain pipe. The principle of operation of a "pipe-in-pipe" exchanger is presented in the figure 2. The efficiency of exchangers is about 60%. For larger installations with the flows that exceed the operating range of a single exchanger, combinations of devices cooperating with each other are provided.

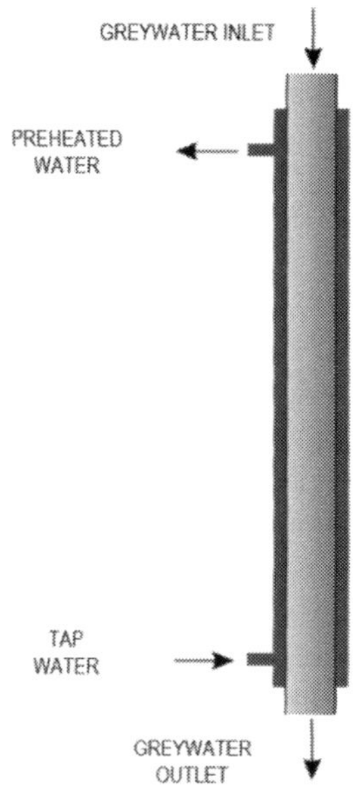

https://doi.org/10.1051/e3sconf/20183003003

Fig. 2. A "pipe in pipe" heat exchanger.

4.2 Heat recovery from sewage systems

Heat from sewage can also be received at the stage of its transportation, i.e. in the collector. The average annual temperature of household waste is about 18-20°C in the summer and about 10-12 °C in the winter. When using special heat exchangers and with cooperation of a system with a heat pump, a large amount of heat energy can be recovered for use. In many European cities such as Paris, Hamburg and Oslo, a sewage recovery system has been applied. Research in this direction was already conducted in the 80's in Switzerland. The first designs of devices were quite expensive to invest in, and the warm pumps used at that time were of poor operational efficiency, but nowadays the solutions are gaining more and more interest and their profitability is increasing.

Heat exchangers are available as built into the structure of sewage pipe, used in the case of new constructed collectors, as well as ones for the use on the existing collectors, at the stage of the renovation, mounted by using a trenchless technology [10].

The first ones are most commonly used for newly built pipes, but also at the stage of replacing the used with new ones. The exchangers mentioned are most often polyethylene or corrosion resistant steel pipes, placed on the bottom of the collector in a profiled half-round gutter of a concrete or reinforced concrete pipe. They are characterized by high structural durability and long heat storage capacity. Their efficiency is in the range of 1-2 kW per tube length meter. Exemplary solutions are shown in the figure 3.

Modern technologies allow the exchanger to be built into existing collectors at the stage of their renovation using trenchless methods. The most commonly used solutions are ready-made modules that can be mounted in the collector. The discretion to combine them in stretches of any length and low investment cost make this technology very popular. The exchanger design is an arc-shaped profile under which two or three polyethylene tubes, in which intermedium circulates, are fixed. The minimum required efficiency of 20 l/s provides thermal efficiency of up to 2 kW/ tube length meter [2]. An example of such a solution is presented in the figure 4.

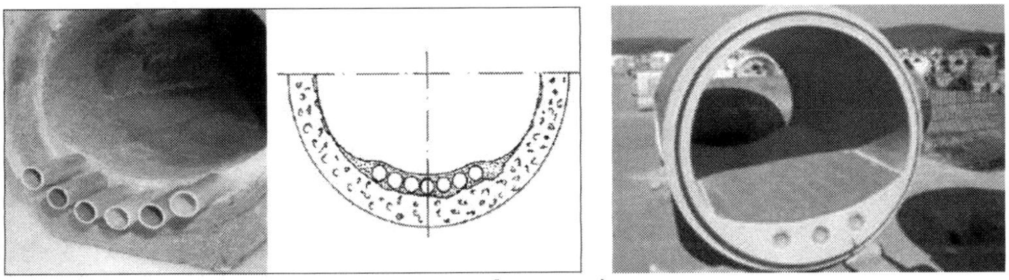

Fig. 3. Heat exchangers built into the structure of sewage pipe.

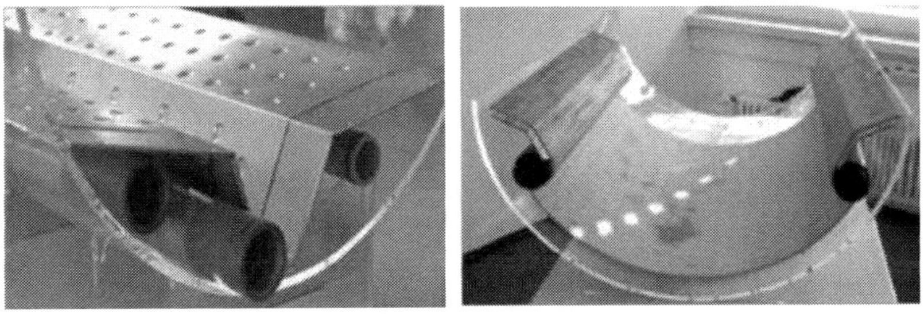

Fig. 4. Ready-made modules of heat exchangers for the use on the existing collectors.

Each heat exchanger used in the collector must be resistant to the aggressive effects of sewage, so they are made of materials such as polyethylene, stainless steel and they are often also additionally enclosed.

In order to use energy from the sewage transported by the collector, the heat recovery is done using heat pumps. Sewage flowing in the sewer pipe gives back the heat on the length of the pipe with the heat exchanger mounted. The process involves taking over heat by intermedium flowing through the heat exchanger placed in the collector. It may be glycol, brine or even water, which is then directed to the heat pump located in the boiler room at the customer's house or in the boiler room serving the larger area (district).

At present, heat recovery technology from sewage collectors is an unknown and very innovative solution in Poland. At the same time, there are numerous examples of its use in Western European countries and continuous development in this field, as well as the growing interest in acquiring such energy, are conducive to the implementation of such solutions in our country.

5 Exemplary waste water heat recovery system

The use of energy from sewage as an alternative energy source to preheat water for buildings can be realized by using the presented heat exchangers cooperating with the heat pump. In order to increase the efficiency of heat recovery system from waste water, especially in larger installations, each heat recovery system should consist of the following components:

- wastewater retention tank - necessary because of the irregularity of waste water discharges and their varied quantity within 24 hours,
- hot tap water tank - to store fresh water for further use,
- sewage filter - is intended to stop pollution, located in front of the retention tank,
- heat recovery system – a heat exchanger and a heat pump.

An exemplary waste water heat recovery system is shown in the figure 5.

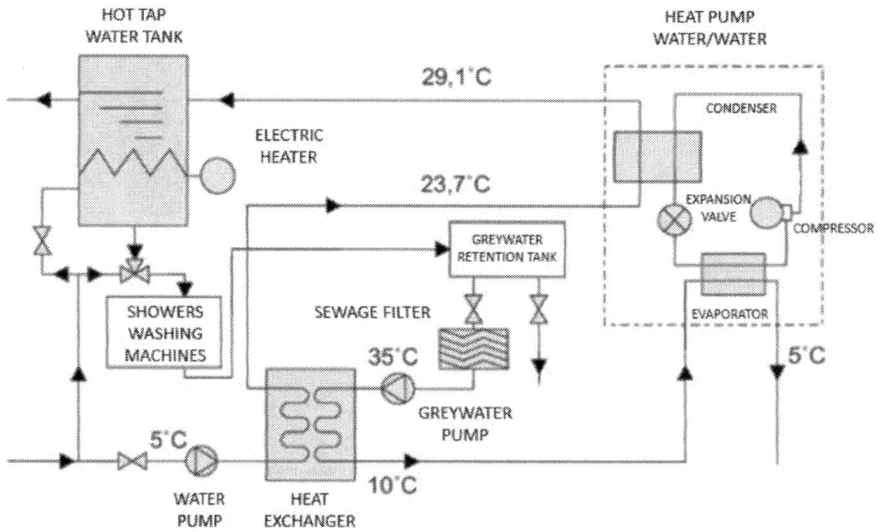

Fig. 5. An exemplary greywater heat recovery system [5].

Particularly noteworthy is the possibility to reduce charges related to the preparation of domestic hot water. In the average household in Poland, the second largest item in fees for energy is the cost of preparing hot water.

For example, when using a spiral 50% efficiency heat exchanger recovering heat from the grey effluents obtained only from a shower (the temperature of waste water is 38°C) in a single-family dwelling inhabited by 5 people, the amount of energy saved for preheating water is about 1100 kWh, which gives cost savings of 580 PLN. Taking into account the total investment costs including the purchase price and the installation of the device as well as the costs resulting from the operation of the system during the warranty period (total: 4 000 PLN), the costs would return within 7 years of using the system.

6 Summary

All over the world new sources of heat as well as modern solutions allowing for waste heat recovery are being looked for. Recovering energy from sewage will not solve the problem of rising energy demand, but it can contribute to some savings in this field.

The presented overview of construction solutions of waste water heat recovery devices shows the multidimensionality of their applications. Universal solutions that will appeal both to homeowners and industrialists are being looked for.

The innovative solution is to obtain energy in the sewage collectors. Studies show that systems of this type can be profitable due to the ease and low costs of acquiring a new heat source.

On the other hand, as for now, no mention is made of the possible effects of the reduction of the temperature of sewage as a result of heat recovery before their inflow to the treatment plant. In addition, due to the lack of data on the quality of greywater, there are concerns about their impact on the efficiency of devices (operation problems - blocking of pipes, pumps), as well as the natural environment.

There is no doubt, however, that the installation of a DWHR exchanger can contribute to saving the energy used to prepare hot tap water. Obviously, in the case of a detached house the benefits are small. Greater savings can be anticipated where large amounts of water are used, and therefore, the amount of waste water discharged is greater or it is discharged more frequently.

It should also be mentioned that the presented solutions are innovative, still rarely found in our country. Currently, the work is being done on improving these devices, increasing their efficiency and it can be expected that their attractiveness will increase over time.

References

[1] J. Chudzicki, *Techniczne możliwości oszczędzania wody i energii w budynkach użyteczności publicznej – wnioski ze STEP-u*, Energia i Budynek, **2**, 22-25 (2011)

[2] P. Kuliczkowski, *Rodzaje wymienników ciepła stosowanych w kolektorach kanalizacyjnych umożliwiających odzysk ciepła ze ścieków*, Instal, **12**, 46-49 (2009)

[3] E. Müller, F. Schmid, W. Stodtmeister, B. Kobel, *Ratgeber für Bauherrn Und Kommunen*, Heizen und Kühlen mit Abwasser (2005)

[4] W. Joniec, *Odzysk ciepła z kanalizacji*, Rynek Instalacyjny, **5**, 72-75 (2007)

[5] S. Kordana, D. Słyś, *Odzysk ciepła odpadowego w instalacjach i systemach kanalizacyjnych*, Wydawnictwo i Handel Książkami „KaBe", Krosno (2013)

[6] Journal of Laws 2006 No. 136, item 964, as amended

[7] A. Kolaszewski, *Odzysk ciepła ze ścieków w krytej pływalni*, Materiały Budowlane, **2**, 48-49 (2000)

[8] A. Kimmels, *Shower Heat Recovery: Overview of Commercially Available DWHR Systems*, Meander Heat Recovery (2011)

[9] Cooperman, J. Dieckmann, J. Brodrick, *Drain Water Heat Recovery*, ASHRAE Journal, **11**, 58-62 (2011)

[10] P. Kuliczkowski, *Alternatywne pozyskiwanie energii z kanałów sanitarnych za pomocą technologii bezwykopowych*, Nowoczesne Budownictwo Inżynieryjne, **3/4**, 55-58 (2010)

Evaluation of selected sewage sludge gasification technological parameters

Grzegorz Gałko[1], *Danuta* Król[1]

[1]Faculty of Energy and Environmental Engineering, Silesian University of Technology, Konarskiego 18, 44-100 Gliwice, Poland

Abstract. Evaluation of selected sewage sludge gasification technological parameters was shown in this paper. Degree of carbon conducted in combustible substance and syngas efficiency (technological readiness coefficient) in accordance with equations were calculated. Enthalpy of individual compounds formation and energy balance were calculated in accordance with rule of Hess.

1 Introduction

One of the main materials, that are formed during the wastewater treatment and production are sewage sludge's. Physical parameters like size of particles and composition of them depends on wastewater treatment conditions [1]. Methods of sewage sludge management should ensure sanitary safety. Disposal solutions of hazardous substances included in sludge's are available by thermal processes like gasification. Combustible gas (syngas) arises after cycle of the reactions between fuel and gasification agent is one of the main products of this process. Technological parameters like: syngas lower heating value, degree of combustible substance conversion and syngas chemical efficiency(technological readiness coefficient) for evaluation of efficiency of the process can be used. In range of temperatures over than 700°C for sewage sludge air gasification, fuel conversion degree is in the range 65-72% and range of syngas efficiency is from 30 to 50% what was shown by authors [2]. However, if temperature range increase from 600 to 900°C the chemical efficiency of gasification process will increase to 47,5%, what has been proved by [3]. The lower heating value of produced syngas achieved by authors was 4,19 MJ/m^3. Sewage sludge may be used for co-gasification process. Then fuel used in gasification is blended from different materials like sewage sludge with wood, biomass waste or coal. Fuel mixtures changes cause increases some technological parameters like syngas efficiency. Increase of this parameter is available by increase of sludge's to wood ratio in fuel mixture what has been found by the authors of [4]. Evaluation of selected technological parameters like: degree of carbon conversion contained in combustible substance and syngas chemical efficiency during sewage sludge air gasification was the aim of this paper.

2 Research material

Dried sewage sludge from municipal wastewater treatment plant in Bialystok was used in the experiment. Fuel properties were prepared in accordance with actual standards and regulations: humidity was determined in accordance with [5], combustible and incombustible parts concentration with [6],volatile matter [7], carbon and hydrogen concentration [8], nitrogen concentration [9], sulphur concentration [10], chlorine [11], higher heating value [12]. Samples were prepared in accordance with [13]. Results of the tests were shown in Table 1.

Table 1. Sewage sludge fuel properties

Parameter	Value
C % $_{(dry\ mass)}$	36.19
H % $_{(dry\ mass)}$	3.63
N % $_{(dry\ mass)}$	1.02
S % $_{(dry\ mass)}$	0.66
Cl % $_{(dry\ mass)}$	0.52
O % $_{(dry\ mass)}$	22.11
A % $_{(dry\ mass)}$	35.87
Volatile matter % $_{(dry\ mass)}$	49.2
Combustible matter % $_{(dry\ mass)}$	64.13
Moisture %	14.78
LHV MJ/kg $_{(dry\ mass)}$	14.68
HHV MJ/kg $_{(dry\ mass)}$	15.47

3 Experiment

Sewage sludge was air-gasified in 5 kW countercurrent gas generator. Gas composition was measured by GAS3100 syngas analyzer. Concentrations of combustible and incombustible gases like CO, CO_2, CH_4, H_2 i O_2 were measured. In the range 700-800°C was the temperature of the process. The measurement of this parameter was taken using nickel type K thermocouples linked with the control panel connected to the computer. The technological parameter of air fuel ratio was calculated in accordance with equation:

$$\Phi=\frac{G_{fuel}}{V_{conv.}}=\frac{1.260}{2.808}=0.45 \quad (1)$$

Where: G_{fuel} – fuel stream (kg/h), V_{conv} –stream of gasification agent (air) (kg/h)
Description of sub processes areas of sludge air gasification in countercurrent pipeline gas generator shown in Fig.1 was prepared in accordance with [15,16].

 https://doi.org/10.1051/e3sconf/20183003004

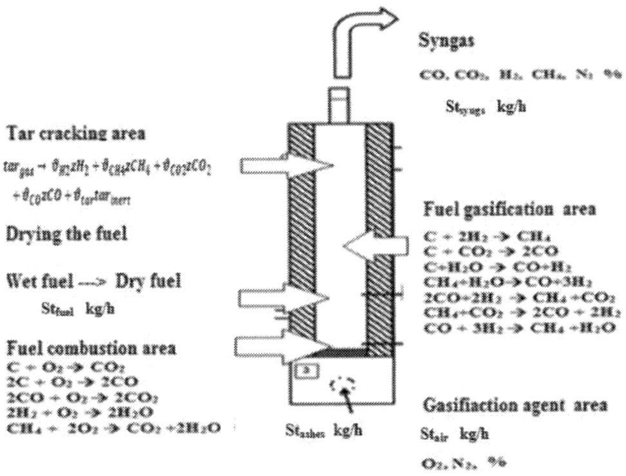

Fig. 1. Graphical concept of sewage sludge gasification in pipeline gas generator

Where: Stfuel – fuel stream [kg/h], Stsyngas – syngas stream [kg/h], ϑ – stoichiometric coefficients [-].

Gasification process consists of different sub processes where specific chemical reactions occurs [14]. Removing of fuel humidity starts at the beginning. Then in primary and secondary reactions gases like: methane, steam, carbon monoxide and also tars are formed. The last product is cracked in the cracking area in accordance with post processing tars [15]. Concentrations of syngas compounds formed in gasification were shown in Fig 2.

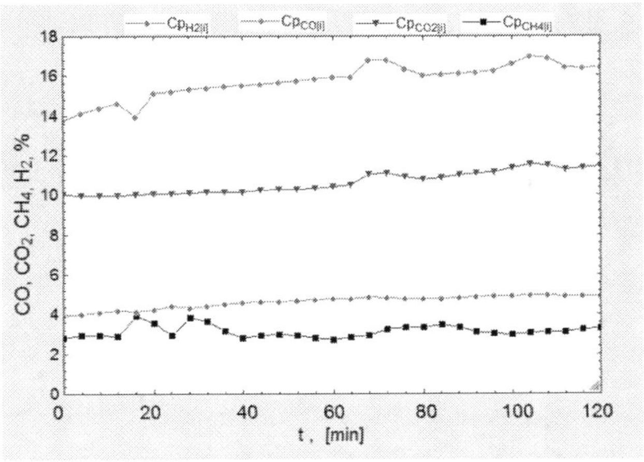

Fig. 2. Syngas composition in percent concentrations while steady state gas generator (without starting and quenching

Combustible gases concentrations like: CO, H_2 and CH_4 were in the ranges 14-17%, 4-5% and 3-4%. Small changes concentrations testify to stability of the process. Descriptions and experimental results of sewage sludge gasification were shown in [14].

4 Energy balance

Syngas lower heating value depends on combustible gases presence like: CO, H_2 and CH_4 Homogenous and heterogeneous reactions are fundamental in gasification process [17]. Energy balance was prepared in accordance with rule of Hess [18].

$$C + O_2 \rightarrow CO_2 \quad (2)$$

$$CO_2 = C + O_2 \quad (3)$$

$$\Delta H^o{}_{CO2} = -393.13 - 0 - 0 = -393.13\ kJ \quad (4)$$

$$2C + O_2 \rightarrow 2CO \quad (5)$$

$$2CO = 2C + O_2 \quad (6)$$

$$\Delta H^o{}_{CO} = -110.53 - 0 - 0 = -110.53\ kJ \quad (7)$$

$$C + 2H_2 \rightarrow CH_4 \quad (8)$$

$$CH_4 = C + 2H_2 \quad (9)$$

$$\Delta H^o{}_{CH4} = -74.78 - (0 - 2 \cdot 0) = -74.78\ kJ \quad (10)$$

$$C + CO_2 \rightarrow 2CO \quad (11)$$

$$2CO = C + CO_2 \quad (12)$$

$$\Delta H^o{}_{CO} = (2 \cdot -110.53) - 0 - (-393.13) = -221.06 + 393.13 = 172.07\ kJ \quad (13)$$

$$2CO + O_2 \rightarrow 2CO_2 \quad (14)$$

$$2CO_2 = 2CO + O_2 \quad (15)$$

$$\Delta H^o{}_{CO2} = (2 \cdot -393.13) - 2 \cdot 110.53 - 0 = -786{,}26 - 221{,}06 = -1007.32\ kJ \quad (16)$$

$$2CO + 2H_2 \rightarrow CH_4 + CO_2 \quad (17)$$

$$CH_4 + CO_2 = 2CO + 2H_2 \quad (18)$$

$$\Delta H^o{}_{CO2+CH4} = (-74.78 - 393.13) - 2 \cdot (-110{,}53) - (2 \cdot 0) = 246.85\ kJ \quad (19)$$

$$CO + 3H_2 \rightarrow CH_4 + H_2O \quad (20)$$

$$CH_4 + H_2O = CO + 3H_2 \quad (21)$$

$$\Delta H^o{}_{CH4+H2O} = (-74.78 - 241.60) - 2 \cdot (-110.53)(3 \cdot 0) = -537.44\ kJ \quad (22)$$

Enthalpy values of individual substrates were taken from thermodynamics tables [19]. If in the syngas there is no presence of higher hydrocarbons, this energy balance is adequate. Otherwise, if there are higher hydrocarbons higher than CH_4 syngas quality analyze is necessary for preparing chemical reactions

5 Evaluation of selected technological parameters

Evaluation of technological parameters like: syngas lower heating value, degree of carbon conversion to combustible substances and chemical efficiency were calculated in accordance with equation [20].

$$LHV_{syng} = \frac{126 \cdot uCO}{1000} + \frac{108 \cdot uH_2}{1000} + \frac{359 \cdot uCH_4}{1000}, [\frac{MJ}{m^3 n}], \quad (23)$$

The syngas lover heating value during the stable work in function of time was shown in Fig. 3.

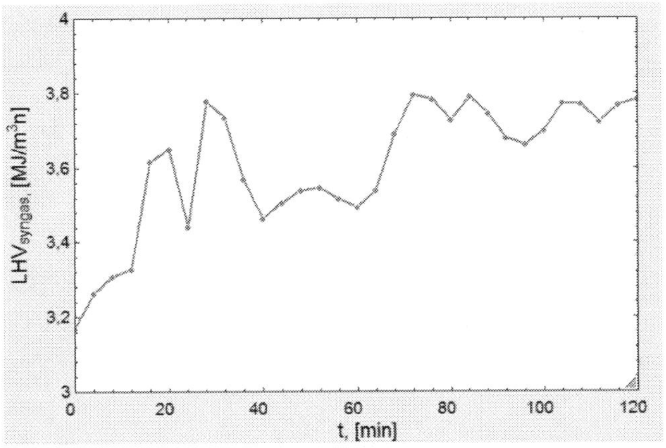

Fig. 3. Syngas lower heating value during steady state conditions in function of time

Average lower heating value of combustible syngas was 3,98 MJ/m^3. Lower heating value range was from 3,1 to 3,8 MJ/m^{3}n. During measurements the LHV value was fluctuated. The technological parameters used for process evaluation like chemical efficiency and carbon degree conversion were calculated in accordance with [2].
Syngas chemical efficency:

$$\eta_{chem} = \left(\frac{St_{gas}\cdot(yH_2\cdot LHV_{H2}+yCH_4\cdot LHV_{CH4}+yCO\cdot LHV_{CO})}{Stfuel\cdot LHV_{fuel}}\right)\cdot 100, [\%], \quad (24)$$

Degree of carbon conversion in fuel:

$$C_{conw.} = \left(1-\frac{St_{gas}(yCO_2\frac{12}{44}+yCO\frac{12}{28}+yCH_4\frac{12}{16})}{Stfuel\cdot yC}\right)\cdot 100, \ [\%], \quad (25)$$

Where: St_{gas} – Syngas content [m^3/h –efficiency, kmol/h – degree of conversion],y – gases content [-], $LHV_{H2,CO,CH4}$ – lower heating value of: hydrogen, carbon monoxide and methane [MJ/m^{3}n], St_{fuel} – fuel stream [kg/h – efficiency, kmol/h –degree of conversion], LHV_{fuel} –Lower heating value of fuel [MJ/kg], yC –mass content of carbon in combustible matter of fuel[-]

6 Balance of the process

Syngas molar stream was calculated in accordance with equation :

$$St_{fuel} + St_{air} = St_{syngas} + St_{ashes}, \left[\frac{kmol}{h}\right] \quad (26)$$

Where: St_{fuel}–fuel stream [kmol/h] (St_{fuel}=0,08012), St_{con}– gasification agent stream [kmol/h] (St_{con}=0,08926), St_{syngs} – syngas stream [kmol/h], St_{ashes}–ashes stream [kmol/h] (St_{ashes}=0,00296)

$$St_{syngas} = St_{fuel} + St_{air} - St_{ashes} = 0.139, \left[\frac{kmol}{h}\right] \quad (27)$$

$$V_{syngas} = St_{syngas} \cdot 22.42 \left[\frac{m3}{kmol}\right] = 3.133, \left[\frac{m^3}{h}\right] \quad (28)$$

Stream volume of syngas was 3.133 m^3/h. While stream fuel delivered to the chamber was 1.256 kg/h (0,00802 kmol/h). The mass grade residue of combustible substance in ashes was 3%. Content of combustible substance from fuel stopped in ashes was calculated in accordance with equation :

$$X_{A1} = \frac{A \cdot X_A}{A1} = \frac{35.87 \cdot 3}{97} = 1.109, \quad [\%], \quad (29)$$

Where: A – ashes content in fuel [%], X_A – combustible matter content in ash [%], A1 –incombustible matter content in ashes [%], X_{A1}- content of combustible substance from fuel stopped in ashes [%],

Mass stream of ashes was calculated in accordance with equation:

$$B_1 = B \cdot (A + x_{A1}) = 0{,}08012 \cdot 36.97\% = 0.464 \left[\frac{kg}{h}\right] \quad (30)$$

Where: B – fuel stream [kmol/h], A – ashes content in fuel [%], X_{A1}– combustible substance content in ashes [%]

7 Results

Results of syngas chemical efficiency and carbon conversion calculations were shown in Fig 4.

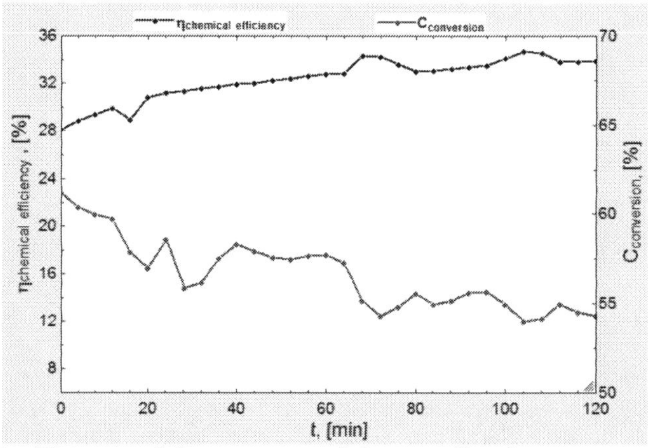

Fig. 4 Chemical efficiency of the process and degree of carbon conversion from sewage sludge in time.

Syngas chemical efficiency produced during municipal sewage sludge air gasification was in the range 33-35%. This value was slightly fluctuated in function of time. and mostly depends on lower heating value of fuel nd produced combustible gas. It also depends on humidity concentration in fuel. Increase of humidity cause decrease of calorific value what has been found by authors [21]. The degree of carbon conversion contained in combustible substance of fuel was in range 53-58%. Conversion changes depends on concentrations of gases in which composition one of the element was carbon. To this gases belongs CO, CH_4 and CO_2. In correlation with carbon element transformation gasification chemical efficiency is increasing while the decrease of carbon conversion. The highest efficiency (33%) was in achieved while the value of carbon conversion was 58%. And inversely 35% syngas efficiency was achieved in 54% carbon conversion. Transformation of coal element contained in combustible to carbon monoxide (incombustible gas) which not increase the lower heating value of syngas justify this dependence.

8 Conclusions

Evaluation of gasification process is possible with using selected technological parameters. Different values of parameters shown in this paper were depended on experimental conditions. Chemical efficiency was in correlation with degree of carbon conversion. Increase of chemical efficiency was increasing with syngas calorific value. One of the main gases that don't have influence on efficiency but on degree of carbon conversion was incombustible CO_2. Combustible syngas from fuel gasification is useful for power in forms of electricity and heat production. This product can be used in combined heat and power and integrated gasification combined cycle (IGCC) technologies. High efficiency gasification in this solution is one of the most desirable parameters. It means that evaluation of gasification processes is necessary to improve and find effective solutions for energy production.

References

1. Smoczynski L., Kosobucka M., Smoczynski M.Ratnaweera H., Pieczulis-Smoczynska K. Water Science and Technology, **75** (4), (2017)
2. Wei-Hsin Chen, Chih-Jung Chen, Chen-I Hung, Cheng-Hsin Shen, Heng-Wen Hsu. Applied Energy,**112**, (2013)
3. Seong-Wan Kang, Jong-In Dong, Jong-Min Kim, Woo-Chan Lee,.Journal of Material Cycles and Waste Management, **13** (2011)
4. Kyung-Won Lee, Woo Chan Lee, Hyuk Jun Lee, Jong In Dong . Journal of Material Cycles and Waste Management, **16**, (2014)
5. PN-EN ISO 18134-2:2015-11, Solid biofuels – Total moisture determination – dryer method: Part 1: Total moisture – reference method (in English)
6. PN-ISO 1171:2002 – Ash determination (in Polish)
7. PN-EN ISO 18123:2016-01 – Solid biofuels – Determination of volatiles by weight method (in Polish)
8. ISO 609:1996 – Solid fuels – Determination of carbon and hydrogen by high-temperature combustion (in Polish)
9. PN-EN 16169:2012E – Sludge, treated biowaste and soil - Determination of Kjeldahl nitrogen (in Polish)
10. PN-ISO 351:1999 –Solid fuels – Determination of total sulfur by high-temperature combustion (in Polish)
11. PN-ISO 587/2000 – Determination of chlorine using Eschka mixture (in Polish)
12. PN-ISO 1928:2002 - Solid fuels – Determination of higher heating value by calorimetric bomb combustion and lower heating value calculation) (in Polish)
13. PN-EN 14778:2011 – Solid biofuels – samples collection
14. Król D., Gałko G., Przemysł Chemiczny (Industrial Chemistry) **2**, 2017 (in Polish)
15. Zhehan Ong, Yongpan Cheng, Thawatchai M., Zhiyi Yao,YenWah Tong, Chi-Hwa Wang (American Institute of Chemical Engineers)AIChE Journal, 61, (2015)
16. Król D., Poskrobko S., Energy & Fuels, **31** (4), (2017)

17. Tomeczek, J. *Coal gasification,* (Central scripts **155** , Silesian University of Technology Publisher, Gliwice 1991)
18. Hołyst R., Poniewierski A., Ciach A. *Thermodynamics for Physics, Chemists and Engineers*, (Warsaw, Physical Chemistry Institute and School of Exact Science Publisher , Polish Academy of Science 2003) (in Polish)
19. R.Lide D. *CRC Handbook of Chemistry and Physics,84thEdition ,* (National Institute of Standards and Technology 2004)
20. Król D., *Biomass and fuels formed from waste in low emission combustion technologies*, (Silesian University of Technology Publisher, Gliwice 2013) (in Polish)
21. Ptasinski J. K., Prins M. J., Pierik A. Energy **32** ,4, (2007)

Sewage sludge conditioning with the application of ash from biomass-fired power plant

Marta Wójcik[1,*], *Feliks* Stachowicz[1] , and *Adam* Masłoń[2]

[1]Rzeszow University of Technology, Department of Materials Forming and Processing, Powstańców Warszawy 8, 35-959 Rzeszow, Poland
[2] Rzeszow University of Technology, Department of Chemistry and Environmental Engineering, Powstańców Warszawy 6, 35-959 Rzeszow, Poland

Abstract. During biomass combustion, there are formed combustion products. Available data indicates that only 29.1 % of biomass ashes were recycled in Poland in 2013. Chemical composition and sorptive properties of ashes enable their application in the sewage sludge treatment. This paper analyses the impact of ashes from biomass-combustion power plant on sewage sludge dewatering and higienisation. The results obtained in laboratory tests proved the possitive impact of biomass ashes on sewage sludge hydration reduction after dewatering and the increase of filtrate volume. After sludge conditioning with the use of biomass combustion by-products, the final moisture content decreased by approximatelly 10÷25 % in comparison with raw sewage sludge depending on the method of dewatering. The application of biomass combustion products in sewage sludge management could provide an alternative method of their utilization according to law and environmental requirements.

1 Introduction

Biomass plays a special importance amongst renewable energy sources. It is estimated that biomass energy constituted approximately 53 % of world renewable energy production in 2010 [1]. Hydrogeological location of Poland prevents a significant share of renewable energy from natural sources, for example: from wind and sun and for this reason, biomass energy is becoming more and more popular in Poland. For heating purposes, it is particularly used forest or agricultural biomass [2]. The main sources of biomass in Poland are energetic crops, for example: straw and willow *Salix viminalis*. The largest share of whole energetic plants has a willow tree. In Poland, there are approximately 780 willow plant plantations with a surface area of 9480 ha which contains approximately 95 % of perennial plants cultivation [3].

Biomass comprises components which are contained in organs, for example: proteins, fat and lignin and structural material of biomass: polymerized carbonohydres, lignin and cellulose. The definite advantage of biomass combustion is "zero emission" of carbon dioxide. During the biomass burning, only so much carbon dioxide is emitted as a plant

* Corresponding author: m.wojcik@prz.edu.pl

takes from the atmosphere by photosynthesis when the plants grow [4]. Therefore, biomass could be treated as a source of clean energy.

The biomass consumption is systematically growing what is confirmed by the available data. In Poland the biomass consumption was 9 317 763 GJ in coal-burning power plants in 2005. For comparison, the use of biomass grew to 42 917 011 GJ in 2013. In 2013 the biomass consumption was 26 275 970 GJ in biomass-burning power plants [2]. During the biomass burning, there are formed combustion by-products, such as fly and bottom ashes. The increasing share of biomass in final energy production causes the problem with the ash utilization. Available data indicates that only 29.1 % of biomass ashes were recycled in Poland in 2012 [5]. Specific properties of biomass combustion products, for example: high losses of ignition and the content of chlorine, prevent their application in many sectors of economy. The main destination of biomass ashes is agriculture which applied them as a valuable fertilizer improving the plant growth [5].

Sewage sludge is a by-product of sewage treatment which contains mineral and organic compounds. Due to the connection of new areas to the sewerage, the amount of sludge is systematically growing. According to the Central Statistical Office Report [6], there were produced 556 Mg dry mass of municipal sewage sludge in Poland in 2014. Before the land application of sewage sludge, there have to be processed in order to decrease the health risk. Additionally, the effective dewatering could reduce the transport cost of sewage sludge. Raw sewage sludge has a stable structure with a low dewatering and sedimentation capacity. For this reason, sludge dewatering is enhanced by conditioning with the use of mineral and organic reagents. Recently, the application of polyelectrolytes is very popular in treatment plants. The disadvantages of aforementioned reagents are their high cost. What is more, high amounts of polyelectrolytes are required for effective dewatering. The alternative solution for popular organic flocculants might be combustion by-products. Biomass ashes might be used as a conditioner [8]. Powdered materials with a diameter below 300 μm are the weights of sewage sludge flocs [9]. In the wake of incorporation of ash in sludge flocs, the shape, dimension and structure of flocs are altered. Ashes are integrated in sludge floc matrix, what could improve sewage sludge dewatering [10].

The aim of this study was to determine the influence of ash from biomass combustion on the effectiveness of sewage sludge dewatering.

2 Materials and Methodology

2.1 Materials

The academic research concerning the influence of biomass ash on sewage sludge dewatering was performed under laboratory conditions. Sewage sludge used in tests was obtained from the inlet of the thickening tank from Świlcza WWTP (Podkarpackie region, Poland). Sewage sludge was taken during summer/autumn and was characterized by grey-brown colour and earthy smell. The physical and chemical properties of unconditioned sludge were as follows: pH = 6.40 ± 0.44; the moisture content = 97.48 ± 0.81%; the capillary suction time (CST) = 130.89 ± 28.71 s.

In laboratory research, ash from "Łężańska" Power Plant in Krosno (Podkarpackie region, Poland) was used. The biomass ash was characterized by particles in the range of 0.399 ÷ 399 μm. Due to the fact that ash was not sieved before laboratory research, there were some coarse particles above 300 μm. The biomass ash contains the following types of particles: irregular sharp, polyhedral and highly porous (Fig. 1).

The chemical composition of ash was determined by means of Energy Dispersive Spectroscopy (EDS). The results of EDS analysis showed that the ash comprises mainly the

 https://doi.org/10.1051/e3sconf/20183003005

following compounds: Ca, O, S, and Cl. The significant amount of Ca and S could classify the tested ash as a calcium and sulphate ash. Due to the biomass combustion from surrounding timber industry, ash from power plant also contains molybdenum. The phase composition of biomass ash was as follows: SiO_2 (α-quartz), Fe_2O_3 (hematite), CaO, Na_2O_2, SO_2, Na_2O, P_2O_5 and KO_2 (Table 2).

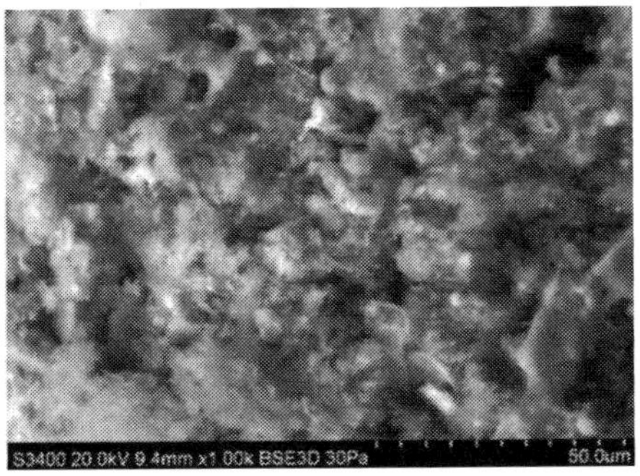

Fig. 1. SEM image of biomass ash.

Table 1. Chemical composition of tested biomass ash.

Element	**Content [%]**
CaO	45.64
SiO_2	15.17
K_2O	9.78
SO_3	7.58
P_2O_5	4.59
MgO	4.25
Al_2O_3	4.14
Fe_2O_3	3.85
MnO	1.61

2.2 Laboratory experiments

Laboratory tests of sewage conditioning and dewatering were carried out in tree series. The sewage sludge conditioning was carried out as follows: five beakers with a volume of 1 dm^3 were filled with 500 cm^3 of raw sewage sludge. To four beakers with sludge, there are applied appropriate dosages of biomass ash: 5; 7.5; 15 and 30 $g \cdot dm^{-3}$. The dosages of biomass ash were expressed as the weight ratio of ash to sewage sludge dry mass: 1:4; 1:3; 1:2 and 1:1, respectively. Firstly, mixtures were rapidly stirred with a speed of 200 rpm for

1 minute and after that, they were stirred with a speed of 50 rpm for 15 minutes. In sewage sludge after conditioning, pH and capillary suction time (CST) were determined.

2.3 Analytical methods

All parameters of sewage sludge after dewatering and conditioning were investigated in line with the applicable research procedure.
CST was measured by means of the CST meter ProLabTech in accordance with PN-EN 147011:2007. In sewage sludge, the pH was analyzed with pH-meter HACH HQ40d in line with PN-EN 15933: 2013-02. After that, for prepared samples of sewage sludge, the dewatering capacity was investigated. Mechanical dewatering process was done in a Büchner funnel and in a centrifuge. The vacuum filtration was done under 0.01 and 0.02 MPa vacuum pressure for approximately 15 minutes. Centrifugation of sewage sludge was done for two rotation speeds: 2000 and 2500 rpm for 5 minutes in a laboratory centrifuge (MPW-350). After mechanical dewatering, the moisture content was determined according to PN-EN 15934: 2013-02.

3 Results and discussion

This study showed different efficiencies in the improvement of sewage sludge dewatering, depending on the amount of biomass ash.
The impact of sludge conditioning on dewatering was assesed by means of CST measurement. The obtained results have shown that the biomass ash influenced the improvement of sewage sludge dewatering in various ways, depending on the amount of ash (Fig. 2). The best result was achieved for a dose of 30 g·dm^{-3} and for this reason, this amount of ash was considered as an optimal dosage. Detailed information concerning the influence of biomass ashes on CST value are presented in [12]. Similar results of sludge conditioning were achieved by Piotrowska-Cyplik and Czarnecki [10] for cationic polyelectrolyte F-410. Wójcik et al. [7] also proved that the addition of willow ash in the dosage of 30 g·dm^{-3} could decrese the CST value of sewage sludge by approximately 55%.

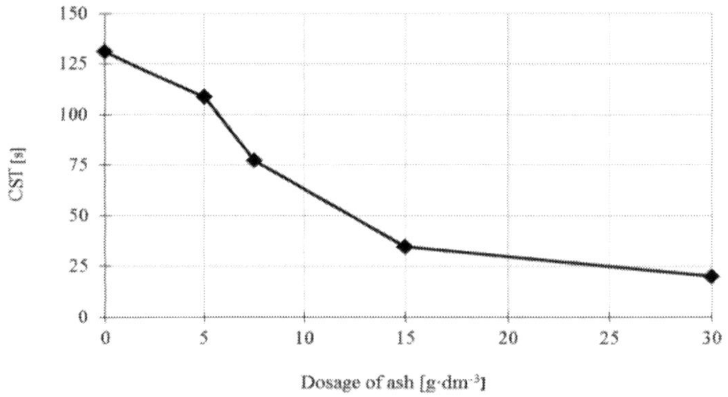

Fig. 2. Influence of sewage sludge conditioning by means of biomass ash on the CST [12]

Biomass ash might be an effective reagent in sewage sludge higienisation. The pH for raw sewage sludge was mean 6.40 and for sewage sludge after conditioning with biomass ash in dosage of 30 g·dm^{-3} was 12.24. The dosages of ash from 5 to 15 g·dm^{-3} could obtain the growth of the pH to the value: 7.38; 8.32 and 11.44, respectively (Fig. 3). The positive impact of biomass ash on the growth of pH is connected with the content of alkaline metals,

https://doi.org/10.1051/e3sconf/20183003005

such as: Ca and Na. In the case of high dosages, biomass ash might be an effective reagent in sewage sludge higienisation.

After conditioning, sewage sludge was dewatered by means of centrifugation. The research was carried out for two rotation speed: 2000 rpm and 2500 rpm. Results showed that the moisture content decreased with an increase of dosage of ash (Fig. 4). The best results were obtained for the highest amount of ash. With 30 g·dm^{-3} dosage of ash, the moisture content decreased from 92.71% (raw sewage sludge) to 84.52% (conditioned sewage sludge) for the rotation speed of 2000 rpm. Low dosages of ash influenced on the sludge moisture content in a lesser degree. Better results were obtained for the higher rotation speed. The application of 30 g·dm^{-3} dosage of biomass ash reduced the moisture content from 92.52% (raw sludge) to 76.21% (conditioned sludge). In comparison with raw sewage sludge, the highest dosage of biomass ash reduced the moisture content of approximately 19% for a rotation speed of 2500 rpm. In any case, low dosages of ash did not influence the reduction of moisture content in a significant way.

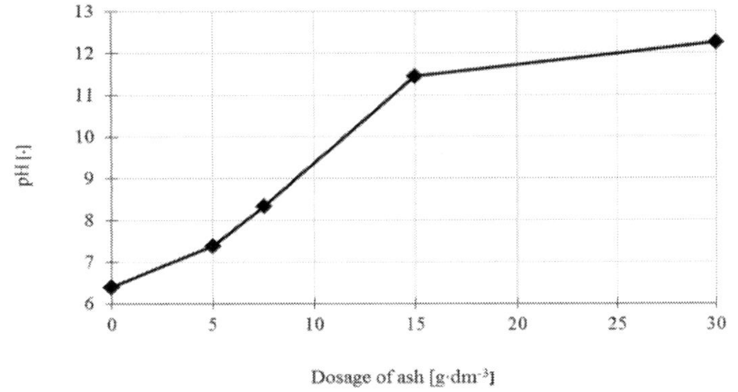

Fig. 3. Influence of sewage sludge conditioning by means of biomass ash on the pH of sewage sludge.

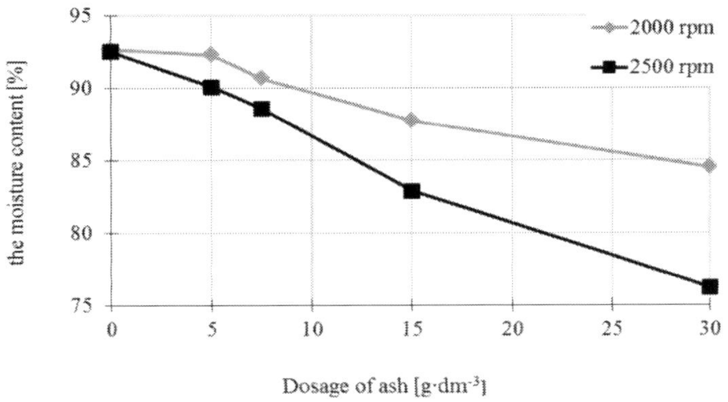

Fig. 4. Influence of biomass ash on the sewage sludge moisture content after centrifugation.

CST increased with the rise of dosage of biomass ash (Fig. 5). It was observed that CST depends on the sewage sludge moisture content. The higher value of CST could mean that sewage sludge has a more consistent structure. In the wake of elimination of excessive water, sludge is resistant to further dewatering. Higher values of CST were obtained for the rotation speed of 2500 rpm what corresponds to the better effectiveness of centrifugation.

Additionally, the centrifugation of conditioned sewage sludge was more efficient in comparison with raw sewage sludge, as witnesses by the liquid volume after dewatering. The water volume was approximately 27 cm^3 for raw sewage sludge and in the range of 31÷34 cm^3 (2000 rpm) and 32÷34 cm^3 (2500 rpm) for conditioned sewage sludge in

https://doi.org/10.1051/e3sconf/20183003005

different dosages of ash (Fig. 6). Although the water volume did not change significantly, the dividing line between sludge and liquid was clearer than for raw sewage sludge.

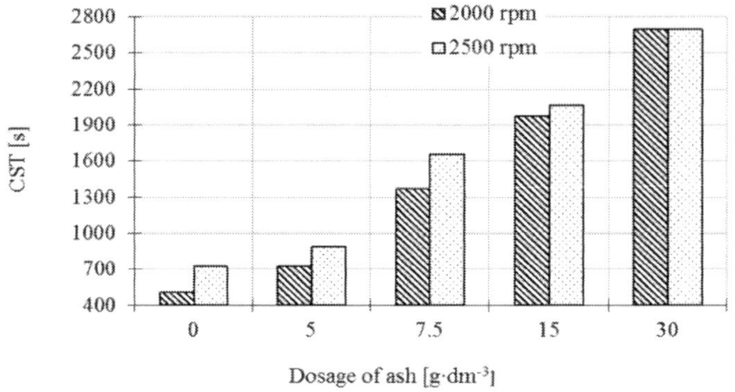

Fig. 5. Influence of biomass ash on the CST after centrifugation.

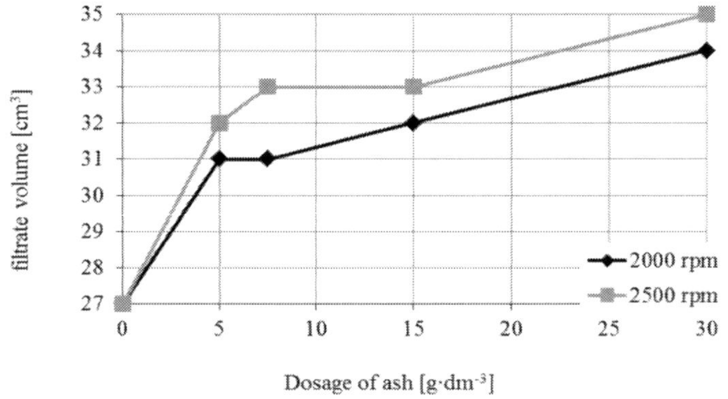

Fig. 6. Influence of biomass ash on the filtrate volume after centrifugation.

The vacuum filtration provided the best results of sewage sludge dewatering (Fig. 7). The hydration of sewage sludge decreased as the dosage of ash and the vacuum pressure increased. The application of biomass ash in a dose of 30 g·dm^{-3} reduced the moisture content of approximately 20% for a vacuum pressure of 0.01 MPa and about 25% for a vacuum pressure of 0.02 MPa. Detailed information concerning the influence of biomass ashes on CST value are presented in [12]. Similar results were achieved by Changya et al. [9] for sewage sludge with the addition of coal fly ash in the dosage of 300% d.m. Likewise, Wójcik et al. [7] investigated that the addition of willow ash in the dosage of 30 g· dm^{-3} could decrease the sewage sludge hydration by approximatelly 13% for the vacuum pressure of 0.01 MPa and 14% for the vacuum pressure of 0.02 MPa.

Increasing both biomass ash dosage and the vacuum pressure caused an increase of CST. In comparison with the results for centrifugation, CST after the vacuum filtration was approximately 50% higher. The higher value of CST is a result from the better effectiveness of sewage sludge dewatering in vacuum filtration (Fig. 8).

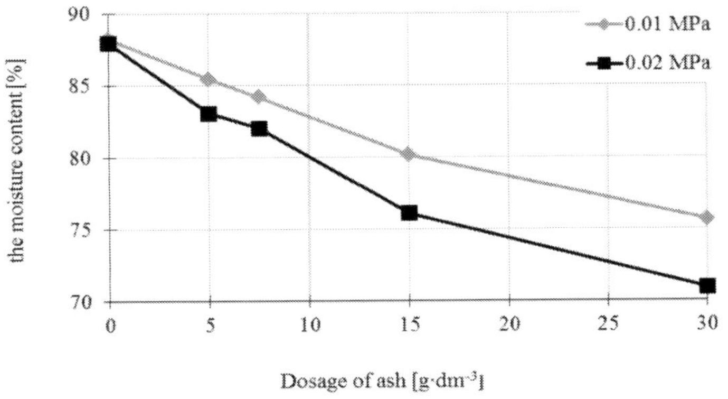

Fig. 7. Influence of biomass ash on the sewage sludge moisture content after vacuum filtration [12].

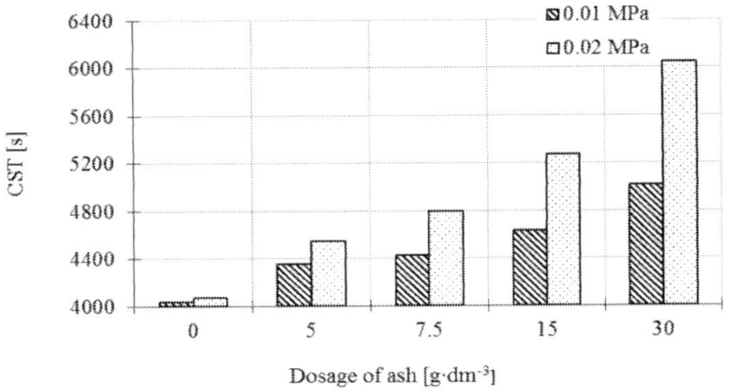

Fig. 8. Influence of biomass ash on the CST after the vacuum filtration.

The effectiveness of vacuum filtration was also done by the measurement of filtrate volume (Fig. 9). Although the filtered water volume did not change significantly, the sludge filtration was more rapid in comparison with raw sewage sludge. With 30 $g \cdot dm^{-3}$ dosage of ash, the filter duration reduced from 200 s to about 60 s [12].

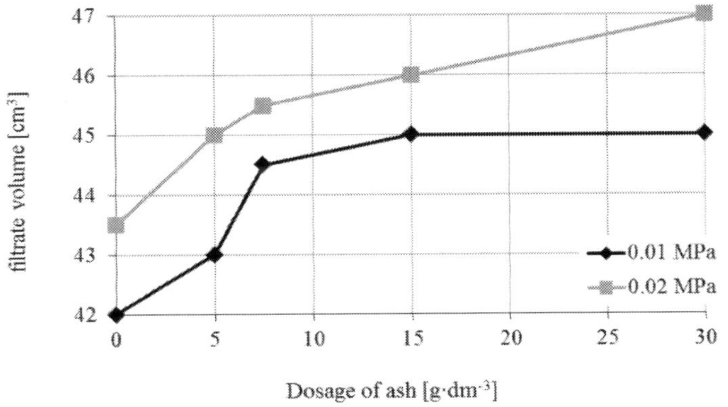

Fig. 9. Influence of biomass ash on the filtrate volume after the vacuum filtration [12].

4 Summary

The biomass ash is problematic waste which has to be managed in line with environmental and law requirements. Stringent requirements prevent biomass ashes from being used as a component in concrete and cement production. For this reason, the new biomass ash utilization methods are developed.

According to the specific physico-chemical properties, biomass ashes could be used as a sewage sludge conditioner. Sewage sludge after conditioning with ash showed a much stronger dewatering capacity than raw sewage sludge. The main sludge conditioning mechanisms with application of biomass ash can improve the floc formation and provide the water transmitting passages by skeleton builder [8, 11].

The obtained results of laboratory tests allow us to draw the following conclusions:

1. The application of biomass ash has a positive impact on sewage sludge dewatering what is confirmed by the measurement of CST. Aforementioned parameter decreased with the increase of dosage of ash.
2. The analysis of results has shown that the addition of biomass ash can improve the capacity of sewage sludge dewatering. The application of higher doses of ash yielded better results which are confirmed by the filtrate volume and moisture content after dewatering. With a 30 $g \cdot dm^{-3}$ dosage of biomass ash, the sewage sludge moisture cake content decreased of approximately 10 ÷ 25 %, depending on the method of dewatering. The best results of sewage sludge dewatering were obtained for vacuum filtration for 0.02 MPa vacuum pressure. The addition of low dosages of biomass ash could slightly improve the sewage sludge dewatering.
3. The application of ash from biomass-fired power plant might raise the pH of conditioned sewage sludge. For this reason, it is possible to reduce the amount of pathogens and achieve the partial higienization of sewage sludge.

The confirmation of the effectiveness of biomass ashes in sewage sludge conditioning might reduce the consumption of polyelectrolytes and contribute to the reduction of operating costs of treatment plants.

References

1. J. Poluszyńska, ICMB **12**, 60 (2013).
2. A. Uliasz-Bocheńczuk, A. Pawluk, J. Skierka, Mineral Resources Management **31**, 145 (2015).
3. K. Rajczyk, E. Giergiczny, M. Szota, ICMB **12**, 73 (2013).
4. T. Czech, T.A. Sobczyk, A. Jaworek, A. Krupa, The Conference POL-EMIS, Sienna 2012, 73 (2012).
5. [online] [access: 10.05.2016] https.://www.kierunekenergetyka.pl/magazyn, popioły-z-biomasy.html
6. Central Statistical Office Report on Environmental Protection, 2015.
7. M. Wójcik, F. Stachowicz, A. Masłoń, E3S Web Conf. **17**, 1 (2017).
8. A. Masłoń, Instal **4**, 51 (2015).
9. C. Changya, Z. Panyue, Z. Guangming, D. Jiuhua, Z., L. Haifeng, CHEM ENG J. **158**, 616 (2010).
10. A. Piotrowska-Cyplik, Z. Czarnecki, J. Res. Appl. Agric. Engng **50**, 21 (2005).
11. S. Wang, T. Viraraghavan, WASTE MANAGE. **17**, 443 (1997).
12. M. Wójcik, F. Stachowicz, A. Masłoń, IOS **20**, 155 (2017).

Numerical analysis of a heat exchanger with differentiated temperatures surface at varying distances from the wall

Magdalena Orłowska[1,*]

[1] Technical University of Koszalin, Faculty of Civil Engineering, Environmental and Geodetic Sciences, al. Śniadeckich 2, 75-453 Koszalin, Poland

Abstract. This article is one of a series of articles by the author. For many years she conducts research on convective heat exchange. The work is mainly concerned on knowing the effect of positioning the heater on the heat output of the device. It turns out that the correct location is very important.

1 Introduction

When installing a radiator, we have to take many considerations into account, such as a wall with a window or interior, the distance from the floor and the sill or wall. It is also important to consider additional decorative covers.

This article is a follow-up to the author's series on airflow for radiators operating under different boundary conditions.

1.1 Convection heat transfer and radiators

Convection occurs when we are dealing with the macroscopic movement of fluid particles (liquid or gas) in the direction of temperature drop. The movement of fluid molecules determines the heat flux and the temperature field. This term refers to the case of the energy flow between the moving fluid and the surface of the barrier at different baffle and plate temperatures. We divide it into natural and forced. With forced velocity fluids are much higher. The movement of heat during convection coincides with conduction [1]. Currently, 90% of all radiators that are installed are plate radiators that are flat. They are characterized by low weight, low water content, nice appearance, large heat transfer surface, wide range of dimensions. The surfaces of the radiators can be smooth or wavy. There are solutions one and several rows. The front of these radiators gives off the heat by radiation and convection while the rear of the radiator mostlyby convection [2].

* Corresponding author: magdalena.orlowska@tu.koszalin.pl

2 Methodology of numerical research

All simulations were performed using the simulation code Ansys Mechanical APDL/CFD Flo Preferences for GUI Filtering Flotran CFD 2D Flotran 141 Library of Elements Types [3].
„The program is based on solving the balance equations of energy, momentum and mass used in CFD- Computational Fluid Dynamic model. The standard version of the simulation code allows you to conduct modeling based on traditional dependence phenomenological equations and additional closures, describing the previously mentioned turbulent momentum and energy streams. The test system was modeled in two-dimensional space. Program calculates after entering the boundary conditions: temperature of the walls of the heat exchanger, room air velocity.The article presents couple of selected systems with different geometry attempted to explain its effects on the intensity of heat exchange (heat transfer coefficient)" [4,5].

3 Results

The numerical analysis include in this issue refers to the proposed three heights of the location of the heater x at the wall (x_1=1cm; x_2=2cm; x_3=2,5cm) Fig.1 with the differentiated temperatures surface set on the walls.

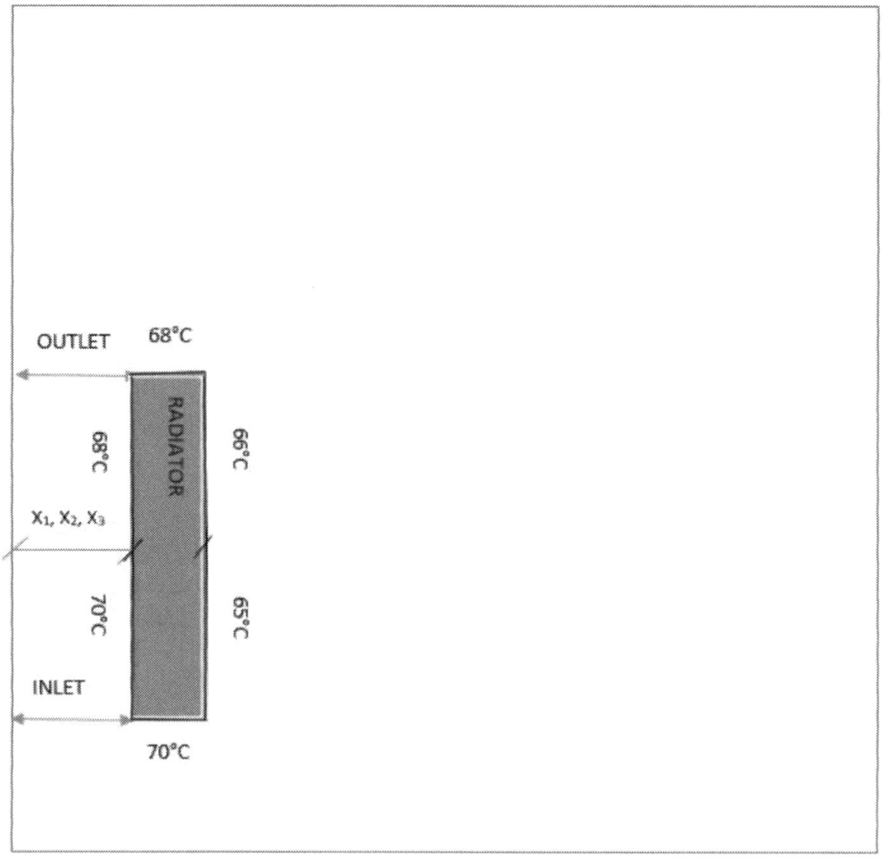

Fig. 1. The scheme of testing stand.

Suggested distances let you determine which of the systems under test is the most efficient in terms of unit efficiency and thus the energy efficiency of the heat exchanger.
The numerical analysis gave the fallowing results Fig. 2,3,4,5,6,7.

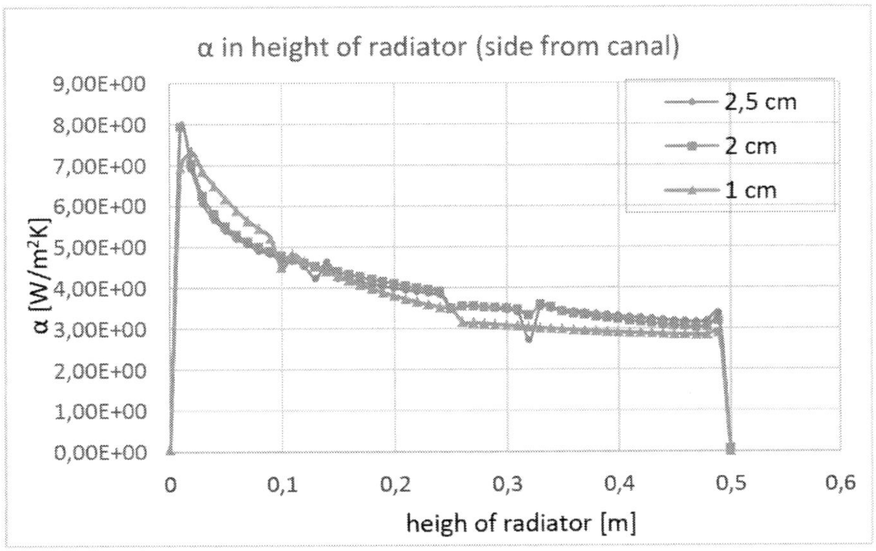

Fig. 2. Heat transfer coefficient α in height of radiator function.

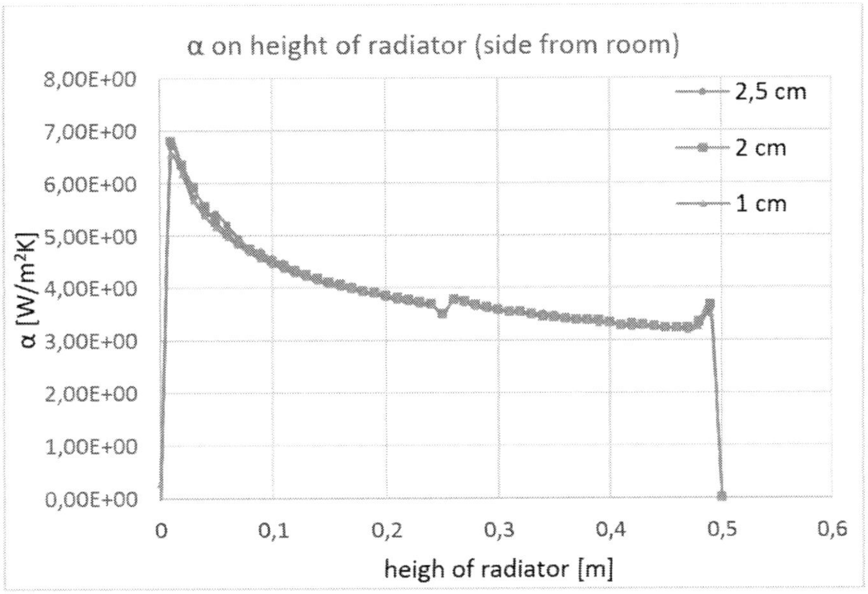

Fig. 3. Heat transfer coefficient α in height of radiator function (side from the room)

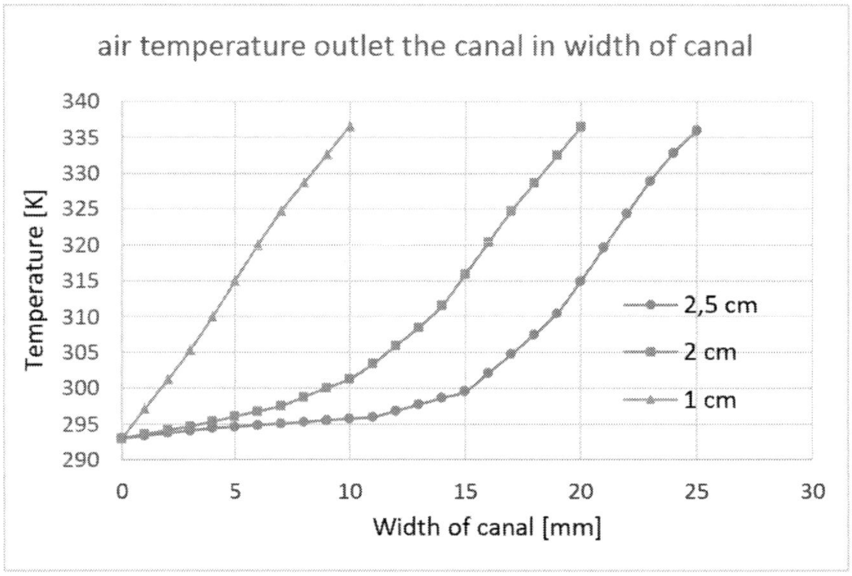

Fig. 4. Air temperature outlet the canal in width of canal function

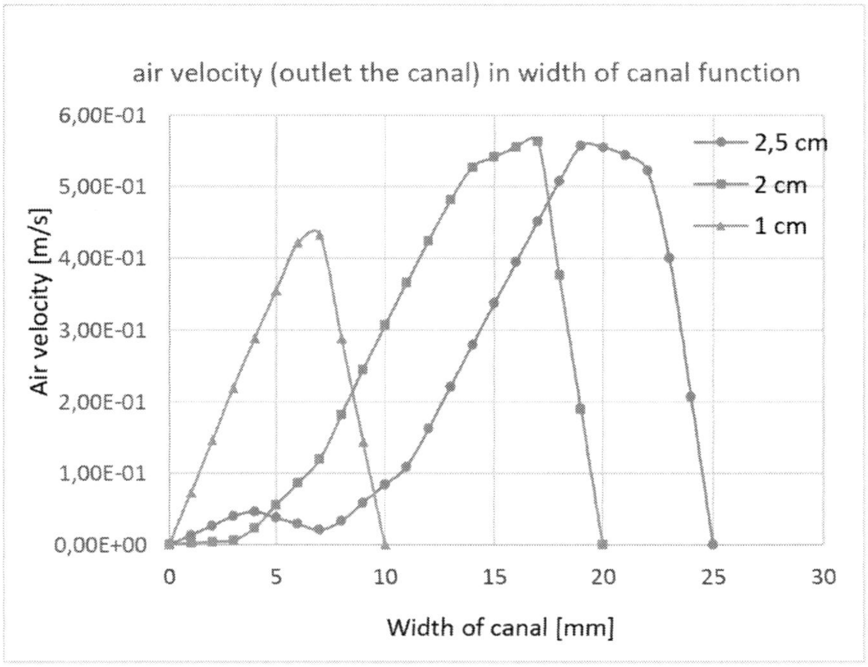

Fig. 5. Air velocity outlet the canal in width of canal function

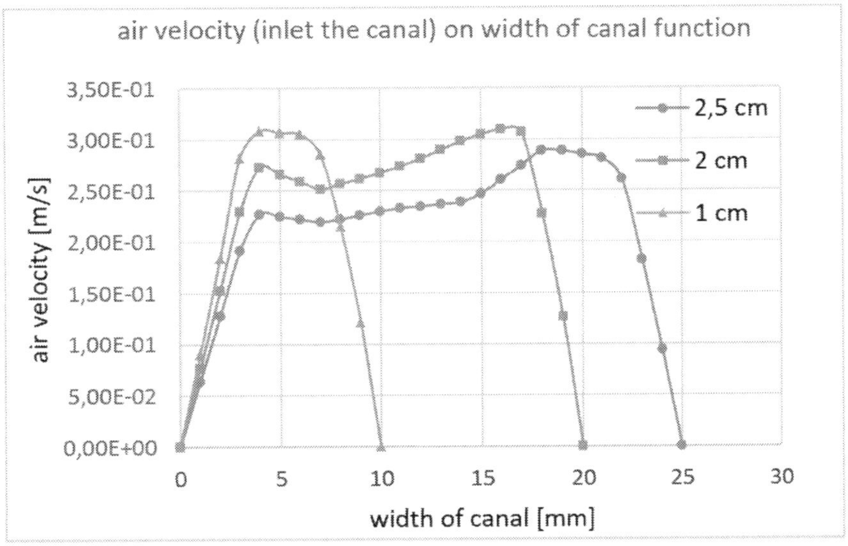

Fig. 6. Air velocity inlet the canal in width of canal function

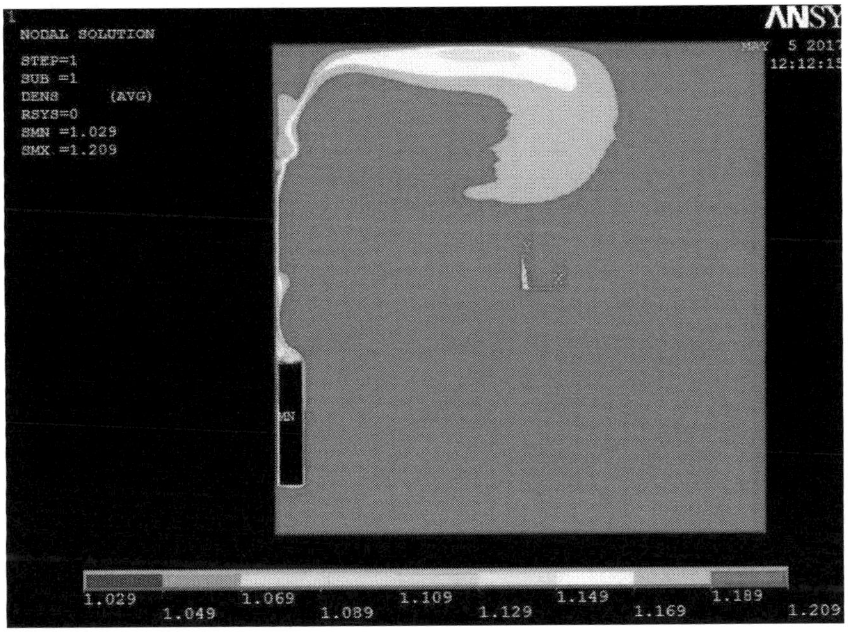

Fig. 7. Air density of testing model, [3]

4 Discussion on the obtained results

The best results in terms of heat output and therefore the heat transfer coefficient for the heater channel side were obtained in a system with 2 cm distance from the heater to the wall and it was α=3,93 [W/m^{2}K], slightly lower in a 2,5 cm distance α=3,89 [W/m^{2}K] and the lowest in a distance of 1 cm α=3,77 [W/m^{2}K]
In terms of the temperature at the outlet of the channel most preferred system was the arrangement of 1 cm distance, where the temperature was 314,92 K, did slightly less variant

with a distance of 2 cm 307,08 K and the lowest value was obtained at a distance of 2,5 cm 304,2 K.
The velocity at the outlet of the duct was the highest because the outlet was w_{wylot}=2,41 m/s reached in a system with a 2 cm distance between the wall and the heater, the lower at 2,5 cm of the slope w_{wylot} =2,17 m/s the lowest slope w_{wylot}=2,15 m/s in a layout with a distance of 1cm.
The highest inlet velocity reached in the system with a 2 cm distance to the inlet w=2,24 m/s, lower in the variant with 2,5 cm inlet distance w=2,06 m/s and lowest in the system with
1 cm distance to the inlet w=1,91 m/s.
The article shows the optimal distance (in terms of the heater's thermal performance) from the radiator's distance traveled. It is important that this performance is maximized.
It turns out that the influence of the environment is affected by it. It is not only about the distance of the radiator from the wall, but also the distance from the floor, from the window sill, the width and shape of the window sill. The radiator elements such as - decorating, and architectural assembly and mounting walls also affects performance. Most often it is a wall under the window and it is certainly better in terms of performance than a wall without a window. So, deciding on the location we should take into account not only the selection of suitable for the room of the heating power, which the radiator should provide but also rethink the mounting conditions, which remain insignificant.

5 Conclusion

Among the systems proposed in this study most preferred option is with a distance of 2 cm between the wall and the heater and this is the optimal distance. The layout with 1 cm distance could cause interference in the boundary layer hence the low values of the obtained parameters. In turn, the 2,5 cm system is already too far away, although it is only 0,5 cm away from the most advantageous variant.
Based on the numerical analysis, it can be said with absolute certainty that the radiator's distance from the wall to the temperature and air velocity fields and the efficiency of the device is affected. So consider the location of the heat exchanger before installing it.

References

1. A. Skoczylas, *Przenoszenie ciepła*, Oficyna Wydawnicza Politechniki Wrocławskiej, p.11, (1999)
2. M.B. Nantka, *Ogrzewnictwo i ciepłownictwo, Volume **II***, Wydawnictwo Politechniki Śląskiej, p.27-28, (2010)
3. Ansys CFD, ver.12,1, *Mechanical APDL Product Launcher*, User Instruction
4. M. Orłowska, *Numerical analysis of the heat exchanger energy efficiency depending on location from the floor,* E3S Web Conf. Volume **17**, 2017 9th Conference on Interdisciplinary Problems in Environmental Protection and Engineering EKO-DOK 2017, (2017)
5. M. Orłowska, *Numerical analysis at a plate heat exchanger with covers-selected variations*, Heat Transfer and Heat Exchangers –Heat Transfer and Renewable Sources of Energy HTRSE 2016, p. 225–228, (2016)

Computer Simulation in Predicting Biochemical Processes and Energy Balance at WWTPs

Jakub Drewnowski[1,*], *Ewa* Zaborowska[1] , and *Carmen* Hernandez De Vega[2]

[1] Gdansk University of Technology, Faculty of Civil and Environmental Engineering, ul. Narutowicza 11/12, 80-233 Gdansk, POLAND

[2] Polytechnic University of Valencia, School of Civil Engineering, Dept. of Hydraulic Engineering and Environment, Camino de Vera, s/n 46-022 Valencia, SPAIN

Abstract. Nowadays, the use of mathematical models and computer simulation allow analysis of many different technological solutions as well as testing various scenarios in a short time and at low financial budget in order to simulate the scenario under typical conditions for the real system and help to find the best solution in design or operation process. The aim of the study was to evaluate different concepts of biochemical processes and energy balance modelling using a simulation platform GPS-x and a comprehensive model Mantis2. The paper presents the example of calibration and validation processes in the biological reactor as well as scenarios showing an influence of operational parameters on the WWTP energy balance. The results of batch tests and full-scale campaign obtained in the former work were used to predict biochemical and operational parameters in a newly developed plant model. The model was extended with sludge treatment devices, including anaerobic digester. Primary sludge removal efficiency was found as a significant factor determining biogas production and further renewable energy production in cogeneration. Water and wastewater utilities, which run and control WWTP, are interested in optimizing the process in order to save environment, their budget and decrease the pollutant emissions to water and air. In this context, computer simulation can be the easiest and very useful tool to improve the efficiency without interfering in the actual process performance.

1 Introduction

Mathematical modelling becomes an integral part of the design and operation of water treatment systems, especially with activated sludge (AS) systems [1]. The use of mathematical models and computer simulation allow analysis of many different technological solutions as well as testing various scenarios in a short time and at low financial budget in order to simulate the scenario under typical conditions for the real system and help to find the best solution in design or operation process [2]. Nowadays, a lot of mathematical models, which describe a wastewater treatment plant (WWTP) in different

* Corresponding author: jdrewnow@pg.gda.pl

ways with good results, had been developed. The biochemical mathematical models called Activated Sludge Model (ASM), which is one of the simplest, describe the conversion of organic compounds, phosphorous and nitrogen. More complex models (such as Mantis2) developed in last years integrate ASM with anaerobic digestion processes, precipitation and/or anammox.

Computer simulation has become a helpful tool in analysis of wastewater treatment systems performance and effectiveness. Using advanced program it is possible to create a mathematical model of a real WWTP, run a simulation and subsequently interpret results under various conclusions. However, the most important to get a successful model is the organization. Firstly, the laboratory tests, full-scale measurements and calculations should be carried out in order to better integrate computer model to the conditions prevailing in the local parameters of each individual parts of WWTP. Secondly, the development of a computer models require a detailed analysis of the whole technological system and taking into account the specific variables to each modelled device/object. The proper estimation of these parameters in model has strong influence on the results of mathematical modelling and computer simulation as well as it consumes a lot of time. Furthermore, the quality of the calibration depends on the objectives of the study which defined the accuracy of the model calibration parameters. For that reason, this step is the hardest and one of the most important of this study.

Water and wastewater utilities, which run and control WWTP, are interested in optimizing the process in order to save environment, their budget and decrease the pollutant emissions to water and air. In this context, computer simulation can be the easiest and very useful tool to improve the efficiency without interfering in the actual process performance. The aim of this study was to evaluate different concepts of the biochemical processes and energy balance modelling using a comprehensive models Mantis2. For this purpose, the sample layout of a WWTP was modelled in a computer program GPS-x ver. 6.4 (Hydromantis, Canada), which was submitted to dynamic simulation.

2 Material and methods

2.1 Study site

This study was based on archives date from real "Wschód" WWTP (presented earlier by Swinarski et al. [3]) as a base-point for newly developed "virtual" plant layout, extended in the range of sewage sludge treatment with the following device:

- primary sludge thickener,
- waste activated sludge dewatering,
- anaerobic digester,
- digested sludge dewatering (centrifuges).

The schematic layout of plant was created in GPS-x and is presented in Figure 1. Additional information about wastewater treatment process line including activated sludge MUCT system with a single bioreactor compartment (volumes of individual compartments, average DO concentrations in the aerobic zone) are listed in Table 1 as well as characteristics of the average operating conditions at "Wschód" WWTP for computer simulation study can be found in Table 2.

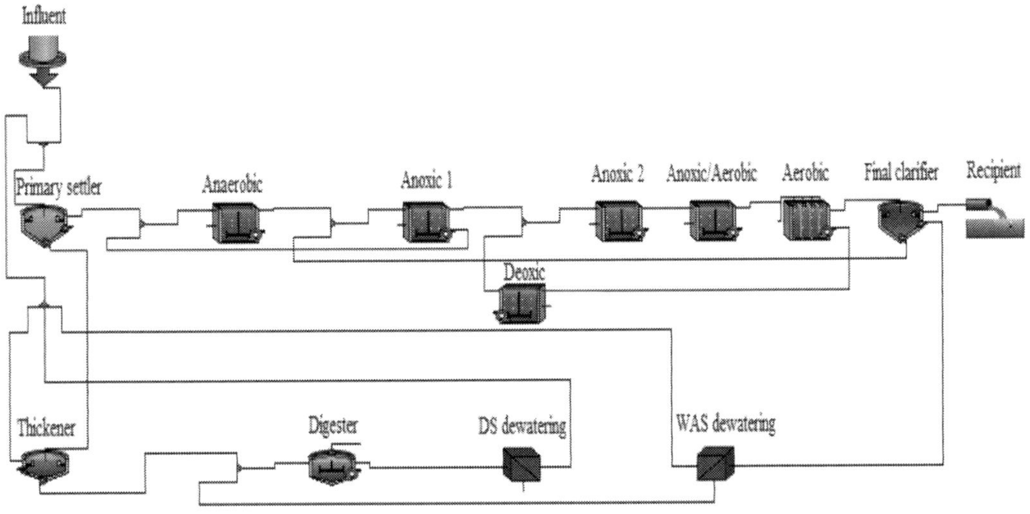

Fig. 1. A schematic layout of the MUCT bioreactor extended in the range of wastewater sludge treatment created in the GPS-x simulator.

The "Wschód" WWTP is located in north of Poland and it is serving a population of Gdansk and adjusting communities (altogether approx. 550,000 inhabitants in 2008), being one of the largest facilities in the Baltic Sea. The virtual plant layout, in which this study is based on, was modified as the following structure. The process wastewater treatment line start with a primary settler followed by a MUCT (Modified University of Cape Town) bioreactor with an additional deoxic zone in the internal recirculation line from the aerobic zone to the anoxic zone and a final clarifier. On the one hand, the sludge produced in the primary settler is carried to the thickener and on the other hand, the sludge produced in the final clarifier is carried to the dewatering process. Afterwards, the thickened sludge from the primary and final clarifiers is mixed and conducted to the digester in which the digestion process will be done in order to produce biogas. Finally, the digested sludge is dewatering in order to get a suitable mud for other possible uses. The water which was separated in the thickened and dewatering processes is carried to header plant with the aim of clean it and do not have any discharge of dirty water. The feedstock for fermentation is a mixture of thickened primary sludge approx. 4% TSS (Total Suspended Solids) and thickened waste activated sludge (6% TSS). Reject water from raw and digested sludge dewatering processes is recirculated to the head of wastewater train.

https://doi.org/10.1051/e3sconf/20183003007

Table 1. Characteristics of the wastewater treatment process line including activated sludge MUCT system with a single bioreactor compartment (volumes, average DO concentrations etc.) used for simulation study in GPS-x.

Parameter	Unit	Primary settler	Anaerobic	Anoxic 1	Anoxic 2	Anoxic / Aerobic	Aerobic 1/2	Aerobic 3/4	Aerobic 5/6	Deoxic	Final clarifier	Digester
Surface area	m^2	1000	-	-	-	-	-	-	-	-	3020	-
Max. Volume	m^3	-	3500	2050	3930	3930	1950	1950	1950	1250	-	5000
Depth	m	3.5	4	4	4	4	4	4	4	4	3	-
DO concentration	gO_2/m^3	-	-	-	-	-	2.93/2.25	1.66/1.21	0.066/0.60	-	-	-

2.2 Extracting experimental data

Laboratory batch test (e.g. nitrogen utilization rate – NUR, P release rate and anoxic/aerobic P utilization rate - PRR and anoxic/aerobic PUR, oxygen/ammonia utilization rate – OUR/AUR) with the process biomass and settled wastewater were carried out in a specially designed and constructed experimental set-up consisting of two parallel batch reactors (max. volume of 4.0 dm3), control system and computer. This batch experiments results were obtained by Drewnowski and Makinia [4] at a large (600,000 PE) BNR "Wschód" WWTP in Gdansk (Poland) provided the experimental database for comparison of the model predictions as well as additional 96-hour measurement campaign in the full-scale MUCT bioreactor at this plant presented earlier by Swinarski et al. [3] was integrated to calibration/validation process. During that time only full-scale MUCT bioreactor was simulated using base model ASM2d. This study was set as a extension to evaluate different concepts of the biochemical processes and energy balance modes using a comprehensive model Mantis2 in order to developed procedure of the whole "virtual" plant layout as a base point and test of complex model for practical use in full-scale object.

Table 2. Characteristics of the average operating conditions from "Wschód" WWTP as a base point for "virtual" plant layout simulation study in GPS-x.

Parameter	Unit	Average value in summer 2008
Concentrations in settled wastewater:		
COD	$gCOD/m^3$	695
BOD_5	$gBOD/m^3$	357
$N_{tot.}$	gN/m^3	65.9
$N\text{-}NH_4^+$	gN/m^3	56.1
$P_{tot.}$	gP/m^3	50.9
$P\text{-}PO_4^-$	gP/m^3	45.6
Concentrations in secondary effluent:		
COD	$gCOD/m^3$	42.1
"Soluble" COD	$gCOD/m^3$	34.5
$N_{tot.}$	gN/m^3	10.3
$N\text{-}NH_4^+$	gN/m^3	0.135
$N\text{-}NO_3^-$	gN/m^3	7.55
$P_{tot.}$	gP/m^3	2.53
$P\text{-}PO_4^-$	gP/m^3	0.812
Operating parameters:		
Q_{INF}	m^3/d	21568
Q_{MLR1} (anox 1 – anaer)	m^3/d	40000
Q_{MLR2} (aer – anox 2)	m^3/d	120000
Q_{RAS}	m^3/d	20110
Process temperature	°C	19
Sludge Retention Time	d	21.3
Biomass characteristics:		
MLSS	g/m^3	5450
MLVSS/MLSS (i_{VT})	-	0.437

2.3 Organization of the modelling study procedure

The comprehensive model Mantis2 incorporates the most commonly observed biological, physical, and chemical processes in WWTPs. This model integrates carbon, nitrogen and phosphorus removal with anaerobic digestion process. The basic structure of the model is based among others on ASM2d and UCTADM1 models. The computer simulations with this model were carried out using the software GPS-x ver. 6.4 simulation platform (Hydromantis, Canada), more can be found in [5]. The model of activated sludge process was calibrated based on results of the batch tests (summer study period) and 96-hour measurement campaign in the full-scale MUCT bioreactor from "Wschód" WWTP in summer 2008. The results of the batch tests with the process biomass and settled wastewater under winter sessions were used to compare model predictions in terms of NH_4-N, NO_3-N, PO_4-P behavior and validated the model. This set of the process stoichiometric and the kinetic parameters in ASM2d was used as a base-point of further process calibration in newly developed comprehensive model Mantis2. The Nelder-Mead simplex method was additionally used in order to numerically optimize and final recalibrated/validate parameters of Mantis2 for simulation in newly developed whole "virtual" plant layout in GPS-x as was presented in Table 3.

Table 3. List of the kinetic/stoichiometric parameters in Mantis2 adjusted during model calibration in the GPS-x simulator.

Symbol	Unit	Default value	Calibrated value
Heterotrophic organisms:			
Aerobic heterotrophic yield on soluble substrate	gCOD/gCOD	0.666	0.625
Anoxic heterotrophic yield on soluble substrate	gCOD/gCOD	0.533	0.625
Maximum specific growth rate on substrate	1/d	3.2	6
Saturation/inivition coefficient for Sac	g/COD/m^3	5	4
Reduction factor for denitrification on nitrate-N	-	0.32	0.8
Aerobic heterotrophic decay rate	1/d	0.62	0.4
Phosphorus Accumulating Organisms:			
Aerobic yield on PAO growth	gCOD/gCOD	0.639	0.625
Anoxic yield on PAO growth	gCOD/gCOD	0.511	0.625
PHA storage yield	gP/gCOD	0.4	0.5
Saturation coefficient of PAO for Sac	gCOD/m^3	4	2
Rate constant for storage of poly-phosphate	gP/gPAO/d	1.5	2.5
Aerobic decay	1/d	0.2	0.18

https://doi.org/10.1051/e3sconf/20183003007

coefficient for PAO			
Anoxic reduction factor for decay rate	-	0.9	0.6
Poly-P lysis coefficient	1/d	0.2	0.1
Ammonia-oxidizing Organisms:			
Oxygen saturation for ammonia oxidizer	gO_2/m^3	0.25	0.2
Ammonia oxidizer aerobic decay rate	1/d	0.17	0.12
Anoxic reduction factor for decay rate	-	0.5	1.5
Nitrite-oxidizing Organisms:			
Nitrite saturation coefficient for nitrite oxidizer	gN/m^3	0.5	0.04
Oxygen saturation for nitrite oxidizer	gO_2/m^3	0.68	0.1
Anoxic reduction factor for decay rate	-	0.5	0.3
Hydrolysis:			
Hydrolysis rate constant for Xs	1/d	3	2.5
Saturation coefficient for particulate COD	-	0.1	0.2
Anoxic hydrolysis reduction factor	-	0.28	0.6
Anaerobic hydrolysis reduction factor	-	0.4	0.1

The model of AS bioreactor was extended by primary settler and anaerobic digestion chamber in order to develop a complex model of a WWTP. Energy balance followed the simulations of wastewater and solids treatment processes. The operating cost model was used to simulate aeration and pumping energy concurrently to wastewater processes simulation. This study focused on energy consumption by blowers supplying air to a nitrification tank and by pumps for recirculation of activated sludge from a secondary clarifier to the head of the biological reactor. Each of the selected objects was set up to estimate electric power and daily energy consumption. The potential for renewable energy generation was calculated based on simulated biogas production in anaerobic digestion process operated at 37°C. For a mixture of primary and secondary sludge digestion a default (Mantis2 model) set of the process parameters was used. The simulations were carried out at steady-state conditions, but dynamic condition was also made in order to calibrate whole "virtual" plant layout for simulation study in GPS-x.

3 Results and discussion

The paper presents the example of application of the WWTP model as a tool to optimize plant operation. Sample results and model predictions of batch experiments are presented in Figure 2 a-b [4, 6]. Set of the process stoichiometric and kinetic parameters in ASM2d was used as a base point of further process calibration in newly developed comprehensive model Mantis2 for simulation study of "virtual" plant layout in GPS-x (Figure 2c). The values of Mantis2 parameters adjusted during dynamic calibration of the full-scale MUCT bioreactor were compared to the default values in Table 3. With these values, both process rates (PRR, anoxic/aerobic PUR, NUR and AUR) in two-phase batch tests (Figure 2 a,b) and behavior of NH_4-N, NO_3-N and PO_4-P concentrations in the full-scale bioreactor (Figure 2c) were matched accurately by the model predictions.

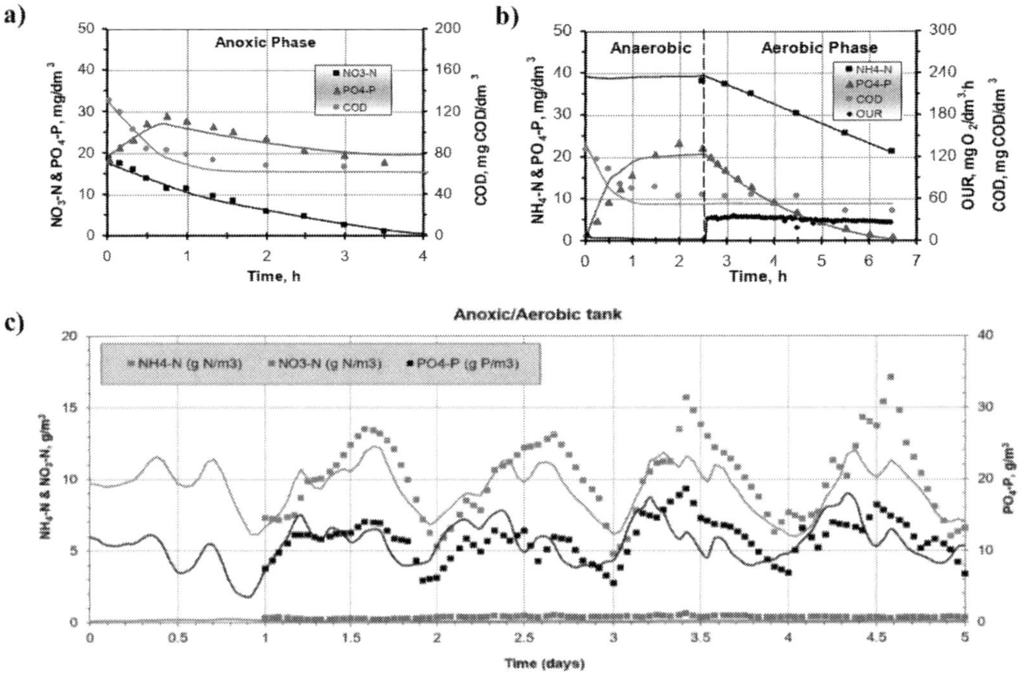

Fig. 2. Sample results of different process rate measurements and model ASM2d/Mantis2 predictions: (a) conventional NUR experiments, (b) PRR/aerobic PUR experiments, (c) anoxic/aerobic zone MUCT bioreactor simulation study of "virtual" plant layout.

During the simulation the changes in the concentration of nitrogen and phosphorus in biological tanks and in the effluent from WWTP was observed. The model predictions were fitted to the measured NURs by adjusting two parameters (based on selected kinetic ASM2d coefficients): the maximum growth rate of heterotrophs (μ_H) and hydrolysis rate constant (k_h) (Figure 2a). In the NURs during anoxic P uptake, no further modifications were needed to calibrate the NUR in the anoxic phase of the PRR/anoxic PUR test (not shown). The PRR and PUR tests were calibrated with the parameters: the rate constant for storage of PHA (q_{PHA}), saturation coefficient for PAOs with respect to S_A ($K_{SA,PAO}$), saturation coefficient for PAOs with respect to polyphosphate (K_{PP}), anaerobic hydrolysis reduction factor (η_{fe}) and saturation coefficient for particulate COD (K_X). The nitrification process based on the measured data from PRR and aerobic PUR batch tests was calibrated with kinetic parameters, including the maximum growth rate of autotrophs (μ_A) and the NH_4-N saturation coefficient ($K_{NH4,A}$) (Figure 2b). The process calibration in newly developed comprehensive model Mantis2 for simulation study of "virtual" plant layout with the full-scale MUCT bioreactors in GPS-x were started with calibrated results of a series of batch tests with the process biomass (ASM2d) in order to match accurately kinetic/stoichiometric parameters. Then previous model of "Wschód" WWTP [3] as a base-point of further process calibration was used in conjunction with optimization techniques Nelder-Mead simplex to estimate kinetic/stoichiometric parameters in Mantis2 as was shown in Table 3.

The simulations run on the calibrated and validated plant-wide Mantis2 model included scenarios that may influence the WWTP process operation and energy balance. For the purpose of this study, in order to show the possibilities of using simulation software as a tool for predicting a WWTP energy output, the following scenarios were considered:

(1) variation in suspended solids removal efficiency from the primary settler,

(2) variation in returned activated sludge (RAS) flow rate,

(3) variation in dissolved oxygen concentration in the activated sludge reactor nitrification zones.

The first scenario predicted variations in the amount of particulate matter removed form the primary settler. The influence on digestion process and biogas production was observed for three values of primary sludge removal efficiency: 20%, 30% and 40%. The potential for electric energy production by biogas driven combined heat and power plants was calculated for assumed efficiency equal to 35%. The results, presented in Table 4, show a positive influence of primary sludge digestion on biogas production. The volume of biogas is nearly 25% higher for suspended solids removal efficiency equal to 40% in comparison to the reference state. That can definitely improve a WWTP energy balance, increase renewable energy production while decreasing energy purchase. On the other hand a maximum volume of raw sludge removed from a settler is limited by carbon demand in biological step of wastewater treatment and restrictions for effluent quality.

Table 4. Simulated results of digestion process and potential for energy production

Suspended solids removal efficiency	Biogas production	Biogas yield per VSS destroyed	Energy production in cogeneration	Change in relation to the reference state
%	m^3/d	$m^3/kgVSS$	kWh/d	%
20	3320	1,06	7185	Reference state
30	3677	1,06	7957	+8.7
40	4293	1,07	9290	+24.7

The second scenario focuses on returned activated sludge flow rate and its influence on energy consumption for pumping. The results of simulations are presented in Table 5. A higher RAS rate may lead to better effluent quality but simultaneously may result in distinctly higher energy demand for pumping.

Table 5. Simulated values of RAS pumping energy

RAS flow rate		Pumping power	Pumping energy required	Change in relation to the reference state
m^3/d	%	kW	kWh/d	%
25000	100	24	584	Reference state
75000	300	73	1752	+200
100000	400	97	2336	+300

The subject of the third scenario predicted variation in oxygen concentration in aerobic zones of the activated sludge reactor. The reference state in Table 6 represents the assumptions close to the calibration terms. The lower oxygen concentration was still enough to remove ammonia but at decreased demand for blower energy. Both examples confirm a significant role of oxygen control in aeration tanks as energy required for blowers often has a dominant share in WWTPs energy balance.

Table 6. Simulated results of aeration energy demand

Oxygen concentration in aerobic zones	Total air flow to aeration tank	Blower wire power	Blower energy required	Change in relation to the reference state
gO_2/m^3	m^3/h	kW	kWh/d	%
1.0-1.5	15420	235	5640	-17.5
2.0-3.0	18690	285	6840	Reference state
3.0-4.5	23090	352	8448	+23,5

The IWA Activated Sludge mathematical models model is accepted as a useful model to describe biological nitrogen and EBPR in a variety of biological nutrient removal process configurations. The ASM models have been proven to be useful tools for evaluating and optimizing the effect of carbon sources for nitrate/nitrite removal, provided that wastewater biodegradability (COD fractions), kinetics and stoichiometric parameters were determined by Onnis-Hayden and Gu [7]. Using a simplified version of the Activated Sludge Model no. 2d (ASM2d), De Lucas et al. [8] modeled the results of nitrate utilization experiments with various wastewaters domestic and/or agro-food industrial. Subsequently, the denitrification potential of each carbon source and values of the most important kinetic and stoichiometric parameters for the denitrification process were evaluated. Takacs et al. [9] accurately simulated four full-scale wastewater treatment plants (WWTPs) with an expanded BioWin model (Envirosim, Canada) under aerobic conditions. Dold et al. [10] used modeling to design optimal batch tests procedures, and to estimate the maximum specific growth rate of heterotrophs under anoxic conditions. Phillips et al. [11] used a dynamic model to optimize the mixed liquor recycle rate and studied factors that affected nitrogen removal, including the influent wastewater readily biodegradable COD/N ratio and acetate addition. The strategies and calculations of the optimum COD/N ratios were investigated by Filali-Meknassi et al. [12] (in sequencing batch reactor (SBR)) and Latimer et al. [13] (in continuous flow reactor). In this study not only biological part of WWTP was simulated but more complex newly developed model of whole plant Mantis2 (which is extension of well known ASM) was calibrated/validated, behaviors of real date NH_4-N and NO_3-N and PO_4-P in the anoxic/aerobic tank of full-scale MUCT bioreactor, as a sample results presented Figure 2c, were matched accurately by the model predictions in the GPS-x simulator. Moreover full-scale simulations during the validation process revealed that addition of winter scenario in Mantis2 compensate with the temperature parameters resulting in a similar NH_4-N and NO_3-N and PO_4-P behavior compared to the reference case (data not shown) presented earlier by Swinarski et al. [3].

4 Conclusions

Process optimization performed based on biological system physiology is the next level of system operations that are required to maximize the use of existing facilities. One of the main reasons that has limited the use of this approach has been the lack of proper tools to simulate the biological processes. In this study the simulation platform GPS-x and Mantis2 model was found as useful tools for this purpose.

During model calibration, kinetic parameters in previous ASM2d as well as new complex Mantis2 model were adjusted until the real date from "Wschód" WWTP match with simulations. With the calibrated ASM2d/Mantis2, the principal process rates (such as NUR, PRR and anoxic/aerobic PUR, OUR/AUR) were accurately predicted in one/two-phase batch tests and for simulation study of "virtual" plant layout with the full-scale MUCT bioreactors in GPS-x. Continual feedback between the model and real-time measurements (e.g. NH_4-N and NO_3-N and PO_4-P behavior) from "Wschód" WWTP were additionally used to confirm initial modeling results. It was shown that a "well" calibrated model within an assessment study could be used to evaluate process capacities, operational changes, upgrades and/or process changes. Therefore, confidence in the model simulations increases with the degree of calibration conducted. Historical periods defining different operating conditions, intensive sampling and other optimization techniques assist in matching the performance of the model to the full-scale facility.

Energy consumption and generation are crucial in determining wastewater treatment plant operating costs. Computer simulations enable to predict energy balance for different treatment strategies simultaneously with biological and chemical processes. The examined scenarios showed the influence of sample operating parameters variations on the WWTP energy performance. Primary sludge removal efficiency was found as a significant factor determining biogas production under mesophilic conditions and the share of renewable energy in the WWTP energy balance. On the other hand energy savings may be found in different facilities with AS aeration systems.

References

1. W. Gujer. Wat. Sci. Tech., **53** (3), 111–119 (2006).
2. M. Henze, W. Gujer, T. Mino, M. van Loosdrecht. (London, IWA Publishing, 2002).
3. M. Swinarski, J. Mąkinia, K. Czerwionka, M. Chrzanowska, J. Drewnowski. Water Environment Research, **84** (8), 646-655 (2012).
4. J. Drewnowski, J. Makinia. IJEST **11**, 7, 1973-1988 (2014).
5. GPS-X 6.4. *User's guide and technical reference*. Hydromantis Environmental Software Solutions, Inc. Hamilton (2014)
6. J. Drewnowski, J. Makinia. Wat. Sci. Tech. **67**, 9, 2067-2074 (2013).
7. A. Onnis-Hayden, A.Z. Gu. *Comparison of Organic Sources for Denitrification: Biodegradability, Denitrification Rates, Kinetic Constants and Practical Implication for Their Application in WWTPs. Proc. the 81st Annual WEF Technical Exhibition and Conference*, Water Environment Federation: Alexandria, Virginia, 253 (2008).
8. A. De Lucas, L. Rodríguez, J. Villasenor, F.J. Fernandez. Water Res., 39, 3715 (2005)
9. I. Takacs, M. O'Shaugnessy, S. Murthy, K. Brian, N. Shirodkar, D. Katehis. *Design of Denitrification Systems Using Methanol. Proc. the WEF-IWA Specialized Conference "Nutrient Removal 2007: State of the Art.* Water Environment Federation, Alexandria, Virginia, 218 (2007).
10. P. Dold, I. Takacs, Y. Mokhayeri, A. Nichols, J. Hinojosa, R. Riffat, W. Bailey, S. Murthy. *Denitrification with Carbon Addition – Kinetic Considerations. Proc. the*

WEF-IWA Specialized Conference "Nutrient Removal 2007: State of the Art", Water Environment Federation, Alexandria, Virginia, 218 (2007)

11. H.M. Phillips, J.L. Barnard, C. de Barbadillo, A.R. Shaw, M.T. Steichen, C. Wallis-Lage. *Denitrification Rates: Sampling, Modelling and Design Consideration. Proc. the WEF Specialty Conference "Nutrient Removal 2009. Sustainable Treatment Solutions*, Water Environment Federation: Alexandria, Virginia, 252 (2009).
12. Y. Filali-Meknassi, M. Auriol, R.D. Tyagi, Y. Comeau, R.Y. Surampall. Environ. Technol., **26**, 1081 (2005)
13. R. Latimer, P. Pitt, A. van Niekerk, T. Bruton, S. Dailey, J. Grandstaff, K. Selock. *Secondary. Impacts of Supplemental Carbon Addition to BBNR/ENR Treatment Process. Proc. the WEF Specialty Conference "Nutrient Removal 2009. Sustainable Treatment Solutions.* Water Environment Federation, Alexandria, Virginia, 849 (2009).

AUTHOR INDEX